# Sustainable Consumer Behaviour and the Environment

This book advances the tourism and hospitality industry's contribution to meeting the United Nations Sustainable Development Goal 12 of responsible consumption and production. It enables a collaboration platform across these sectors in pursuit of common goals for promoting sustainable consumption and environmental protection.

Sustainable consumer behavior is a principal topic in the current tourism and hospitality industry as many types of unsustainable consumptions pose a threat to society and the natural environment. Sustainable consumer behavior is a vital facet of protecting the environment that ultimately benefits the entire society. Individuals' irresponsible consumption activities are undeniably considerable elicitors of harmful environmental, social, economic, and economic impacts throughout the world. Comprehending sustainable consumer behavior is of utmost importance for the tourism and hospitality industry to design innovative and responsible strategies to minimize the negative consequences of tourism.

The scope of this book includes various sustainable consumptions, productions, and consumer behaviors in a variety of tourism and hospitality sectors and will be of great value to students, scholars, and researchers interested in areas such as sustainable consumer behaviour, hospitality, sustainable development, and tourism management.

The chapters in this book were originally published as a special issue of the *Journal of Sustainable Tourism*.

**Heesup Han** is Professor in the College of Hospitality and Tourism Management at Sejong University, Korea. His research interests include sustainable consumer behaviour, green tourism and hospitality products, sustainable development, slow tourism, and destination marketing. His papers have been selected as the most downloaded and read articles in many top-tier tourism and hospitality journals. Heesup Han is a 2019 and 2020 highly cited researcher (HCR) of the world in social science (identified by the Web of Science Group - Clarivate).

# Sustainable Consumer Behaviour and the Environment

*Edited by*
**Heesup Han**

**Routledge**
Taylor & Francis Group

LONDON AND NEW YORK

First published 2022
by Routledge
2 Park Square, Milton Park, Abingdon, Oxon, OX14 4RN

and by Routledge
605 Third Avenue, New York, NY 10158

*Routledge is an imprint of the Taylor & Francis Group, an informa business*

© 2022 Taylor & Francis

*British Library Cataloguing-in-Publication Data*
A catalogue record for this book is available from the British Library

ISBN13: 978-1-032-18794-5 (hbk)
ISBN13: 978-1-032-18795-2 (pbk)
ISBN13: 978-1-003-25627-4 (ebk)

DOI: 10.4324/9781003256274

Typeset in Myriad Pro
by codeMantra

**Publisher's Note**
The publisher accepts responsibility for any inconsistencies that may have arisen during the conversion of this book from journal articles to book chapters, namely the inclusion of journal terminology.

**Disclaimer**
Every effort has been made to contact copyright holders for their permission to reprint material in this book. The publishers would be grateful to hear from any copyright holder who is not here acknowledged and will undertake to rectify any errors or omissions in future editions of this book.

# Contents

*Citation Information*                                                    vii
*Notes on Contributors*                                                   ix

1   Consumer behavior and environmental sustainability in tourism and
    hospitality: a review of theories, concepts, and latest research        1
    *Heesup Han*

2   Green-Induced tourist equity: the cross-level effect of regional
    environmental performance                                              23
    *IpKin Anthony Wong, Wenjia Jasmine Ruan, Xiaomei Cai and
    GuoQiong Ivanka Huang*

3   Hotels' sustainability practices and guests' familiarity, attitudes
    and behaviours                                                         43
    *Hossein Olya, Levent Altinay, Anna Farmaki, Ainur Kenebayeva
    and Dogan Gursoy*

4   The anchoring effect of aviation green tax for sustainable
    tourism, based on the nudge theory                                     62
    *Haeok Liz Kim and Sunghyup Sean Hyun*

5   Application of internal environmental locus of control to
    the context of eco-friendly drone food delivery services              78
    *Jinsoo Hwang, Jin-soo Lee, Jinkyung Jenny Kim and Muhammad
    Safdar Sial*

6   The impact of the Middle East Respiratory Syndrome coronavirus on
    inbound tourism in South Korea toward sustainable tourism             97
    *Yunseon Choe, Junhui Wang and HakJun Song*

7   Exploring preferences and sustainable attitudes of Airbnb green users in
    the review comments and ratings: a text mining approach              114
    *Laura Serrano, Antonio Ariza-Montes, Martín Nader,
    Antonio Sianes and Rob Law*

8  An application of Delphi method and analytic hierarchy process in understanding hotel corporate social responsibility performance scale        133
   *Antony King Fung Wong, Seongseop (Sam) Kim,*
   *Suna Lee and Statia Elliot*

9  Comparing resident and tourist perceptions of an urban park: a latent profile analysis of perceived place value        160
   *Hwasung Song and Changsup Shim*

10  Understanding backpacker sustainable behavior using the tri-component attitude model        173
   *Elizabeth Agyeiwaah, Frederick Dayour, Felix Elvis Otoo and Ben Goh*

   *Index*        195

# Citation Information

The chapters in this book were originally published in the *Journal of Sustainable Tourism*, volume 29, issue 7 (2021). When citing this material, please use the original page numbering for each article, as follows:

**Chapter 1**
*Consumer behavior and environmental sustainability in tourism and hospitality: a review of theories, concepts, and latest research*
Heesup Han
*Journal of Sustainable Tourism*, volume 29, issue 7 (2021) pp. 1021–1042

**Chapter 2**
*Green-Induced tourist equity: the cross-level effect of regional environmental performance*
IpKin Anthony Wong, Wenjia Jasmine Ruan, Xiaomei Cai and GuoQiong Ivanka Huang
*Journal of Sustainable Tourism*, volume 29, issue 7 (2021) pp. 1043–1062

**Chapter 3**
*Hotels' sustainability practices and guests' familiarity, attitudes and behaviours*
Hossein Olya, Levent Altinay, Anna Farmaki, Ainur Kenebayeva and Dogan Gursoy
*Journal of Sustainable Tourism*, volume 29, issue 7 (2021) pp. 1063–1081

**Chapter 4**
*The anchoring effect of aviation green tax for sustainable tourism, based on the nudge theory*
Haeok Liz Kim and Sunghyup Sean Hyun
*Journal of Sustainable Tourism*, volume 29, issue 7 (2021) pp. 1082–1097

**Chapter 5**
*Application of internal environmental locus of control to the context of eco-friendly drone food delivery services*
Jinsoo Hwang, Jin-soo Lee, Jinkyung Jenny Kim and Muhammad Safdar Sial
*Journal of Sustainable Tourism*, volume 29, issue 7 (2021) pp. 1098–1116

**Chapter 6**
*The impact of the Middle East Respiratory Syndrome coronavirus on inbound tourism in South Korea toward sustainable tourism*
Yunseon Choe, Junhui Wang and HakJun Song
*Journal of Sustainable Tourism*, volume 29, issue 7 (2021) pp. 1117–1133

## Chapter 7

*Exploring preferences and sustainable attitudes of Airbnb green users in the review comments and ratings: a text mining approach*
Laura Serrano, Antonio Ariza-Montes, Martín Nader, Antonio Sianes and Rob Law
*Journal of Sustainable Tourism*, volume 29, issue 7 (2021) pp. 1134–1152

## Chapter 8

*An application of Delphi method and analytic hierarchy process in understanding hotel corporate social responsibility performance scale*
Antony King Fung Wong, Seongseop (Sam) Kim, Suna Lee and Statia Elliot
*Journal of Sustainable Tourism*, volume 29, issue 7 (2021) pp. 1153–1179

## Chapter 9

*Comparing resident and tourist perceptions of an urban park: a latent profile analysis of perceived place value*
Hwasung Song and Changsup Shim
*Journal of Sustainable Tourism*, volume 29, issue 7 (2021) pp. 1180–1192

## Chapter 10

*Understanding backpacker sustainable behavior using the tri-component attitude model*
Elizabeth Agyeiwaah, Frederick Dayour, Felix Elvis Otoo and Ben Goh
*Journal of Sustainable Tourism*, volume 29, issue 7 (2021) pp. 1193–1214

For any permission-related enquiries please visit:
http://www.tandfonline.com/page/help/permissions

# Notes on Contributors

**Elizabeth Agyeiwaah** Faculty of Hospitality and Tourism Management, Macau University of Science and Technology, Avenida Wai Long, Taipa, China.

**Levent Altinay** Strategy and Entrepreneurship, Oxford Brookes Business School, Oxford Brookes University, Headington Campus, UK.

**Antonio Ariza-Montes** Social Matters Research Group, Universidad Loyola Andalucıa, Spain.

**Xiaomei Cai** School of Tourism Management, South China Normal University, Higher Education Mega Center, Guangzhou, China.

**Yunseon Choe** School of Community Resources and Development, The Hainan University-Arizona State University International Tourism College, Arizona State University, Phoenix, USA.

**Frederick Dayour** Department of Community Development, Simon Diedong Dombo University of Business and Integrated Development Studies, Ghana School of Tourism and Hospitality, University of Johannesburg, South Africa.

**Statia Elliot** School of Hospitality and Tourism Management, University of Guelph, Canada.

**Anna Farmaki** Department of Hotel and Tourism Management, Cyprus University of Technology, Limassol, Cyprus.

**Ben Goh** Professor and Dean, Faculty of Hospitality and Tourism Management, Macau University of Science and Technology, Avenida Wai Long, Taipa, China.

**Dogan Gursoy** Tourism Management, School of Hospitality Business Management, College of Business, Washington State University, Pullman, USA.

**Heesup Han** College of Hospitality and Tourism Management, Sejong University, Seoul, Korea.

**GuoQiong Ivanka Huang** School of Tourism Management, Sun Yat-Sen University, Zhuhai, China.

**Jinsoo Hwang** Department of Food Service Management, The College of Hospitality and Tourism Management, Sejong University, Seoul, Korea.

**Sunghyup Sean Hyun** School of Tourism, Hanyang University, Seoul, Republic of Korea.

**Ainur Kenebayeva** Management and Business Department, University of International Business, Almaty, Kazakhstan.

**Haeok Liz Kim** School of Tourism, Hanyang University, Seoul, Republic of Korea.

**Jinkyung Jenny Kim** Department of Hotel Management, School of Hotel and Tourism Management, Youngsan University, Busan, South Korea.

**Seongseop (Sam) Kim** School of Hotel & Tourism Management, The Hong Kong Polytechnic University, Kowloon, Hong Kong SAR, China.

**Rob Law** School of Hotel and Tourism Management, The Hong Kong Polytechnic University, Hong Kong SAR, China.

**Jin-soo Lee** The School of Hotel and Tourism Management, The Hong Kong Polytechnic University, Hung Hom, China.

**Suna Lee** Hotel and Tourism Event Management Tourism College, Macau Institute for Tourism Studies, Colina de Mong-Ha, Macao, China.

**Martın Nader** Department of Psychological Studies, Universidad ICESI, Colombia.

**Hossein Olya** Sheffield University Management School, UK; School of Tourism and Hospitality, University of Johannesburg, South Africa.

**Felix Elvis Otoo** School of Hotel and Tourism Management, The Hong Kong Polytechnic University, Tsim Sha Tsui East, China.

**Wenjia Jasmine Ruan** School of Tourism Management, Sun Yat-Sen University, Zhuhai, China.

**Laura Serrano** Social Matters Research Group, Universidad Loyola Andalucıa, Spain.

**Changsup Shim** Department of Tourism Management, Gachon University, Seongnam-si, Gyeonggi-do, South Korea.

**Muhammad Safdar Sial** Department of Management Sciences, COMSATS University Islamabad, Pakistan.

**Antonio Sianes** Research Institute on Policies for Social Transformation, Universidad Loyola Andalucıa, Spain.

**HakJun Song** Department of Hotel & Convention Management, Pai Chai University, Daejeon, South Korea.

**Hwasung Song** Research Fellow, Department of Urban Management, Suwon Research Institute, Suwon-si, Gyeonggi-do, South Korea.

**Junhui Wang** Department of Hotel & Convention Management, Pai Chai University, Daejeon, South Korea.

**Antony King Fung Wong** School of Hotel & Tourism Management, The Hong Kong Polytechnic University, Kowloon, China.

**IpKin Anthony Wong** School of Tourism Management, Sun Yat-Sen University, Zhuhai, China.

# Consumer behavior and environmental sustainability in tourism and hospitality: a review of theories, concepts, and latest research

Heesup Han (iD)

**ABSTRACT**

Diverse forms of environmental problems pose a serious threat to the natural environment. Environmental sustainability is the foremost topic in the contemporary tourism and hospitality industry. Environmentally-sustainable consumer behavior is an important aspect of environmental protection, which eventually benefits the society. In order to better understand environmentally-sustainable consumption and promote environmentally responsible consumer behavior, this research provides a sound conceptualization of environmentally-sustainable consumer behavior, and presents a systematic review and perspective on theories (theory of reasoned action, norm activation theory, theory of planned behavior, model of goal-directed behavior, and value-belief-norm theory) established in tourism and environmental psychology. In addition, this study introduces the essential drivers of environmentally-sustainable consumer behavior (green image, pro-environmental behavior in everyday life, environmental knowledge, green product attachment, descriptive social norm, anticipated pride and guilt, environmental corporate social responsibility, perceived effectiveness, connectedness to nature, and green value). Lastly, this paper provides the values of the latest studies on the special issue of environmental sustainability and consumer behavior in tourism and hospitality. This study as an introductory paper along with other articles in this special section help enable a collaboration platform across tourism and hospitality fields in pursuit of universal goals for promoting pro-environmental consumption and environmental sustainability.

## Introduction

Diverse problems (e.g. greenhouse effect, air/water/soil pollution, extinction/loss of species, and exhaustion of natural resources) pose a serious threat to the environment and its sustainability (Wang et al., 2020; Xu et al., 2020). The problems are mostly relevant to environmentally irresponsible human behaviors (Hopkins, 2020; Steg & Vlek, 2009; Xu et al., 2020; Wu et al., 2020). Researchers agree that the problematic issues can be therefore managed and resolved by correcting the human behaviors to be an environmentally-sustainable way (Han, 2020; Steg & Vlek, 2009). Particularly, the change of individuals' consumption behaviors (approaching, buying, and consuming the products in a pro-environmental manner) is considered to be an important requisite for environmental sustainability (Halder et al., 2020; Wang et al., 2020).

Therefore, for the last few decades, eliciting environmentally-sustainable consumer behavior has been a vital topic that has an increasing attention in the consumer marketplace and academia (Dong et al., 2020; Garvey & Bolton, 2017). Environmentally-sustainable consumer behavior whose alternative term is environmentally-responsible consumer behavior is hard to pin down when defining it, but the term is broadly utilized and employed as an umbrella concept that highlights an individual's various actions saving natural resources (e.g. water, energy), reducing environmental harm (e.g. waste decrease), meeting the green needs of society, and improving his/her life quality (Bridges & Wilhelm, 2008; Dong et al., 2020; Han, 2020). Similarly, according to Krajhanzl (2010), 'pro-environmental behavior is such behavior which is generally (or according to knowledge of environmental science) judged in the context of the considered society as a protective way of environmental behavior or a tribute to the healthy environment' (p. 252). Therefore, such terms as environmentally responsible behaviors (Kaiser et al., 1999), environmentally sustainable behaviors (Clayton & Myers, 2009), environment-protective/preserving behaviors (Krajhanzl, 2010), ecological behaviors (Kaiser et al., 1999), and green behaviors (Han, 2020) are often utilized as equivalents for pro-environmental behaviors. Environmentally-sustainable consumer behavior is irrefutably an important facet of pro-social consumption activities (Black & Cherrier, 2010; Halder et al., 2020), which eventually benefits the environment and the entire society (Park et al., 2018; Steg & Vlek, 2009).

Particularly, sustainable consumption is becoming an emerging issue in tourism and hospitality (Kiatkawsin & Han, 2017; Wang et al., 2020). As an increasing number of people in the marketplace recognize that many serious environmental deteriorations are rooted in tourism activities/development (Trang et al., 2019; Wang et al., 2020), the issue of eco-friendly consumption and sustainable product development is becoming more important than ever in the contemporary tourism and hospitality industry. Nowadays, customers in this sector increasingly demand green products (e.g. environmentally responsible hotels, restaurants, cruises, airlines, destinations, resorts, conventions, casinos) and often show a willingness for sustainable consumption (Chen et al., 2012; Ramkissoon et al., 2013; Trang et al., 2019; Wang et al., 2018). Due to this demand and eco-conscious market environment, many tourism and hospitality companies are increasingly showing the tendency to become pro-active in greening their operations and products (Afifah & Asnan, 2015; Hopkins, 2020; Lee et al., 2013). Concurrently, tourism and hospitality customers are becoming acquainted with the inevitability of environmentally-sustainable behaviors in their product-consumption situations as well as in their everyday life (Choi et al., 2015; Wang et al., 2018; Xu et al., 2020).

The present research aims to enrich the extant tourism and hospitality literature in sustainable consumption and purchase behaviors by filling the gaps described subsequently. While scholars in environmental behavior and consumer behavior have described environmentally-sustainable consumer behavior (Bridges & Wilhelm, 2008; Chan, 2001; Joshi & Rahman, 2015), it has not been soundly conceptualized in the tourism and hospitality sector. In addition, although considerable efforts on applying and expanding existing theories in social/environmental psychology have been made (Choi et al., 2015; Han, 2015; Young et al., 2020), the extended discussions on environmentally-sustainable consumer behavior and its related theories are lacking. Moreover, comprehending the drivers of sustainable behavior is undoubtedly crucial to design effective strategies to substantially minimize the negative environmental impacts of tourism (Ramkissoon et al., 2013; Xu et al., 2020). Yet, the thorough reviews and extensive discussions about key concepts that induce pro-environmental consumer behavior in tourism and hospitality have not been sufficiently provided.

This research discusses the concept of environmentally-sustainable consumer behavior and its merits for the environment. In addition, this paper provides a systematic review and perspective on social psychology and environmental psychology theories that are popular and often used for explicating pro-environmental behaviors among consumers. This research also discusses main variables that drive environmentally-sustainable consumer behavior and provides the conceptualization of the variables. Subsequently, in the last section, this paper introduces nine papers included in the special issue of 'Sustainability and consumer behavior' and discusses the value of the papers.

## Environmentally-sustainable consumer behavior

At present, there exists no agreement regarding the definition of environmentally-sustainable consumer behavior. The environmental psychology literature often discusses environmentally-sustainable behavior using the terms such as pro-environment behavior and green behavior (Bridges & Wilhelm, 2008; Han, 2020). That is, environmentally-sustainable behavior is one's behavior that helps environmental sustainability (Halder et al., 2020; Steg & Vlek, 2009). Steg and Vlek (2009) provided one of the clearest definitions that environmentally-sustainable behavior is one's specific behavior that is not harmful (or even beneficial) for the environment. In a coherent manner, environmentally-sustainable consumer behavior in the present study refers to an individual's behavior that hardly harms the environment or even brings the benefit to the environment in a consumption situation of a product/service. In the consumer behavior literature, environmentally-sustainable behaviors are frequently described as green consumption activities (Black & Cherrier, 2010; Chan, 2001; Joshi & Rahman, 2015). These green behaviors have been broken down into eco-friendly product post-purchase/pre-purchase behavior, energy-efficient action, natural resource saving, eco-friendly buying, water saving, garbage sorting behavior, food waste reduction, solid waste minimization, organic/green product use, and public transportation use in the literature (Dong et al., 2012; Leary et al., 2014; Minton et al., 2018; Singh & Verma, 2017). Yet, environmentally-sustainable consumption and its scope cover all forms of consumer behaviors that are helpful for the reduction of the environmental impacts (Han, 2020).

The most frequently employed types of environmentally-sustainable behaviors among consumers are green purchasing, recycling/reusing, and natural resource saving (Dong et al., 2020; Garvey & Bolton, 2017; Zhao et al., 2014). In the tourism and hospitality industry, environmentally-sustainable consumption activities comprising water saving, towel reuse, energy saving, eco-product purchasing, local product use, reuse plastic bottles/bags, and food waste reduction at tourist places/sites has been extensively researched (Choi et al., 2015; Kiatkawsin & Han, 2017; Untaru et al., 2016). Particularly, a considerable number of studies have dealt with eco-friendly/ green product purchase behaviors (Untaru et al., 2016). The most frequently researched products in the tourism and hospitality context can be green hotels (Choi et al., 2015; Wang et al., 2018), green restaurants (Moon, 2021), green cruises (Paiano et al., 2020), eco-friendly/sustainable destinations (Kiatkawsin & Han, 2017; Werner et al., 2020), green cafés (Jeong et al., 2014), green museums (Byers, 2008; Han et al., 2018), and green conventions (Han & Hwang, 2017). These forms of green products partly/wholly fulfill travelers' and the market's increasing needs for going green, and therefore, the firm/destination that offers such green products grow its competitiveness in the tourism and hospitality marketplace.

Pro-environmental purchase as one major form of environmentally-sustainable consumer behavior in tourism and hospitality refers to travelers' eco-friendly choice activity for green products/services for environmental preservation (Han, 2020). Such environmentally responsible choice is derived from intricate pro-environmental decision-making processes (Chan, 2001; Joshi & Rahman, 2015). Individuals who practice environmentally-sustainable consumption activities generate the minimized environment impact and even benefit the natural environment (Chan, 2001; Dong et al., 2020; Joshi & Rahman, 2015; Minton et al., 2018; Singh & Verma, 2017). Likewise, in a tourism product consumption situation, travelers' environmentally-sustainable consumption activities are indisputably key contributors to environmental preservation at a tourist place/destination whereas their environmentally irresponsible consumption behaviors are essential triggers of negative environmental impacts on the tourist place/destination (Kiatkawsin & Han, 2017).

## Theories underpinning environmentally-sustainable consumer behavior

This section provides a review on social psychology and environmental psychology theories (i.e. theory of reasoned action, theory of planned behavior, model of goal-directed behavior, norm activation theory, and value-belief-norm theory). These are well-established theories in environmental sustainability, which have been successfully applied and extensively used in explaining pro-environmental behaviors among consumers (Han, 2020; Manosuthi et al., 2020; Megeirhi et al., 2020; Onwezen et al., 2013). The theories view environmentally-sustainable consumer behaviors as being stimulated by pro-social motives or self-interest motives (Ajzen & Kruglanski, 2019; Han, 2015; Meng et al., 2020).

### *Theory of reasoned action and theory of planned behavior*

In their endeavor to explicate human pro-environmental behavior, academics and industry practitioners typically center on a specific behavior of interest, such as water saving, recycling, energy saving, public transportation use, avoidance of disposable product use, or environmental protection (Moon, 2021; Paiano et al., 2020; Untaru et al., 2016). These behaviors are largely of interest as comprehending the determining factors of the behaviors provides the base for planning/making interventions to promote such eco-friendly behaviors and lessen environmental problems. For the past few decades, academics in environmental psychology, social behavior, and traveler behavior have relied considerably on the theory of reasoned action (Fishbein & Ajzen, 2010) and the theory of planned behavior (Ajzen, 1991) as their conceptual framework for explaining customer pro-environmental behaviors (Garay et al., 2019; Han, 2020). In diverse behavioral domains, the prediction capability of the reasoned action theory and the planned behavior theory has been demonstrated.

According to the reasoned action theory, an individual's behavioral intention, which is an immediate precursor of actual action, is built by his/her attitude toward the behavior and subjective norm (Fishbein & Ajzen, 2010). In other words, the reasoned action theory recognizes the usefulness of attitudinal and social factors but applies the variables to the specific behavior of interest (Meng et al., 2020). Hence, the key determinants of one's intention and behavior within this theory are attitude toward the behavior and subjective norm concerting the action under consideration (Ajzen & Kruglanski, 2019). Attitude toward the behavior is the degree to which practicing a certain behavior is favorably/unfavorably valued (Ajzen, 1991). It forms based on the combination of outcome beliefs and subjective value of the anticipated outcomes (Manosuthi et al., 2020). The outcome beliefs whose alternative term is behavioral beliefs indicate an individual's perceived probability that the action will generate given outcomes (Manosuthi et al., 2020).

Subjective norm indicates an individual's perceived social pressure to practice or not to practice a certain behavior (Ajzen, 1991). Jacobson et al. (2011) similarly described the term as the degree of one's perception about social approval or disapproval for a specific behavior. Subjective norm is assumed to be a critical function of normative beliefs of important referents and motivation to comply with the referents (Meng et al., 2020; Moon, 2021). Normative beliefs indicate certain behavioral expectations of the important referents whereas motivation to comply is about how critical the referents' expectations are. These referents can be family/relatives, teachers, supervisors, co-workers, or peers (co-workers peers) (Meng et al., 2020). Ajzen (2012) and Moon (2021) described attitude toward the behavior and subjective norm including their predictors as a volitional process.

The theory of planned behavior is an extended version of the planned behavior theory (Ajzen, 1991). Within this theory, one's behavioral intention is a proximal determinant of his/her actual behavior, and this intention forms through both volitional process and non-volitional process (Garay et al., 2019). Ajzen (1991) differentiated the planned behavior theory from the reasoned action theory by integrating the non-volitional process into his theoretical framework. The key aspect of the non-volitional process is perceived behavioral control that refers to an individual's level of perception regarding his/her capability to carry out a given behavior (Ajzen, 2012). Perceived behavioral control forms based on the combination of control beliefs and perceived power (Manosuthi et al., 2020; Moon, 2021). While control belief refers to one's perception about the presence/absence of the factors that facilitate/hinder his/her performance of the behavior, perceived power is about how important the factors are to him/her (Ajzen, 1991; Meng et al., 2020).

### Model of goal-directed behavior

In their effort of understand individuals' environmentally-sustainable behavior, researchers often examine the achievement of certain goals, such as green product use, green healthy food consumption, environmentally responsible traveling, or energy-efficient lifestyle, which minimize the possible harms to the environment (Ajzen & Kruglanski, 2019; Han & Hwang, 2014; Perugini & Bagozzi, 2001). Undoubtedly, these kinds of goals are of criticality in that a clear understanding of their drivers helps designing/developing tactics to motivate/uphold pro-environmental behaviors and alleviate environmental deteriorations. Due to its stronger prediction power and anticipation ability as compared to the planned behavior theory (Kim et al., 2020), the model of goal-directed behavior has been broadly applied and expanded for explicating goal-centered behaviors in a variety of domains. The theory of planned behavior, utilized broadly for the anticipation/amendment of individual activities, has a behavior-focused approach as its base (Ajzen & Kruglanski, 2019). However, within the model of goal-directed behavior, one' behavior in general serves as a means to his/her certain goal (Ajzen & Kruglanski, 2019; Perugini & Bagozzi, 2001). That is, the model of goal-directed behavior includes a goal-centered approach as its basis.

According to the goal-directed behavior theory, an individual's behavioral intention, which is the most proximal and single direct determinant of actual action, is generated through motivation process (desire toward the behavior), volitional process (attitude toward the behavior and subjective norm), non-volitional process (perceived behavioral control), emotional process (positive and negative anticipated emotions), and habitual process (frequency/infrequency of past behavior). For more comprehensive explanation of an individual's intention/behavior, the model of goal-directed behavior incorporated such critical factors as desire toward the behavior, positive and negative anticipated emotions, and frequency/infrequency of past behavior into the planned behavior theory (Perugini & Bagozzi, 2004). That is, the goal-directed behavior theory made three key improvements on the planned behavior theory.

First, the goal-directed behavior theory includes desire as a proximal driver of intention and a direct determining force of the intention. Desire toward the behavior is one's state of mind where he/she has motivations to carry out the behavior for attaining the goal related to the behavior (Perugini & Bagozzi, 2001; Bagozzi & Dholakia, 2006). The motivations form based on the combination of attitude, subjective norm, perceived behavioral control, and anticipated emotions (Carrus et al., 2008; Han & Hwang, 2014; Kim et al., 2020). Second, the goal-directed behavior theory includes anticipated emotions that refer to anticipated post-behavioral affective responses, which are either positive or negative (Kim et al., 2020; Thomson et al., 2008). The theory relates such expected emotional reactions to goal desire parallel to volitional and non-volitional factors (Perugini & Bagozzi, 2004). Third, the model of goal-directed behavior considers the influence of past behavior. The inclusion of an individual's frequency and recency of past behavior allows the integration of essential information relating to a habitual/experiential aspect of goal-centered behaviors, which is not considered in the planned behavior theory (Han & Hwang, 2014). Past behavior is particularly crucial when a certain goal-centered action is performed in an unstable sector (or is not sufficiently learned) (Carrus et al., 2008; Ouellette & Wood, 1998). In this circumstance, past behavior acts as a direct driving force of intention/behavior (Bagozzi & Dholakia, 2006).

### Norm activation theory

Individuals' behaviors in their daily life including consumption activities cause diverse environmental harm, and they can practice pro-social/pro-environmental behaviors for the mitigation of the harm (Rosenthal & Ho, 2020). Over the last four decades, researchers have relied heavily on the norm activation theory (Han, 2020; Klöckner, 2013; Schwartz, 1977) when explaining such pro-social/pro-environmental behaviors as their theoretical framework (Denley et al., 2020; Klöckner, 2013; Rosenthal & Ho, 2020; Shi et al., 2017). The alternative term of the norm activation theory is norm activation model. The norm activation theory encompasses awareness of consequences, ascription of responsibility, and personal norm as its constituents (Bamberg & Möser, 2007; Denley et al., 2020; Schwartz, 1977; Stern et al., 1999).

Personal norm is the direct precursor that maximizes and mediates the effect of awareness of consequences and ascription of responsibility on altruistic behavior (Bamberg & Möser, 2007; Shi et al., 2017). Personal norm is increasingly becoming the core concept when explicating environmentally responsible behaviors (Klöckner, 2013). This concept refers to a person's sense of moral obligation to carry out a certain action derived from his/her awareness of the harmful consequences of not practicing the action and his/her feeling of responsibility to act (Han, 2014; Rosenthal & Ho, 2020). The key aspect of personal norm is one's feeling of moral obligation to engage (or not to engage) in the behavior (Schwartz, 1977). Thus, personal norms are alternatively used with the terms, such as moral norm, moral obligation, and sense of moral obligation (Denley et al., 2020; Han, 2014). This personal norm guides pro-social/pro-environmental behavior (Shi et al., 2017). The norm activation arises when a person is aware of the possible negative consequences of their socially/environmentally irresponsible behaviors and perceives/admits personal responsibility (De Groot & Steg, 2009).

The norm activation theory is interpreted in two main ways (Han, 2014; Steg & De Groot, 2010). The first interpretation is a sequential model where personal norm, which is a direct determinant of altruistic behavior, is activated by awareness of consequences indirectly through ascription of responsibility (Onwezen et al., 2013; Steg & De Groot, 2010). The second way of the interpretation implies that norm activation is a progression where awareness of consequences together with ascription of responsibility as direct antecedents elicit personal norm, which leads to a certain pro-environmental behavior (Bamberg & Möser, 2007; Rosenthal & Ho, 2020). The first interpretation (awareness of consequences → ascription of responsibility → personal norm → altruistic behavior) is more generally accepted and applied in environmental psychology and consumer behavior (Onwezen et al., 2013; Steg & De Groot, 2010). Han (2014) and Onwezen et al. (2013) empirically demonstrated its effectiveness. In addition, this sequential model is also coherent with the proposition of Schwartz and Howard (1981).

### Value-belief-norm theory

Schwartz's (1977) norm activation model is developed to explicate general pro-social/altruistic behavior whereas Stern et al.'s (1999) value-belief-norm theory is specifically designed to explain one's pro-environmental behavior. The value-belief-norm theory is an extended version of the norm activation model (Choi et al., 2015; Han, 2015; Megeirhi et al., 2020; Young et al., 2020).

Value orientations and ecological worldview were linked to the norm activation framework. It theorizes that an individual's environmentally responsible action is formed based on the associations among normative factor (sense of obligation to take the pro-environmental action), value orientations (biospheric, altruistic, and egoistic), and belief factors (ecological worldview, adverse consequences for valued objects, ascribed responsibility) (Choi et al., 2015; De Groot et al., 2007; Stern, 2000). The value-belief-norm theory is a sequential model where pro-environmental behavior is activated by the relationship chain through value orientations, ecological worldview, adverse consequences for valued objects, ascribed responsibility, and sense of obligation to take pro-environmental actions in sequence (Klöckner, 2013; Young et al., 2020).

Within the value-belief-norm theory, value indicates 'a desirable trans-situational goal varying in importance, which serves as a guiding principle in the life of a person or other social entity" (Schwartz, 1992, p. 21). Among the constituents of value orientations, biospheric value embraces the concept of environmental value (Stern, 2000; Stern et al., 1999). Thus, this value as a term of biospheric/environmental value is most frequently employed and applied for explicating pro-environmental behaviors (De Groot et al., 2007; Kiatkawsin & Han, 2017). Biospheric/environmental value is also often utilized as a sole dimension of value in environmental psychology and tourism (Han, 2015; Kiatkawsin & Han, 2017). Biospheric value is to personal value accentuating the biosphere and the natural environment (Han, 2015; Stern, 2000). Meanwhile, altruistic value is pertinent to the welfare of people, and egoistic value is about maximization of an individual's benefits (Kiatkawsin & Han, 2017; Stern, 2000).

Ecological worldview refers to individuals' tendency to engage in a specific behavior with eco-friendly intent (Stern et al., 1999). The terms 'adverse consequences for valued objects' and 'awareness of consequences' are interchangeably used (De Groot & Steg, 2009). Adverse consequences for valued objects indicated that individuals' consciousness level about undesirable outcomes/consequences for the things that they value when not practicing a behavior in a pro-social/pro-environmental manner (Schwartz, 1977). Ascribed responsibility indicates individuals' feeling of personal responsibility for undesirable outcomes or consequences of not behaving in a pro-social/pro-environmental manner (Schwartz & Howard, 1981). In line with the concept of personal norm within the norm activation theory, sense of obligation to take pro-environmental actions refers to one's personal moral obligation to carry out a certain pro-environmental action in a given situation (Han, 2015; Young et al., 2020). The value-belief-norm theory encompassing these concepts has been largely validated in the extant environmental behavior and tourism literature (Choi et al., 2015; Kiatkawsin & Han, 2017).

## Key factors affecting environmentally-sustainable consumer behaviors

Considerable research indicated the criticality of the core variables within the above mentioned social and environmental psychology theories (attitude toward the behavior, subjective norm, perceived behavioral control, anticipated emotions, desire toward the behavior, past behavior, awareness of consequences, ascription of responsibility, personal norm, value orientations, ecological worldview) (Ajzen & Kruglanski, 2019; Denley et al., 2020; Han, 2015; Ramkissoon, 2020; Ramkissoon et al., 2020; Rosenthal & Ho, 2020; Shi et al., 2017; Untaru et al., 2016; Young et al., 2020). Indeed, these variables and their roles in explicating diverse consumer environmentally responsible behaviors have been extensively applied and tested in many empirical studies (Bamberg & Möser, 2007; Klöckner, 2013; Onwezen et al., 2013; Paiano et al., 2020; Steg & De Groot, 2010; Untaru et al., 2016).

Besides these crucial concepts, academics and empirical evidence in their studies across many consumer behavior and environmental psychology sectors support that green image (Lee et al., 2010), pro-environmental behaviors in everyday life (Han & Hyun, 2018; Untaru et al., 2016), environmental knowledge (Chan et al., 2014; Laroche et al., 1996), attachment (Rosenthal & Ho,

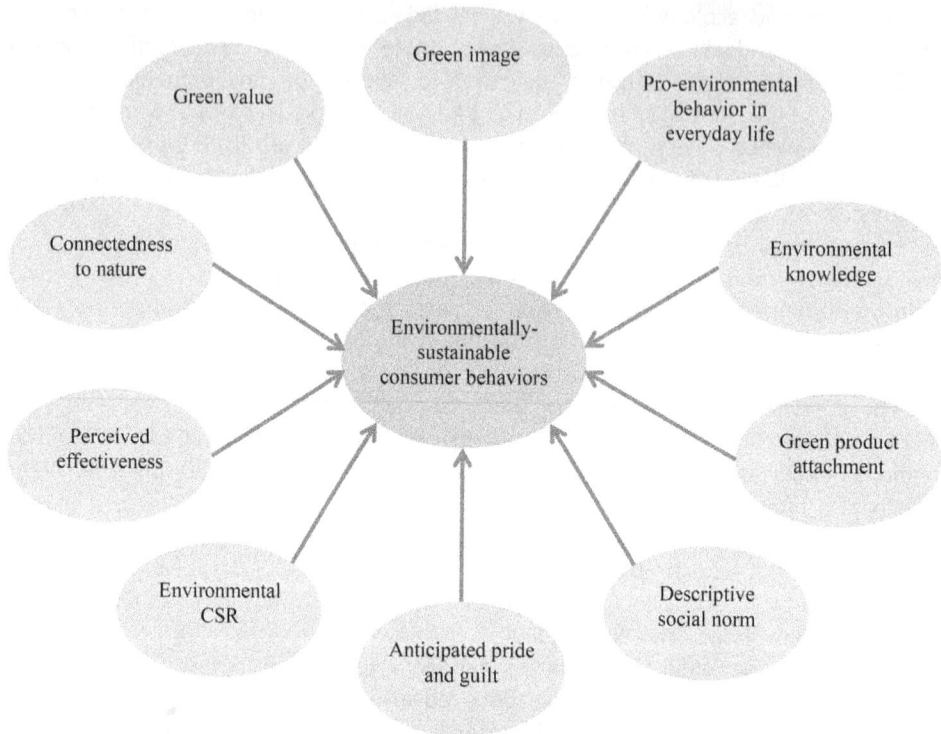

**Figure 1.** Key drivers of environmentally-sustainable consumer behaviors.

2020; Yuksel et al., 2010), descriptive social norm (Matthies et al., 2012), anticipated pride and guilt (Klöckner & Matthies, 2004; Steg & Vlek, 2009), environmental corporate social responsibility (Afifah & Asnan, 2015; Lee et al., 2013), perceived effectiveness (Han et al., 2017), connectedness to nature (Dutcher et al., 2007), and green value (Halder et al., 2020) are essential constituents for more comprehensive and clear comprehension of environmentally-sustainable decision-making processes and behavior among consumers. These variables are the crucial aspects of cognitive/perceptual process (green image, environmental knowledge, environmental corporate social responsibility, perceived effectiveness, green value), affective process (anticipated pride and guilt), conative process (attachment, connectedness to nature), normative process (descriptive social norm), and habitual process (pro-environmental behaviors in everyday life), which are fundamental in environmentally-sustainable behavior. These concepts are therefore extensively adopted and used in the broad range of the environmental consumption literature. The variables and definitions are presented in Figure 1 and Table 1.

## Green image

Image is often conceptualized as the set of beliefs/impressions that an individual has about a product/service/place and its attributes (Kotler et al., 1993). Similarily, Han et al. (2018) described green image as patrons' overall perceptions/ideas of a green product/service and its features. This image develops based on information/knowledge about the green object and its attributes, which are obtained and processed over time (Assael, 1984; Lee et al., 2010). Some researchers asserted that image is constituted of multiple phases (cognitive image → affective image → conative image → overall image) (Baloglu & McCleary, 1999). Yet, a general view is that image is cross-sectional (Assael, 1984; Han et al., 2018; Kotler et al., 1993; Kotler & Gertner, 2002). A favorable image of a green product is a fundamental requisite for the long-term success of every

**Table 1.** Definitions of key concepts driving environmentally-sustainable consumer behaviors.

| Variable | Definition |
|---|---|
| Green image | Green image refers to a consumer's overall perceptions/ideas of a green product/service and its features. |
| Pro-environmental behavior in everyday life | Pro-environmental behaviors in everyday life refer to a consumer's eco-friendly daily activities for environmental protection, which are habitual when they frequently practiced. |
| Environmental knowledge | Environmental knowledge refers to a consumer's ability to know/understand the environmental concepts, issues, problems, and model his/her activities. |
| Green product attachment | Green product attachment refers to a consumer's strong sense of belonging to a specific green product. |
| Descriptive social norm | Descriptive social norm refers to the extent to which an action is commonly perceived in a particular consumption situation. |
| Anticipated pride and guilt | Anticipated pride and guilt refer to a consumer's anticipated feeling states (feeling of either pride or guilt) that he/she would experience by practicing a certain pro-environmental behavior in the future. |
| Environmental corporate social responsibility | Environmental corporate social responsibility refers to a company's business activities in a manner, which is coherent with environment protection and complies with local regulations and governmental policies. |
| Perceived effectiveness | Perceived effectiveness refers to a consumer's beliefs about the influence of his/her environmental preservation endeavors/activities in decreasing the harm to nature. |
| Connectedness to nature | Connectedness to nature refers to the extent to which a consumer thinks he/she is a part of nature. |
| Green value | Green value refers to a consumer's cognitive appraisal of the efficacy of an eco-friendly product/service on the basis of his/her perception of what is obtained and what is sacrificed. |

business (Han et al., 2018; Wang et al., 2018). Because of its criticality, green image has long been a crucial topic in the extant literature of consumer behavior (Kotler & Gertner, 2002), environmental psychology (Wu et al., 2016), and tourism (Lee et al., 2010; Wang et al., 2018).

## Pro-environmental behaviors in everyday life

An eco-friendly consumer who makes approach decision/choice/action for a green product/service often engages in environmentally-sustainable behaviors in their daily life and has a belief about the effectiveness of such daily eco-friendly practices in protecting the natural environment (Laroche et al., 2001; Paco & Rapose, 2009; Untaru et al., 2016). Such conservation activities become habitual when they are frequently practiced (Han & Hyun, 2018; Untaru et al., 2016). These pro-environmental behaviors in everyday life as a form of habitual behaviors (Steg & Vlek, 2009) are regarded as one of the fundamental elements for clear explanation of environmentally-sustainable consumer behavior as the eco-friendly daily activities directly trigger sustainable consumer decision and behavior (Han & Hyun, 2018; Laroche et al., 2001). An individual's awareness/concern for the environmental deteriorations is frequently displayed in their actions in daily life (e.g. avoidance of disposable product use, water conservation, energy saving, towel reuse, recycling) and in their consumption behaviors (e.g. green product purchase) (Laroche et al., 2001; Paco & Rapose, 2009). The effectiveness of promoting pro-environmental behaviors in everyday life in boosting eco-friendly consumption has been largely proven in the tourism and hospitality sector (Han & Hyun, 2018; Untaru et al., 2016).

## Environmental knowledge

Due to its critical role, environmental knowledge is considered as a main cognitive dimension when explaining environmentally-sustainable consumption behavior (Boo & Park, 2013; Han &

Hyun, 2017; Kaiser et al., 1999). According to Chan et al. (2014), environmental knowledge is a vital precondition of customer sustainable intention/behavior for a hospitality/tourism product. Environmental knowledge indicates a patron's ability to know/understand the environmental concepts, issues, problems, and model their activities (Han & Hyun, 2017; Laroche et al., 1996). A customer tends to avoid engaging in a certain consumption activity where his/her knowledge/understanding to direct the behavior in that given situation is not sufficient (Chan et al., 2014; Kaiser et al., 1999). Such tendency of not practicing the behavior is for the minimization of probable uncertainty (Han & Hyun, 2017; Laroche et al., 1996). A consumer in general perceives that his/her knowledge level about certain consumption is high when believing that he/she knows a product/service/behavior related to the consumption better than others (Boo & Park, 2013; Kaiser et al., 1999). This knowledge in an eco-friendly product consumption situation often contributes to activating one's moral obligation for taking a pro-environmental behavior (Han & Hyun, 2017).

### Green product attachment

Customers, who feel attached to a specific product, show strong commitment to continue purchasing the product (Fedorikhin et al., 2008). In addition, patrons, who are attached to a certain brand/place, exhibit willingness/intention to sustain an existing relation with the brand/place (Thomson et al., 2005). Likewise, when an individual feels attached to a green/eco-friendly product, they are likely to continue buying and consuming the product (Jang et al., 2015). Green product attachment indicates a patron's strong sense of belonging to a specific green product (Ailawadi et al., 2001). It is also described as an emotional bond between a company's green product and its consumer (Jang et al., 2015; Ramkissoon et al., 2018; Yuksel et al., 2010). The high level of a consumer's attachment to a green product generates his/her favorable decision for the product and elicits environmentally-sustainable consumption behavior (Jang et al., 2015). Green product attachment often leads to the repeat purchase of a sustainable product over non-green alternatives and to the high loyalty for the product (Fedorikhin et al., 2008; Yuksel et al., 2010). Therefore, for every environmentally responsible company, the ability of generating attachment between its green product and its consumer has been suggested to be a key priority in sustaining the relationship with its customers and in running a successful business (Pedeliento et al., 2016; Thomson et al., 2005).

### Descriptive social norm

In the environmental behavior literature, social norm contains two dimensions (i.e. injunctive social norm and descriptive social norm) (Matthies et al., 2012; Steg & Vlek, 2009). Injunctive social norm is interchangeably used with the term 'subjective norm' (Han & Hwang, 2017) that refers to one's social pressure to engage or not to engage in a particular action (Ajzen, 1991; Ajzen & Kruglanski, 2019). That is, the concept of injunctive social norm is mostly about what important referents approve/disapprove (Smith et al., 2012). Yet, descriptive social norm indicates the extent to which an action is commonly perceived in a particular consumption situation (Han & Hwang, 2017; Matthies et al., 2012). In other words, the concept of descriptive social norm is mainly about what most people are doing in a given situation (Matthies et al., 2012). Unlike personal norm encompassing an intrinsic nature (intrinsic motivation), descriptive social norm comprises an extrinsic nature (extrinsic motivation) (Steg & Vlek, 2009). That is, when social norm is high, individuals are behaving with extrinsic motivation when a pro-environmental action is elicited (Smith et al., 2012; Steg & Vlek, 2009). Descriptive social norm is believed to include the power to affect diverse environmentally-sustainable activities (e.g. green hotel choice, green food consumption, sustainable transportation use, natural resource saving) (Han & Hwang, 2017; Steg & Vlek, 2009).

### Anticipated pride and guilt

A consumer experiences positive/negative affect when performing a specific consumption behavior, and also anticipates good/bad feeling that he/she would experience by practicing the behavior in the future (Bamberg & Möser, 2007; Klöckner & Matthies, 2004; Steg & Vlek, 2009). Perugini and Bagozzi (2001) conceptualized these affective states/responses as anticipated emotions. Yet, the scope of positive and negative anticipated emotions is too broad and less effective in explicating pro-environmental behavior (Han et al., 2017). Hence, anticipated pride and guilt as narrower terms are more prevalently used in the context of environmentally-sustainable consumer behavior (Han et al., 2017). Of many self-conscious emotional components, anticipated pride and guilt have largely proven to be essential and relevant for the apparent understanding of a consumer's sustainable decision formation and behaviors (Bamberg & Möser, 2007; Han, 2014; Lerner & Keltner, 2001; Onwezen et al., 2014). Different from other self-conscious emotions, anticipated pride and guilt arise specifically when a patron feels responsible for his/her action and evaluates the action with respect to his/her own ethical and social standards (Harth et al., 2013; Onwezen et al., 2014). Anticipated pride often has a feeling of proud, accomplished, confident, and worthwhile as its constituents whereas anticipated guilt includes a feeling of guilty, remorseful, sorry, and bad as its components (Han et al., 2017).

### Environmental corporate social responsibility

An ethical firm that is pro-active in corporate social responsibility activities for the environmental preservation has a better change for high customer retention and high reputation in the increasingly eco-conscious marketplace (Afifah & Asnan, 2015; Han et al., 2019; Lee et al., 2013). Many tourism and hospitality companies make continuous efforts to practice and promote environmental corporate social responsibility activities in order to maximize their competitiveness (Han et al., 2019). A firm's environmental corporate social responsibility endeavors make a feasible contribution to the eco-friendly development of a society and business environment while preserving the environment (Chen et al., 2012). Environmental corporate social responsibility helps a company do its business in a manner, which is coherent with environment protection and complies with local regulations and governmental policies (Montgomery & Stone, 2009). Irrefutable, a company's environmental corporate social responsibility practices influence its customer decision-making process and consumption behaviors (Afifah & Asnan, 2015; Lee et al., 2013). According to Han et al. (2019), when travelers perceive that a firm practices/promotes environmental corporate social responsibility activities, they form a better image of the firm and make approach behaviors to the firm and its eco-friendly products.

### Perceived effectiveness

Many studies rooted in social/environmental psychology theories with one's pro-social motives revealed that the efficient activation of moral norm needs cognitive elicitors (Han & Yoon, 2015; Roberts, 1996; Straughan & Roberts, 1999). Perceived effectiveness is irrefutably one of the essential constituents of such cognitive triggers (Judge et al., 2007; Roberts, 1996). Perceived effectiveness refers to a patron's beliefs about the influence of their environmental preservation endeavors/activities in decreasing the harm to nature (Han & Yoon, 2015; Straughan & Roberts, 1999). Perceived effectiveness is alternatively described as self-efficacy in the environmental consumer behavior literature (Han et al., 2017; Judge et al., 2007). An eco-friendly person is more sensitive to environmental problems, feel jointly responsible for the problems, recognize the value of environmental protection, and believe the efficacy of sustainable efforts/behaviors in resolving the problems (Judge et al., 2007; Roberts, 1996). A consumer pro-environmental consumption decision/behavior for a tourism product is indisputably influenced by his/her belief about the effectiveness of their conservation effort and activity (Han & Yoon, 2015).

**Table 2.** Articles in this special issue.

| Number | Article title |
| --- | --- |
| (1) | Green-induced tourist equity: The cross-level effect of regional environmental performance |
| (2) | Hotels' sustainability practices and guests' familiarity, attitudes and behaviors |
| (3) | The anchoring effect of aviation green tax for sustainable tourism, based on the nudge theory |
| (4) | Application of internal environmental locus of control to the context of eco-friendly drone food delivery services |
| (5) | The impact of the Middle East Respiratory Syndrome coronavirus on inbound tourism in South Korea toward sustainable tourism |
| (6) | Exploring preferences and sustainable attitudes of Airbnb green users in the review comments and ratings: A text mining approach |
| (7) | **An application of Delphi method and analytic hierarchy process in understanding hotel corporate social responsibility performance scale** |
| (8) | Comparing resident and tourist perceptions of an urban park: A latent profile analysis of perceived place value |
| (9) | Understanding backpacker sustainable behavior using the tri-component attitude model |

### Connectedness to nature

A customer who feels a strong connectedness to the natural environment is generally more active in engaging in an environmentally-sustainable consumption behavior (Dutcher et al., 2007; Mayer & Frantz, 2004). Indeed, academics asserted that patrons behave in an environmentally responsible manner more often when they feel that they are linked to the natural world (Gosling & Williams, 2010; Han & Hyun, 2017; Poon et al., 2015). Connectedness to nature refers to the extent to which a patron thinks he/she is a part of nature (Schultz, 2001). Undoubtedly, a person's willingness to preserve the natural environment enlarges when he/she has a feeling of connection to the nature environment increases (Mayer & Frantz, 2004; Poon et al., 2015; Schultz, 2001). Coherently, patrons' feeling of closeness to the natural world boosts their feeling of attachment to an eco-friendly product/service and leads to environmentally responsible behavior in a consumption situation (Han & Hyun, 2017; Mayer & Frantz, 2004).

### Green value

Academics have developed and applied a variety of conceptual frameworks to explicate customer environmentally responsible decision-making processes and behaviors (Halder et al., 2020; Monroe, 1991). Green value is often integrated in such frameworks as a focal constituent contributing to their explanatory power (Halder et al., 2020; Han et al., 2018). According to Zeithaml (1988), value is one's evaluation of the efficiency of a product based on his/her perception regarding what is gained and what is lost. Coherently, Han et al. (2018) described green value as patrons' cognitive appraisal of the efficacy of an eco-friendly product/service on the basis of their perception of what is obtained and what is sacrificed. When consumers believe that their gains (e.g. excellent green product performance, healthy consumption experience) are greater than their losses in a consumption situation (e.g. financial expenses, time/effort costs), their green value perception becomes high (Monroe, 1991; Oh, 2000; Zeithaml, 1988). In contrast, if they feel that what is given is greater than what is received, their perceived value decreases (Monroe, 1991; Oh, 2000).

## Value of the papers in the special issue of environmental sustainability and consumer behavior

The articles of this special issue deal with a variety of vital topics in tourism and hospitality and provide crucial insights to academics and practitioners about environmental sustainability and consumer behavior. The titles of the articles are displayed in Table 2. Given the ongoing environmental crisis and its influence on the global tourism and hospitality industry, understanding and promoting environmentally-sustainable consumer behaviors are more important than ever. This special section including a total of nine articles contributes to enriching the extant tourism and hospitality literature, broadening the range of methods, and expanding practical knowledge in the sector of environmentally-sustainable consumer behavior.

(1) While a large body of the literature has discussed various mechanisms of specific greening practices that could render powerful forces that influence tourists' attitudes, perceptions, and behaviors, they have largely relied on individual dispositions, ignoring the broader environmental context that gives shape to their cognitions and actual behaviors. As a result, sustainable tourism investigations thus far have focused primarily on tourists' in-destination perceptions and behaviors, without taking consideration of the environmental influence from the source market. Wong, Ruan, Cai, and Huang (2020) develop the concept of green equity to denote tourist perceptions of a destination's environmental programs, which should ultimately bring an array of benefits that could foster favorable value, brand, and relationship equities (i.e. tourist equity) of the travel locale. They further synthesize a multilevel green-induced tourist equity model to underscore a process in which a destination's greening practices ultimately influence tourist revisit propensity through tourist equity that is underpinned by value, brand, and relationship travel appeals. The model highlights a subliminal priming mechanism in which destination greening helps activate tourists' unconscious choices that are embedded within as a natural habitat based on the source market environment cues (i.e. regional ecological performance). Their study is a rather unique multilevel inquiry in the sustainable tourism literature, as it opens a new avenue of research on how unconscious beliefs that are embedded in a place of origin could ultimately play a role in guiding destination evaluations and choices.

(2) Sustainability is a complex process that involves social, economic and environmental dimensions. Although hotels are not among frontier industries adopting the sustainability, hospitality scholars studied a separate dimension of sustainability such as the environmental or social aspect (e.g. Namkung & Jang, 2017; Martinez & del Bosque, 2013). Olya, Altinay, Farmaki, Kenebayeva, and Gursoy (2020) extend the current knowledge of hotels' sustainability by inclusion of three dimension of sustainability (social, economic, and environmental) in a predictive model to understand the interplay between guest familiarity with the three dimensions of sustainability and their satisfaction and loyalty. Using data obtained from hotel guests in Kazakhstan and a multi-methods approach, they found that guests' familiarity increases their satisfaction but has no impact on their loyalty. A combination of social and economic dimensions led to loyalty. An alternative recipe for loyalty is a configuration of environmental and social dimensions. All three dimensions of sustainability are necessary conditions to achieve satisfaction and loyalty of hotel guests. Necessary conditions are those factors where outcome (e.g. loyalty) cannot be attained in their absence. Although familiarity is not necessary, its combination with the social dimension can boost guest satisfaction and loyalty.

(3) The aviation industry's significant impact on climate change is causing increasing concern on environment sustainability. The aviation industry comprises the largest share (40%) of total $CO_2$ emissions, followed by car transport (32%) (Scott et al., 2010). Kim and Hyun (2020) discuss the legitimacy of imposing aviation green taxes on passengers in countries that recognize the importance of the environment and climate change. It examines the anchoring effect of the nudge theory that affects the decisions of consumers who are voluntarily willing to pay environmental tax. The results show that the anchor had a significant effect. Their study contributes to the literature on sustainable tourism by investigating the factors that induce voluntary change in consumer behavior. The practical implications are to improve transparency and to minimize tax resistance by positively influencing converting travelers' perceptions of aviation green taxes. Further, in countries where an aviation green tax is not introduced, these results can serve as a

guide to encourage travelers' voluntary participation to pay aviation green taxes. As such, tourists' voluntary participation will contribute to sustainable tourism in the long-term.

(4) Drones in food delivery services make consuming much more environmentally friendly compared to the current delivery methods. Today, more consumers are interested in participating in sustainability through their food consumption choices. The group of people who possess a sense of internal environmental locus of control (INELOC) devote themselves to creating eco-friendly environment by conducting related activities. Hwang et al. (2020) apply the concept of multifaceted INELOC, namely, green consumers, activists, advocates, and recyclers, to the context of eco-friendly drone food delivery services and they endeavored to explain the intricate associations among INELOC, anticipated emotions, and intention to use eco-friendly drone food delivery services which have not been discovered to date. Their study provides rich theoretical originality and the findings suggest various ways to increase consumers' pro-environmental intentions.

(5) Sustainable tourism consumption can be impeded by epidemics, the impact of which may be greater than expected in modern societies where exchanges between countries are high (Jamal & Budke, 2020). The spread of epidemics is closely related to living in a globalized world and not only limits individual's sustainable tourism behaviors, but also results in a country's economic losses (Jung & Sung, 2017). Despite the declaration of the end of the MERS (Middle East Respiratory Syndrome) Coronavirus outbreak in South Korea in December 2015, the depressed domestic economy and tourism sector did not immediately recover. Little estimation has been conducted related to the impact of epidemics on inbound tourism demand. Choe et al. (2020) investigate the impact of MERS on the tourism demand in South Korea as well as estimated tourism data for the period affected by MERS using a variety of rigorous forecasting methods (e.g. Autoregressive Integrated Moving Average Model, Winters Exponential Smoothing Model, Stepwise Autoregressive Model). The authors' results show that MERS was statistically significant at $p < 0.05$ with a negative sign, indicating that MERS had a strong negative impact on Korea's inbound tourism industry in 2015. Employing three popular time-series models, they found the impact of MERS on inbound tourism and evidence of the total effect was estimated to be 1,968,765 tourists with a loss of 3.1 billion USD in receipts from June 2015 to September 2015.

(6) The irruption of the sharing economy in the Short-Term Rental (STR) tourist accommodation industry is linked to the development of technologies, social networks and the internet. The sharing economy breaks into the lodging sector modifying not only the tourist's consumption patterns, but also the conditions of sale, the source of information and the value that is generated. The STR platform Airbnb emerges as the benchmark for the business model of this new type of economy, whose rise was originally motivated largely by consumers' desire for greater sustainability in their consumer actions. Indeed, numerous studies have investigated this shift in consumer behaviour towards green choice intention, but only a few studies have attempted to explore different variables related to green consumption and green consumers using unstructured data and big data approaches. Even scarcer are articles that combine approaches to big data with unstructured data, such as online reviews from users of specific platforms. Thus, Serrano et al. (2020) approach this research with the aim of exploring the preferences and attitudes of Airbnb users characterized as green through an sentimental analysis and text mining of online comments through a set of data from online reviews published on Airbnb listings worldwide. The study suggests, on the one hand, that there is a positivity bias in the online reviews of Airbnb green users, and on the other, the results reveal that the latent aspect of 'sustainability' predominates in the online opinions of Airbnb green users among the six latent identified aspects.

(7) The fundamental indicators of hotel Corporate Social Responsibility (CSR) performance measurement with a standardized and composite CSR performance measurement index for the hotel industry have not been explored. Wong, Kim, Lee, and Elliot (2020) incorporate both Delphi and Analytic Hierarchy Process (AHP) methods through surveying with three stakeholder groups including academicians, hotel managers, and hotel customers. Their findings imply that

three traditional CSR domains (legal, ethical, and social/philanthropic) are primary contributors to generating CSR performance, followed by two new environmental domains (room and restaurant; other general areas), and financial/economic domains as secondary contributors. Since the high level of consistency in the responses from stakeholder groups is found, this study espouses the effectiveness of the scale as a valuable tool to measure hotel CSR performance. Interestingly, now that domain weighted scores do differ slightly by respondent characteristic, it means that the impacts of CSR are sensitive to respondent diversity. The results are very useful in manifesting indicators of CSR and domains in measuring CSR performance.

(8) In this era of the visitor economy, urban parks have become popular recreational spaces for tourists visiting cities and urban residents alike. City governments, therefore, have paid much attention to making urban parks more attractive for both residents and tourists. Previous studies have shown that residents and tourists are likely to assign different values to the same place, implying that sustainable management of urban parks can only be achieved by understanding the different perceptions of these two groups of park visitors. However, little research has empirically analyzed these differences. Song and Shim (2021) address this research gap in the context of sustainable management of urban parks by examining visitors at Gwanggyo Lake Park (GLP) in South Korea to compare tourist and resident perceptions. Employing Latent Profile Analysis (LPA), the authors' study identifies three valid sub-groups for visitors of GLP, namely Relationship Seekers, Activity Seekers, and Environment Seekers. Their study further reveals that residents most value the environmental aspects of urban parks while tourists assign the most value to urban parks as spaces for recreational activities. They also identify significant variations in demographic and behavioral characteristics of residents and tourists visiting the park. This study provides a valuable starting point in identifying the changing functions of urban parks in contemporary society as tourists increasingly occupy an essential and ever-present role as participants in urban life.

(9) Research on sustainable practices of backpackers has gained scholarly attention over the past three decades; however, a comprehensive model for understanding their sustainable behavior is yet to be developed in tourism literature. While backpacker tourism contributes to sustainability through a wide range of economic, cultural, social, and environmental activities associated with it, it also represents an important context where unsustainable behaviors occur. However, the potential underlying explanation for such attitudinal and behavioral patterns of (un)sustainability among these travelers are rarely examined as part of directing backpacker's behaviors towards responsible consumption in line with the United Nations Sustainable Development Goal 12. Agyeiwaah et al. (2021) address this inherent research gap by examining the relationships among backpacker motivation, perceived impacts of backpacking, backpacker (un)sustainable behaviors, and their satisfaction using the tri-component attitude model. Their results reveal two cognitive factors (i.e. backpacker motivations and perceived positive impacts of backpacking) predict backpacker sustainable behavior while one cognitive factor (i.e. backpacker motivations) predicts unsustainable behavior. Both cognitive (perceived positive impacts of backpacking) and behavioral (backpacker sustainable behavior) factors predict backpacker satisfaction. Backpacker motivation is key to understanding backpacker's (un)sustainable behavior.

## Conclusion

Customers contribute considerably to attaining the long-term environmental sustainability, looking beyond the short-term gains, when adopting environmentally-sustainable consumption patterns and practicing eco-friendly consumption behaviors (Dong et al., 2020; Kiatkawsin & Han, 2017; Laroche et al., 2001; Wang et al., 2018). Academics in tourism and hospitality can have a crucial role in reducing the environmental impacts of consumer behaviors by promoting customer behavioral changes to be environmentally responsible in diverse consumption situations.

The promotion of pro-environmental consumption activity is in general more efficient when it is planned and implemented in a systematic manner and evaluated in a continuous manner (Manosuthi et al., 2020; Steg & Vlek, 2009; Untaru et al., 2016). To do so, it is a fundamental requisite to clearly understand what environmentally-sustainable consumer behavior is, to know the theories that are effective for explicating the behavior, to recognize the factors triggering it, and to know contemporary tourism and hospitality studies dealing with the behavior, which were not wholly uncovered in the extant literature. The present study made a novel contribution by bridging the voids in the extant literature. This research successfully addresses these four issues. Specially, this study (1) provided the conceptualization of environmentally-sustainable consumer behavior, (2) presented the thorough review on social psychology and environmental psychology theories, (3) provided the discussion on key drivers of environmentally-sustainable consumer behavior, and (4) introduced the new tourism and hospitality studies about sustainability and consumer behavior.

The present research as an introduction to the special issue of environmental sustainability and consumer behavior is to advance the contribution of the tourism and hospitality research to comprehending environmentally responsible consumer behavior, supporting it, and achieving the environmental sustainability. The sustainable development goals can be described as the blueprint to attain a better and more sustainable future for humanity. Taking actins for the sustainable development goals is of necessity for all. An environmental dimension is one of the core aspects of such goals. The range of topics covered in this special section and the methodology applied would offer encouragement across the hospitality, tourism, consumer behavior, and environmental psychology fields to work in close cooperation in pursuit of common goals for promoting pro-environmental consumption, environmental sustainability, and environmentally sustainable development. The present special issue including this introductory paper and other important articles helps enable a platform for such close work across the fields and future studies on sustainable customer behavior, meeting the needs of the eco-conscious marketplace and the society and the goals of sustainable development.

## Acknowledgement

The reviewers' constructive comments and insightful suggestions help improve the quality of the papers in this special issue. The guest editor and all authors of the papers appreciate all reviewers' timely and thorough reviews of the manuscripts selected for this special issue.

## Disclosure statement

No potential conflict of interest was reported by the author.

## ORCID

*Heesup Han* http://orcid.org/0000-0001-6356-3001

# References

Afifah, N., & Asnan, A. (2015). The impact of corporate social responsibility, service experience and intercultural competence on customer company identification, customer satisfaction and customer loyalty (case study: PDAM Tirta Khatulistiwa Pontianak West Kalimantan). *Procedia - Social and Behavioral Sciences, 211*, 277–284. https://doi.org/10.1016/j.sbspro.2015.11.035

Agyeiwaah, E., Dayour, F., Otoo, F. E., & Goh, B. K. (2021). Understanding backpacker sustainable behavior using the tri-component attitude model. *Journal of Sustainable Tourism*. https://doi.org/10.1080/09669582.2021.1875476

Ailawadi, K. L., Neslin, S. A., & Gedenk, K. (2001). Pursuing the value-conscious consumer: Store brands versus national brand promotions. *Journal of Marketing, 65*(1), 71–89. https://doi.org/10.1509/jmkg.65.1.71.18132

Ajzen, I. (1991). The theory of planned behavior. *Organizational Behavior and Human Decision Processes, 50*(2), 179–211. https://doi.org/10.1016/0749-5978(91)90020-T

Ajzen, I. (2012). The theory of planned behavior. In P. A. M. Lange, A. W. Kruglanski, & E. T. Higgins (Eds.), *Handbook of theories of social psychology* (Vol. 1, pp. 438–459). Sage.

Ajzen, I., & Kruglanski, A. W. (2019). Reasoned action in the service of goal pursuit. *Psychological Review, 126*(5), 774–786. https://doi.org/10.1037/rev0000155

Assael, H. (1984). *Consumer behavior and marketing action*. Kent.

Bagozzi, R. P., & Dholakia, U. M. (2006). Antecedents and purchase consequences of customer participation in small group brand communities. *International Journal of Research in Marketing, 23*(1), 45–61. https://doi.org/10.1016/j.ijresmar.2006.01.005

Baloglu, S., & McCleary, K. W. (1999). A model of destination image formation. *Annals of Tourism Research, 26*(4), 868–897. https://doi.org/10.1016/S0160-7383(99)00030-4

Bamberg, S., & Möser, G. (2007). Twenty years after Hines, Hungerford, and Tomera: A new meta-analysis of psycho-social determinants of pro-environmental behavior. *Journal of Environmental Psychology, 27*(1), 14–25. https://doi.org/10.1016/j.jenvp.2006.12.002

Black, I. R., & Cherrier, H. (2010). Anti-consumption as part of living a sustainable lifestyle: Daily practices, contextual motivations and subjective values. *Journal of Consumer Behaviour, 9*(6), 437–453. https://doi.org/10.1002/cb.337

Boo, S., & Park, E. (2013). An examination of green intention: The effect of environmental knowledge and educational experiences on meeting planners' implementation of green meeting practices. *Journal of Sustainable Tourism, 21*(8), 1129–1147. https://doi.org/10.1080/09669582.2012.750327

Bridges, C. M., & Wilhelm, W. B. (2008). Going beyond green: The "why and how" of integrating sustainability into the marketing curriculum. *Journal of Marketing Education, 30*(1), 33–46. https://doi.org/10.1177/0273475307312196

Byers, R. (2008). *Green museums & green exhibits: Communicating sustainability through content and design* [Unpublished master's thesis]. University of Oregon.

Carrus, G., Passafaro, P., & Bonnes, M. (2008). Emotions, habits and rational choices in ecological behaviours: The case of recycling and use of public transportation. *Journal of Environmental Psychology, 28*(1), 51–62. https://doi.org/10.1016/j.jenvp.2007.09.003

Chan, E. S. W., Hon, A. H. Y., Chan, W., & Okumus, F. (2014). What drives employees' intentions to implement green practices in hotels? The role of knowledge, awareness, concern and ecological behavior. *International Journal of Hospitality Management, 40*, 20–28. https://doi.org/10.1016/j.ijhm.2014.03.001

Chan, R. Y. (2001). Determinants of Chinese consumers' green purchase behavior. *Psychology and Marketing, 18*(4), 389–413. https://doi.org/10.1002/mar.1013

Chen, F.-Y., Chang, Y.-H., & Lin, Y.-H. (2012). Customer perceptions of airline social responsibility and its effect on loyalty. *Journal of Air Transport Management, 20*, 49–51. https://doi.org/10.1016/j.jairtraman.2011.11.007

Choe, Y., Wang, J., & Song, H. (2020). The impact of the Middle East Respiratory Syndrome coronavirus on inbound tourism in South Korea toward sustainable tourism. *Journal of Sustainable Tourism*. https://doi.org/10.1080/09669582.2020.1797057

Choi, H., Jang, J., & Kandampully, J. (2015). Application of the extended VBN theory to understand consumers' decisions about green hotels. *International Journal of Hospitality Management, 51*, 87–95. https://doi.org/10.1016/j.ijhm.2015.08.004

Clayton, S., & Myers, G. (2009). *Conservation psychology: Understanding and promoting human care for nature*. Wiley-Blackwell.

De Groot, J. I. M., & Steg, L. (2009). Morality and prosocial behavior: The role of awareness, responsibility, and norms in the norm activation model. *The Journal of Social Psychology, 149*(4), 425–449. https://doi.org/10.3200/SOCP.149.4.425-449

De Groot, J. I. M., Steg, L., & Dicke, M. (2007). Morality and reducing car use: Testing the norm activation model of prosocial behavior. In F. Columbus (Ed.), *Transportation research trends*. NOVA Publishers.

Denley, T. J., Woosnam, K. M., Ribeiro, M. A., Boley, B. B., Hehir, C., & Abrams, J. (2020). Individuals' intentions to engage in last chance tourism: Applying the value-belief-norm model. *Journal of Sustainable Tourism, 28*(11), 1860–1881. https://doi.org/10.1080/09669582.2020.1762623

Dong, X., Liu, S., Li, H., Yang, Z., Liang, S., & Deng, N. (2020). Love of nature as a mediator between connectedness to nature and sustainable consumption behavior. *Journal of Cleaner Production, 242,* 1–12.

Dong, X., Yang, Z., & Li, Y. (2012). Influencing factors of urban residents' SCB. *Urban Problem, 10,* 55–61.

Dutcher, D. D., Finley, J. C., Luloff, A. E., & Johnson, J. B. (2007). Connectivity with nature as a measure of environmental values. *Environment and Behavior, 39*(4), 474–493. https://doi.org/10.1177/0013916506298794

Fedorikhin, A., Park, C. W., & Thomson, M. (2008). Beyond fit and attitude: The effect ofemotional attachment on consumer responses to brand extensions. *Journal of Consumer Psychology, 18*(4), 281–291. https://doi.org/10.1016/j.jcps.2008.09.006

Fishbein, M., & Ajzen, I. (2010). *Predicting and changing behavior: The reasoned action approach.* Psychology Press.

Garay, L., Font, X., & Corrons, A. (2019). Sustainability-oriented innovation in tourism: An analysis based on the decomposed theory of planned behavior. *Journal of Travel Research, 58*(4), 622–636. https://doi.org/10.1177/0047287518771215

Garvey, A. M., & Bolton, L. E. (2017). Eco-product choice cuts both ways: How proenvironmental licensing versus reinforcement is contingent on environmental consciousness. *Journal of Public Policy & Marketing, 36*(2), 284–298. https://doi.org/10.1509/jppm.16.096

Gosling, E., & Williams, K. J. H. (2010). Connectedness to nature, place attachment and conservation behavior: Testing connectedness theory among farmers. *Journal of Environmental Psychology, 30*(3), 298–304. https://doi.org/10.1016/j.jenvp.2010.01.005

Halder, P., Hansen, E. N., Kangas, J., & Laukkanen, T. (2020). How national culture and ethics matter in consumers' green consumption values. *Journal of Cleaner Production, 265,* 121754. https://doi.org/10.1016/j.jclepro.2020.121754

Han, H. (2014). The norm activation model and theory-broadening: Individuals' decision-making on environmentally-responsible convention attendance. *Journal of Environmental Psychology, 40,* 462–471. https://doi.org/10.1016/j.jenvp.2014.10.006

Han, H. (2015). Travelers' pro-environmental behavior in a green lodging context: Converging value-belief-norm theory and the theory of planned behavior. *Tourism Management, 47,* 164–177. https://doi.org/10.1016/j.tourman.2014.09.014

Han, H. (2020). Theory of green purchase behavior (TGPB): A new theory for sustainable consumption of green hotel and green restaurant products. *Business Strategy and the Environment, 29*(6), 2815–2828. https://doi.org/10.1002/bse.2545

Han, H., & Hwang, J. (2014). Investigation of the volitional, non-volitional, emotional, motivational, and automatic processes in determining golfers' intention: Impact of screen golf. *International Journal of Contemporary Hospitality Management, 26*(7), 1118–1135. https://doi.org/10.1108/IJCHM-04-2013-0163

Han, H., & Hwang, J. (2017). What motivates delegates' conservation behaviors while attending a convention? *Journal of Travel & Tourism Marketing, 34*(1), 82–98. https://doi.org/10.1080/10548408.2015.1130111

Han, H., Hwang, J., & Lee, S. (2017). Cognitive, affective, normative, and moral triggers of sustainable intentions among convention-goers. *Journal of Environmental Psychology, 51,* 1–13. https://doi.org/10.1016/j.jenvp.2017.03.003

Han, H., & Hyun, S. (2017). Fostering customers' pro-environmental behavior at a museum. *Journal of Sustainable Tourism, 25*(9), 1240–1256. https://doi.org/10.1080/09669582.2016.1259318

Han, H., & Hyun, S. (2018). What influences water conservation and towel reuse practices of hotel guests? *Tourism Management, 64,* 87–97. https://doi.org/10.1016/j.tourman.2017.08.005

Han, H., & Yoon, H. (2015). Hotel customers' environmentally responsible behavioral intention: Impact of key constructs on decision in green consumerism. *International Journal of Hospitality Management, 45,* 22–33. https://doi.org/10.1016/j.ijhm.2014.11.004

Han, H., Yu, J., Jeong, E., & Kim, W. (2018). Environmentally responsible museums' strategies to elicit visitors' green intention. *Social Behavior and Personality: An International Journal, 46*(11), 1881–1894. https://doi.org/10.2224/sbp.7310

Han, H., Yu, J., & Kim, W. (2019). Environmental corporate social responsibility and the strategy to boost the airline's image and customer loyalty intentions. *Journal of Travel & Tourism Marketing, 36*(3), 371–383. https://doi.org/10.1080/10548408.2018.1557580

Harth, N. S., Leach, C. W., & Kessler, T. (2013). Guilt, anger, and pride about in-group environmental behavior: Different emotions predict distinct intentions. *Journal of Environmental Psychology, 34,* 18–26. https://doi.org/10.1016/j.jenvp.2012.12.005

Hopkins, D. (2020). Sustainable mobility at the interface of transport and tourism. *Journal of Sustainable Tourism, 28*(2), 129–143. https://doi.org/10.1080/09669582.2019.1691800

Hwang, J., Lee, J., Kim, J., & Sial, M. S. (2020). Application of internal environmental locus of control to the context of eco-friendly drone food delivery services. *Journal of Sustainable Tourism.* https://doi.org/10.1080/09669582.2020.1775237

Jacobson, R. P., Mortensen, C. R., & Cialdini, R. B. (2011). Bodies obliged and unbound: Differentiated response tendencies for injunctive and descriptive social norms. *Journal of Personality and Social Psychology, 100*(3), 433–448. https://doi.org/10.1037/a0021470

Jamal, T., & Budke, C. (2020). Tourism in a world with pandemics: Local–global responsibility and action. *Journal of Tourism Futures, 6*(2), 181–188. https://doi.org/10.1108/JTF-02-2020-0014

Jang, Y. J., Kim, W. G., & Lee, H. Y. (2015). Coffee shop consumers' emotional attachment and loyalty to green stores: The moderating role of green consciousness. *International Journal of Hospitality Management, 44*, 146–156. https://doi.org/10.1016/j.ijhm.2014.10.001

Jeong, E., Jang, S., Day, J., & Ha, J. (2014). The impact of eco-friendly practices on green image and customer attitudes: An investigation in a café setting. *International Journal of Hospitality Management, 41*, 10–20. https://doi.org/10.1016/j.ijhm.2014.03.002

Joshi, Y., & Rahman, Z. (2015). Factors affecting green purchase behaviour and future research directions. *International Strategic Management Review, 3*(1-2), 128–143. https://doi.org/10.1016/j.ism.2015.04.001

Judge, T. A., Jackson, C. L., Shaw, J. C., Scott, B. A., & Rich, B. L. (2007). Self-efficacy and work-related performance: The integral role of individual differences. *The Journal of Applied Psychology, 92*(1), 107–127.

Jung, E., & Sung, H. (2017). The influence of the Middle East Respiratory syndrome outbreak on online and offline markets for retail sales. *Sustainability, 9*(3), 411–412. https://doi.org/10.3390/su9030411

Kaiser, F. G., Ranney, M., Hartig, T., & Bowler, P. (1999). Ecological behavior, environmental attitude, and feelings of responsibility for the environment. *European Psychologist, 4*(2), 59–74. https://doi.org/10.1027//1016-9040.4.2.59

Kaiser, F. G., Wölfing, S., & Fuhrer, U. (1999). Environmental attitude and ecological behavior. *Journal of Environmental Psychology, 19*(1), 1–19. https://doi.org/10.1006/jevp.1998.0107

Kiatkawsin, K., & Han, H. (2017). Young travelers' intention to behave pro-environmentally: Merging the value-belief-norm theory and the expectancy theory. *Tourism Management, 59*, 76–88. https://doi.org/10.1016/j.tourman.2016.06.018

Kim, H., & Hyun, S. (2020). The anchoring effect of aviation green tax for sustainable tourism, based on the nudge theory. *Journal of Sustainable Tourism*. https://doi.org/10.1080/09669582.2020.1820017

Kim, M. J., Lee, C., Petrick, J. F., & Kim, Y. S. (2020). The influence of perceived risk and intervention on international tourists' behavior during the Hong Kong protest: Application of an extended model of goal-directed behavior. *Journal of Hospitality and Tourism Management, 45*, 622–632. https://doi.org/10.1016/j.jhtm.2020.11.003

Klöckner, C. A. (2013). A comprehensive model of the psychology of environmental behavior – A meta-analysis. *Global Environmental Change, 23*(5), 1028–1038. https://doi.org/10.1016/j.gloenvcha.2013.05.014

Klöckner, C. A., & Matthies, E. (2004). How habits interfere with norm directed behavior: A normative decision-making model for travel mode choice. *Journal of Environmental Psychology, 24*(3), 319–327. https://doi.org/10.1016/j.jenvp.2004.08.004

Kotler, P., & Gertner, D. (2002). Country as brand, product, and beyond: A place marketing and brand management perspective. *Journal of Brand Management, 9*(4), 249–261. https://doi.org/10.1057/palgrave.bm.2540076

Kotler, P., Haider, D. H., & Rein, I. (1993). *Marketing places: Attracting investment, industry, and tourism to cities, states, and nations*. The Free Press.

Krajhanzl, J. (2010). Environmental and pro-environmental behavior. In E. Řehulka (Ed.), *School and health 21* (pp. 251–274). Masarykova Univerzita, MSD.

Laroche, M., Bergeron, J., & Barbaro-Forleo, G. (2001). Targeting consumers who are willing to pay more for environmentally friendly products. *Journal of Consumer Marketing, 18*(6), 503–520. https://doi.org/10.1108/EUM0000000006155

Laroche, M., Toffoli, R., Kim, C., & Muller, T. E. (1996). The influence of culture on pro-environmental knowledge, attitudes, and behavior: A Canadian perspective. In K. P. Corfman & J. G. Lynch, Jr. (Eds.), *Advances in consumer research* (Vol. 23, pp. 196–202). Association for Consumer Research.

Leary, R. B., Vann, R. J., Mittelstaedt, J. D., Murphy, P. E., & Sherry, J. F., Jr. (2014). Changing the marketplace one behavior at a time: Perceived marketplace influence and sustainable consumption. *Journal of Business Research, 67*(9), 1953–1958. https://doi.org/10.1016/j.jbusres.2013.11.004

Lee, E. M., Park, S., & Lee, H. J. (2013). Employee perception of CSR activities: Its antecedents and consequences. *Journal of Business Research, 66*(10), 1716–1724. https://doi.org/10.1016/j.jbusres.2012.11.008

Lee, J., Hsu, L., Han, H., & Kim, Y. (2010). Understanding how consumers view green hotels: How a hotel's green image can influence behavioural intentions. *Journal of Sustainable Tourism, 18*(7), 901–914. https://doi.org/10.1080/09669581003777747

Lerner, J. S., & Keltner, D. (2001). Fear, anger, and risk. *Journal of Personality and Social Psychology, 81*(1), 146–159. https://doi.org/10.1037//0022-3514.81.1.146

Manosuthi, N., Lee, J., & Han, H. (2020). Predicting the revisit intention of volunteer tourists using the merged model between the theory of planned behavior and norm activation model. *Journal of Travel & Tourism Marketing, 37*(4), 510–532. https://doi.org/10.1080/10548408.2020.1784364

Martinez, P., & del Bosque, I. R. D. (2013). CSR and consumer loyalty: The roles of trust, consumer identification with the company and satisfaction. *International Journal of Hospitality Management, 35*, 89–99.

Matthies, E., Selge, S., & Klöckner, C. A. (2012). The role of parental behaviour for the development of behaviour specific environmental norms – The example of recycling and re-use behaviour. *Journal of Environmental Psychology, 32*(3), 277–284. https://doi.org/10.1016/j.jenvp.2012.04.003

Mayer, F. S., & Frantz, C. M. P. (2004). The connectedness to nature scale: A measure of individuals' feeling in community with nature. *Journal of Environmental Psychology, 24*(4), 503–515. https://doi.org/10.1016/j.jenvp.2004.10.001

Megeirhi, H. A., Woosnam, K. M., Ribeiro, M. A., Ramkissoon, H., & Denley, T. J. (2020). Employing a value-belief-norm framework to gauge Carthage residents' intentions to support sustainable cultural heritage tourism. *Journal of Sustainable Tourism, 28*(9), 1351–1370. https://doi.org/10.1080/09669582.2020.1738444

Meng, B., Chua, B., Ryu, B., & Han, H. (2020). Volunteer tourism (VT) traveler behavior: Merging norm activation model and theory of planned behavior. *Journal of Sustainable Tourism, 28*(12), 1947–1969. https://doi.org/10.1080/09669582.2020.1778010

Minton, E. A., Spielmann, N., Kahle, L. R., & Kim, C. (2018). The subjective norms of sustainable consumption: A cross-cultural exploration. *Journal of Business Research, 82*, 400–408. https://doi.org/10.1016/j.jbusres.2016.12.031

Monroe, K. B. (1991). *Pricing: Making profitable decisions*. McGraw-Hill.

Montgomery, C., & Stone, G. (2009). Revisiting consumer environmental responsibility: A five nation cross-cultural analysis and comparison of consumer ecological opinions and behaviors. *International Journal of Management and Marketing Research, 2*(1), 35–58.

Moon, S. (2021). Investigating beliefs, attitudes, and intentions regarding green restaurant patronage: An application of the extended theory of planned behavior with moderating effects of gender and age. *International Journal of Hospitality Management, 92*, 102727. https://doi.org/10.1016/j.ijhm.2020.102727

Namkung, Y., & Jang, S. (2017). Are consumers willing to pay more for green practices at restaurants? *Journal of Hospitality & Tourism Research, 41*(3), 329–356. https://doi.org/10.1177/1096348014525632

Oh, H. (2000). The effect of brand class, brand awareness, and price on customer value and behavioral intentions. *Journal of Hospitality & Tourism Research, 24*(2), 136–162. https://doi.org/10.1177/109634800002400202

Olya, H., Altinay, L., Farmaki, A., Kenebayeva, A., & Gursoy, D. (2020). Hotels' sustainability practices and guests' familiarity, attitudes and behaviours. *Journal of Sustainable Tourism*. https://doi.org/10.1080/09669582.2020.1775622

Onwezen, M. C., Antonides, G., & Bartels, J. (2013). The norm activation model: An exploration of the functions of anticipated pride and guilt in pro-environmental behavior. *Journal of Economic Psychology, 39*, 141–153. https://doi.org/10.1016/j.joep.2013.07.005

Onwezen, M. C., Bartels, J., & Antonides, G. (2014). Environmentally friendly consumer choices: Cultural differences in the self-regulatory function of anticipated pride and guilt. *Journal of Environmental Psychology, 40*, 239–248. https://doi.org/10.1016/j.jenvp.2014.07.003

Paco, A., & Rapose, M. (2009). Green segmentation: An application to the Portuguese consumer market. *Marketing Intelligence and Planning, 27*(3), 364–379.

Paiano, A., Crovella, T., & Lagioia, G. (2020). Managing sustainable practices in cruise tourism: The assessment of carbon footprint and waste of water and beverage packaging. *Tourism Management, 77*, 104016. https://doi.org/10.1016/j.tourman.2019.104016

Park, E., Lee, S., Lee, C. K., Kim, J. S., & Kim, N. J. (2018). An integrated model of travelers' pro-environmental decision-making process: The role of the New Environmental Paradigm. *Asia Pacific Journal of Tourism Research, 23*(10), 935–948. https://doi.org/10.1080/10941665.2018.1513051

Pedeliento, G., Andreini, D., Bergamaschi, M., & Salo, J. (2016). Brand and product attachment in an industrial context: The effects on brand loyalty. *Industrial Marketing Management, 53*, 194–206. https://doi.org/10.1016/j.indmarman.2015.06.007

Perugini, M., & Bagozzi, R. P. (2001). The role of desires and anticipated emotions in goal-directed behaviors: Broadening and deepening the theory of planned behavior. *British Journal of Social Psychology, 40*(1), 79–98. https://doi.org/10.1348/014466601164704

Perugini, M., & Bagozzi, R. P. (2004). The distinction between desires and intentions. *European Journal of Social Psychology, 34*(1), 69–84. https://doi.org/10.1002/ejsp.186

Poon, K., Teng, F., Chow, J. T., & Chen, Z. (2015). Desiring to connect to nature: The effect of ostracism on ecological behavior. *Journal of Environmental Psychology, 42*, 116–122. https://doi.org/10.1016/j.jenvp.2015.03.003

Ouellette, J. A., & Wood, W. (1998). Habit and intention in everyday life: The multiple processes by which past behavior predicts future behavior. *Psychological Bulletin, 124*(1), 54–74. https://doi.org/10.1037/0033-2909.124.1.54

Ramkissoon, H. (2020). Perceived social impacts of tourism and quality-of-life: A new conceptual model. *Journal of Sustainable Tourism*. https://doi.org/10.1080/09669582.2020.1858091.

Ramkissoon, H., Mavondo, F., & Sowamber, V. (2020). Corporate social responsibility at LUX* resorts and hotels: Satisfaction and loyalty implications for employee and customer social responsibility. *Sustainability, 12*(22), 9745. https://doi.org/10.3390/su12229745

Ramkissoon, H., Mavondo, F., & Uysal, M. (2018). Social involvement and park citizenship as moderators for quality-of-life in a national park. *Journal of Sustainable Tourism, 26*(3), 341–361. https://doi.org/10.1080/09669582.2017.1354866

Ramkissoon, H., Smith, L. D. G., & Weiler, B. (2013). Relationships between place attachment, place satisfaction and pro-environmental behaviour in an Australian national park. *Journal of Sustainable Tourism, 21*(3), 434–457. https://doi.org/10.1080/09669582.2012.708042

Roberts, J. A. (1996). Green consumers in the 1990: Profile and implications for advertising. *Journal of Business Research, 36*(3), 217–231. https://doi.org/10.1016/0148-2963(95)00150-6

Rosenthal, S., & Ho, K. L. (2020). Minding other people's business: Community attachment and anticipated negative emotion in an extended norm activation model. *Journal of Environmental Psychology, 69*, 101439. https://doi.org/10.1016/j.jenvp.2020.101439

Schultz, P. W. (2001). The structure of environmental concern: Concern for self, other people, and the biosphere. *Journal of Environmental Psychology, 21*(4), 327–339.

Schwartz, S. H. (1977). Normative influences on altruism. *Advances in Experimental Social Psychology, 10*, 221–279.

Schwartz, S. H. (1992). Universals in the content and structure of values: Theoretical advances and empirical tests in 20 countries. *Advances in Experimental Social Psychology, 25*(1), 1–65.

Schwartz, S. H., & Howard, J. A. (1981). A normative decision-making model of altruism. In J. P. Rushton (Ed.), *Altruism and helping behaviour: Social, personality and developmental perspectives* (pp. 189–211). Erlbaum.

Scott, D., Peeters, P., & Gössling, S. (2010). Can tourism deliver its "aspirational" greenhouse gas emission reduction targets? *Journal of Sustainable Tourism, 18*(3), 393–408. https://doi.org/10.1080/09669581003653542

Serrano, L., Ariza-Montes, A., Nader, M., Sianes, A., & Law, R. (2020). Exploring preferences and sustainable attitudes of Airbnb green users in the review comments and ratings: A text mining approach. *Journal of Sustainable Tourism.* https://doi.org/10.1080/09669582.2020.1838529

Shi, H., Fan, J., & Zhao, D. (2017). Predicting household PM2.5-reduction behavior in Chinese urban areas: An integrative model of theory of planned behavior and norm activation theory. *Journal of Cleaner Production, 145*, 64–73. https://doi.org/10.1016/j.jclepro.2016.12.169

Singh, A., & Verma, P. (2017). Factors influencing Indian consumers' actual buying behavior towards organic food products. *Journal of Cleaner Production, 167*, 473–483. https://doi.org/10.1016/j.jclepro.2017.08.106

Smith, J. R., Louis, W. R., Terry, D. J., Greenaway, K. H., Clarke, M. R., & Cheng, X. (2012). Congruent or conflicted? The impact of injunctive and descriptive norms on environmental intentions. *Journal of Environmental Psychology, 32*(4), 353–361. https://doi.org/10.1016/j.jenvp.2012.06.001

Song, H., & Shim, C. (2021). Comparing resident and tourist perception of an urban lake park: A latent profile analysis. *Journal of Sustainable Tourism.* https://doi.org/10.1080/09669582.2021.1872586

Steg, L., & De Groot, J. (2010). Explaining prosocial intentions: Testing causal relationships in the norm activation model. *British Journal of Social Psychology, 49*(4), 725–743. https://doi.org/10.1348/014466609X477745

Steg, L., & Vlek, C. (2009). Encouraging pro-environmental behaviour: An integrative review and research agenda. *Journal of Environmental Psychology, 29*(3), 309–317. [Database] https://doi.org/10.1016/j.jenvp.2008.10.004

Stern, P. C. (2000). Toward a coherent theory of environmentally significant behavior. *Journal of Social Issues, 56*(3), 407–424. https://doi.org/10.1111/0022-4537.00175

Stern, P. C., Dietz, T., Abel, T., Guagnano, G. A., & Kalof, L. (1999). A value-belief-norm theory of support for social movements: The case of environmentalism. *Research in Human Ecology, 6*(2), 81–97.

Straughan, R. D., & Roberts, J. A. (1999). Environmental segmentation alternatives: A look at green consumer behavior in the new millennium. *Journal of Consumer Marketing, 16*(6), 558–575. https://doi.org/10.1108/07363769910297506

Thomson, J. A., Shaw, D., & Shiu, E. M. K. (2008). An application of the extended model of goal directed behavior within smoking cessation: An examination of the role of emotions. *European Advances in Consumer Research, 8*, 73–79.

Thomson, M., MacInnis, D. J., & Park, W. (2005). The ties that bind: Measuring the strength of consumers' emotional attachments to brands. *Journal of Consumer Psychology, 15*(1), 77–91. https://doi.org/10.1207/s15327663jcp1501_10

Trang, H., Lee, J., & Han, H. (2019). How do green attributes elicit guest pro-environmental behaviors? The case of green hotels in Vietnam. *Journal of Travel & Tourism Marketing, 36*(1), 14–28. https://doi.org/10.1080/10548408.2018.1486782

Untaru, E., Ispas, A., Candrea, A. N., Luca, M., & Epuran, G. (2016). Predictors of individuals' intention to conserve water in a lodging context: The application of an extended theory of reasoned action. *International Journal of Hospitality Management, 59*, 50–59. https://doi.org/10.1016/j.ijhm.2016.09.001

Wang, J., Wang, S., Xue, H., Wang, Y., & Li, J. (2018). Green image and consumers' word-of-mouth intention in the green hotel industry: The moderating effect of Millennials. *Journal of Cleaner Production, 181*, 426–436. https://doi.org/10.1016/j.jclepro.2018.01.250

Wang, S., Wang, J., Li, J., & Yang, F. (2020). Do motivations contribute to local residents' engagement in pro-environmental behaviors? Resident-destination relationship and pro-environmental climate perspective. *Journal of Sustainable Tourism, 28*(6), 834–852. https://doi.org/10.1080/09669582.2019.1707215

Werner, K., Griese, K., & Bosse, C. (2020). The role of slow events for sustainable destination development: A conceptual and empirical review. *Journal of Sustainable Tourism.* https://doi.org/10.1080/09669582.2020.1800021

Wong, A. K. F., Kim, S., Lee, S., & Elliot, S. (2020). An application of Delphi method and analytic hierarchy process in understanding hotel corporate social responsibility performance scale. *Journal of Sustainable Tourism.* https://doi.org/10.1080/09669582.2020.1773835

Wong, I., Ruan, W., Cai, X., & Huang, G. (2020). Green-induced tourist equity: The cross-level effect of regional environmental performance. *Journal of Sustainable Tourism*. https://doi.org/10.1080/09669582.2020.1851700

Wu, H., Ai, C., & Cheng, C. (2016). Synthesizing the effects of green experiential quality, green equity, green image and green experiential satisfaction on green switching intention. *International Journal of Contemporary Hospitality Management*, *28*(9), 2080–2107. https://doi.org/10.1108/IJCHM-03-2015-0163

Wu, J., Font, X., & Liu, J. (2020). Tourists' pro-environmental behaviors: Moral obligation or disengagement? *Journal of Travel Research*. https://doi.org/10.1177/0047287520910787

Xu, F., Huang, L., & Whitmarsh, L. (2020). Home and away: Cross-contextual consistency in tourists' pro-environmental behavior. *Journal of Sustainable Tourism*, *28*(10), 1443–1459. https://doi.org/10.1080/09669582.2020.1741596

Young, H., Yin, R., Kim, J.-H., & Li, J. (2020). Examining traditional restaurant diners' intention: An application of the VBN theory. *International Journal of Hospitality Management*, *85*, 1–12.

Yuksel, A., Yuksel, F., & Bilim, Y. (2010). Destination attachment: Effects on consumer satisfaction and cognitive, affective and conative loyalty. *Tourism Management*, *31*(2), 274–284. https://doi.org/10.1016/j.tourman.2009.03.007

Zeithaml, V. A. (1988). Consumer perceptions of price, quality, and value: A means-end model and synthesis of evidence. *Journal of Marketing*, *52*(3), 2–22. https://doi.org/10.1177/002224298805200302

Zhao, H. H., Gao, Q., Wu, Y. P., Wang, Y., & Zhu, X. D. (2014). What affects green consumer behavior in China? A case study from Qingdao. *Journal of Cleaner Production*, *63*(2), 143–151. https://doi.org/10.1016/j.jclepro.2013.05.021

# Green-Induced tourist equity: the cross-level effect of regional environmental performance

IpKin Anthony Wong(iD), Wenjia Jasmine Ruan, Xiaomei Cai and GuoQiong Ivanka Huang (iD)

**ABSTRACT**

This study extends the customer equity paradigm to propose a green induced tourist equity model in the tourism context. Drawing on the attitude literature pertaining to subliminal persuasion and priming, we argue that the environment performance of the source market plays a pivotal role in inadvertently stimulate tourists to grave from a greener place to visit and hence, greater revisit desire for the destination. We develop a series of multilevel models to test individual-level direct effects leading from green equity to destination loyalty through value equity, brand equity, and relationship equity germane to a destination. We then tested the cross-level moderating effect emanating from source-market's environmental performance to the proposed direct relationships. This study contributes to the literature by bridging the micro and macro perspective of sustainable tourism. It integrates both individual-level and source market regional-level factors into a multilevel framework to better assess how source-market environmental situations could unconsciously influence tourists greening perceptions and behaviors during their excursions. Applications of subliminal persuasion and priming in the current study also broaden the theoretical understanding of the role of attitude at an aggregate level, and they also help strengthen the linkage between regional environmental performance and tourists predispositions.

## Introduction

Greening has often been accentuated to bring a constellation of positive impacts to tourists (Han et al., 2018; Xu et al., 2020), service providers (Collins & Cooper, 2017; Jones et al., 2016; Rosenbaum & Wong, 2015), and hosting destinations (Corte & Aria, 2016; Fok & Law, 2018). For example, implementation of greening and energy-efficient practices has beneficial impacts on tourism development, such as enhancing the comfort of tourist experience, improving the aesthetic value of destinations, and more (Cingoski & Petrevska, 2018). There is no doubt that ecological initiatives could galvanize tourists to react favorably to greening providers and destinations. Greening also facilitates lower operation costs and helps reconcile "risks spoiling the equilibrium of local ecosystems, causing the degradation of the environmental beauty in the long term" (Satta et al., 2019, p. 267). Yet, there is still limited understanding about how destinations' greening programs could ultimately foster favorable tourist responses to functional, symbolic, and relational benefits that comprise the three drivers of customer equity: value, brand, and relationship equities (Lee & Park, 2019; Vogel et al., 2008).

While a large body of the literature has discussed various mechanisms of specific greening practices that could render powerful forces that influence tourists' attitudes, perceptions, and behaviors (Han et al., 2018; Wolf et al., 2017; Wong et al., 2015), these prior studies have largely relied on individual dispositions, ignoring the broader environmental context that gives shape to their cognitions and actual behaviors. As such, with a few exceptions (Xu et al., 2020), sustainable tourism investigations thus far have focused merely on tourists' in-destination perceptions and behaviors, without taking consideration of the environmental influence from the source market. This dearth of sustainable research has prevented the field from using the systems thinking/ theory tradition (Senge, 1994) to embark on synthesizing the linkage between source and destination markets.

This study addresses the above knowledge gaps as follows. First, it improvises the concept of green equity to denote tourist perceptions of a destination's environmental programs, which should ultimately bring an array of benefits that could foster favorable value, brand, and relationship equities (i.e., tourist equity) of the travel locale. Second, we synthesized a multilevel green-induced tourist equity model to underscore a process in which a destination's greening practices ultimately influence tourist revisit propensity through tourist equity that is underpinned by value, brand, and relationship travel appeals. It also highlights a subliminal priming mechanism in which destination greening helps activate tourists' unconscious choices that are embedded within as a natural habitat based on the source market environment cues (i.e., regional ecological performance) (Cooper & Cooper, 2002; Karremans et al., 2006).

In essence, this study investigates and answers the following questions: (1) What is the role of a destination's greening program (i.e., green equity) in tourist equity? (2) How do a source market's environmental conditions subsequently influence tourists' greening dispositions while they sojourn in a foreign place? The key contributions of this study lie in articulation of the green-induced tourist equity model as well as theorizing and examining the linkage between source and host destinations pertinent to sustainable development. This study is a rather unique multilevel inquiry in the sustainable tourism literature, as it opens a new avenue of research on how unconscious beliefs that are embedded in a place of origin could ultimately play a role in guiding destination evaluations and choices.

## Theoretical background

### Green initiatives in the tourism context

Green marketing is a marketing strategy to design, promote, price, and distribute products that are not harmful to the environment (Rahman et al., 2015; Rosenbaum & Wong, 2015). Growing environmental concerns have impelled the tourism industry to embark on more sustainable tourism practices (Satta et al., 2019; Wolf et al., 2017). In the context of tourism, green marketing has been implemented by reducing the operating costs of tourism facilities (through measures such as water saving and energy saving) as well as to create value for tourists (Knezevic Cvelbar et al., 2020). The main purpose of implementing greening has been to respond to ecological concern through minimizing environmental harm from tourism activities and ensuring the quality of the natural environment is diminished or jeopardized (Han et al., 2018). The tourism industry, including hotels, has been a pioneer in implementing green initiatives. These green endeavors can be regarded as hotels' intangible assets, which can improve the corporate image, gain a competitive advantage, and enhance market attractiveness (Gupta et al., 2019; Tanford & Malek, 2015).

In recent years, there is a growing attention towards sustainability and ecological concerns from the tourist's point of view. Thus, the tourism industry has been gradually responding to environmental concern of tourists and societies with respect to tourism facility operations and management through green marketing (Erdogan & Baris, 2007; Mensah, 2006; Merli et al., 2019). For example, there is a sustainable certification initiative called "Green Tourism Business Scheme" in Western England, which has played a crucial role in helping businesses raise environmental awareness, protect sensitive environment areas, reduce water consumption and improve waste management (Jarvis et al., 2010). Green initiatives have become a non-negligible factor in attracting tourists (Han & Yoon, 2015). Several studies have reported a positive relationship between an individual's perception of green initiatives and behavioral responses (Gao et al., 2016; Han et al., 2018), such as satisfaction (Gao & Mattila, 2014; Martínez García de Leaniz et al., 2018; Xu & Gursoy, 2015), loyalty, and willingness to pay a premium price (Teng et al., 2012; Wong et al., 2015). Considering the role of green initiatives in marketing, their adoption can be seen as a rational strategy to improve tourists' appraisals of a destination and to instill confidence in them as well as favorable responses (Ham & Han, 2013; Yusof et al., 2016).

### Conceptualization of the green-induced tourist equity model

The physical environment is a rich repertoire of stimuli that can influence people's attitudes and behaviors consciously and unconsciously (Albarracin & Vargas, 2010). This premise has widely been acknowledged in various disciplines such as environmental psychology (Mehrabian & Russell, 1974; Palanica et al., 2019), management/marketing (Bitner, 1992; Menguc et al., 2016), and tourism and hospitality (Ji et al., 2018; Wong, 2017) to conjecture how the environment could ultimately shape actors' course of actions. The essence of this line of inquiry lies in following the systems thinking/theory paradigm (Senge, 1994) to investigate how human actors are predisposed to certain confined situations (i.e., person-in-situation). These situations could ultimately impel actors to follow suit with certain normative forces (Hirst et al., 2011), such as being more ecologically cautious and in pursuit of more eco-friendly excursion experiences (e.g., consuming less and recycling). The present study draws on this tradition to highlight how the individual tourist's green perceptions of a destination and the impacts of those perceptions on other destination attributes (e.g., tourist equity drivers and destination loyalty) are conditioned based on the source market's environmental forces that ultimately give shape to the tourist's predispositions.

We further draw on the attitude literature pertaining to subliminal priming and motivation (Cooper & Cooper, 2002; Wenke et al., 2010) to argue that the environment performance (i.e., energy consumption and carbon emission) of the source market plays a pivotal role in inadvertently stimulating tourists to crave a greener destination to visit and hence, greater revisit desire for the place. In particular, subliminal persuasion and priming point to a phenomenon in which stimuli are undetected by the receiver yet render as an environmental influence on one's attitudes, choices, and future behaviors inadvertently (Cooper & Cooper, 2002; Karremans et al., 2006). Such a view provides the guiding light on why consumers and tourists alike are affected by the physical environment unconsciously. More importantly, Albarracin and Vargas (2010) note that "subliminal persuasion seems to be most effective when a related motive is already aroused" (p. 402). In other words, our proposed multilevel green-induced tourist equity model argues that perceived greening of a destination (i.e., green equity) evokes a higher level of tourist equity with elevated value, brand, and relational benefits that could ultimately improve a tourist's propensity to revisit the place (i.e., destination loyalty intention) (see Figure 1). Yet, this chain of positive relationships is amplified for tourists whose source markets embark on greening endeavors to improve energy efficiency and reduce carbon emissions.

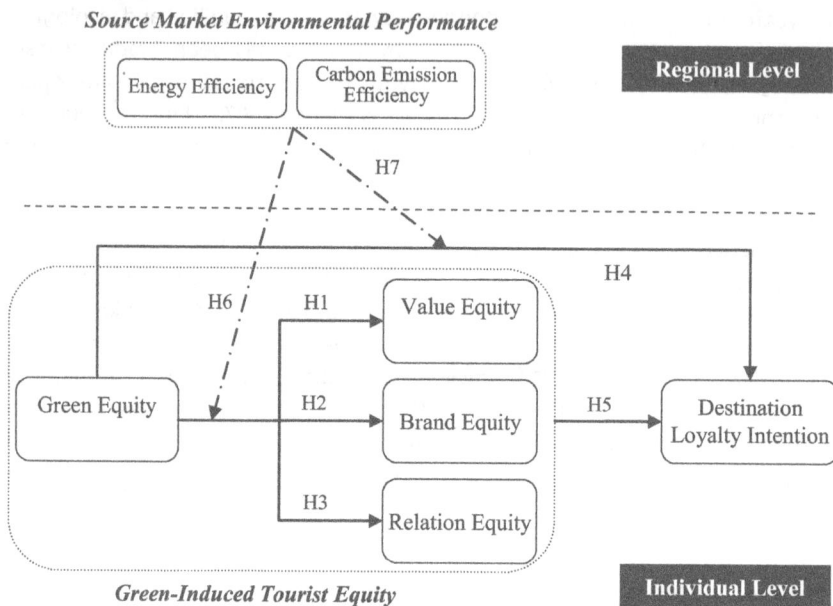

**Figure 1.** Research Framework: Green-Induced Tourist Equity.
*Note:* indicates moderating effect.

That is, these initiatives could inadvertently emanate subliminal persuasions to out-bound tourists.

### Green equity as a driver in the tourist equity model

The section above briefly highlights our logic for the conceptualization of tourist equity and for the proposed model, which we further elaborate in this and the following sections. Although customers' perceptions of an organization's value equity, brand equity and relationship equity can directly influence their loyalty, most customers thoroughly evaluate a firm's marketing programs before considering their patronage options (Gao et al., 2020; Ou et al., 2017). As awareness of ecological issues continues to increase, destination authorities have devoted a substantial amount of efforts to carry out environment-friendly operations along with firms' green programs to build a better habitat for both residents and tourists (Martínez García de Leaniz et al., 2018). Indeed, these greening effects have widely been reported in recent studies. For example, Chen (2016) stated that green perceived value was positively associated with users' green loyalty to public bicycles. Rosenbaum and Wong (2015) investigated guests' subjective appraisal of a hotel's green equity, along with value, brand and relationship equities, on guest loyalty. Their work showed that green equity plays a significant role in customers' overall assessment of a hotel's marketing programs. It further clarifies the role of green marketing programs in hotel management and shows how hotels can benefit from enhanced customer loyalty and decreased operating costs through the implementation of greening programs.

In this study, we propose the green tourist equity model by adding green equity as a driver of tourist loyalty, and reveal how green initiatives may influence other equity drivers that may ultimately impact tourist loyalty intentions. Green equity is described as the tourist's subjective appraisal of green initiatives implemented by the tourism destination in response to environmental concerns such as energy saving, water reuse, pollution reduction, energy consumption reduction, and recycling (Trvst, 2020).

## Hypotheses development

### Tourist equity drivers

The term *tourist equity* was conceptualized based on customer equity. Customer/tourist equity serves as an institution's (i.e., a firm's or a destination's) key strategic initiative to achieve long-term marketing success, and it represents values generated from the institution's current and potential customers/tourists by maintaining a prolonged relationship with them (Ou et al., 2017). It is calculated as the total of the discounted lifetime value summed over the institution's current and potential clienteles (Rust et al., 2004). Yet, it "remains a pipe dream for most firms" (Vogel et al., 2008, p. 98) because it is difficult for all institutions to calculate customer/tourist equity correctly by obtaining accurate customer/tourist lifetime value measures. However, Vogel et al. (2008) pointed out another way to gauge customer equity through its underlying drivers – value equity, brand equity, and relationship equity – as a means to tout loyalty and expenditures. Drawing on this logic, the study conceptualizes tourist equity to entail these properties – value, brand, and relationship equities – as drivers for the success of destination marketing programs. In particular, it has been acknowledged that customer equity drivers can promote relevant managerial outcomes, such as loyalty, sales, and profitability (Gao et al., 2020). Lemon et al. (2001) argued that it was important to reveal which driver is the most critical in customer/tourist equity and will be most effective in promoting loyalty behavior and ultimately increasing their expenditures. These three equity drivers are rather distinct, and they are further detailed in the sections that follow.

Value equity is customers'/tourists' perception of what is sacrificed and what is received during a marketplace exchange (Lemon et al., 2001). The common attributes of value equity in the context of tourism are convenience, price, and quality of tourism products (Lemon et al., 2001; Priporas et al., 2017). Brand equity is customers'/tourists' subjective appraisal towards the brand, which is beyond their objective assessment (Rust et al., 2000, 2004). Evaluation of brand equity often lies in consumer-based assessment (i.e., customer-based brand equity), which is a relative perception of a given brand in reference to other similar labels (San Martín et al., 2019). Relationship equity is the tendency of the customer to keep connection with a brand or tourism product (Ou et al., 2017). The influencing factors of relationship equity in the context of tourism consists of tourists' tendency to know and to stay connected with the destination. Tourist equity is essential to enhance the competitive advantage of a touristic place or product (Lee & Park, 2019; Wong, 2013). Each of these may work interdependently as a conduit to enhance favorable tourists' responses as well as providing an appropriate strategy for destinations to respond to their changing needs.

### The relationship between green equity and tourist equity drivers

Green initiatives have become a key competitive advantage to firms, as they help improve product attractiveness through added eco-related benefits to the product value. Specifically, green marketing could improve product appeals by enhancing customer preferences (Han, 2015; Satta et al., 2019). For example, Wong et al. (2015) empirically illustrated the role greening plays in tourist value creation in the context of food festivals. Destinations and service providers may leverage greening to improve cost control, increase attractiveness and profitability, and ultimately foster stronger tourist interests and preferences (Martínez García de Leaniz et al., 2018). In essence, destination greening renders a means to improve tourist perceptions of existing offerings, thus leading to an accentuated level of travel value (Gupta et al., 2019; Satta et al., 2019). Accordingly, the first hypothesis was proposed as follows:

**Hypothesis 1**: Green equity is positively related to value equity.

With increasing environmental awareness, green practices of tourism destination are considered as an effective way to improve the brand appeal of the place (Fok & Law, 2018). The purpose of implementing green initiatives is to improve marketing competitiveness by elevating the image and reputation of tourism destinations, and ultimately building a strong connection with tourists (Fok & Law, 2018; Wong et al., 2015). As such, DMOs around the globe have embarked on development of sustainable tourism (Corte & Aria, 2016; Wolf et al., 2017) with a constellation of green practices implemented, mainly involving the 4Rs: **R**ecycling paper, metals, and plastics; **R**educing and reusing waste water; using **R**enewable energy through building solar panels; and **R**educing energy consumption through LEDs and other energy-saving instruments (Trvst, 2020); along with other ecological measures (Collins & Cooper, 2017; Satta et al., 2019).

Utility of these green marketing programs contributes to adding tremendous brand benefits to an institution (e.g., tourism products or destinations) and their stakeholders (Koller et al., 2011). Environmental efforts are often demonstrated by the literature to emanate positive influence not only on an institution's branding (Chen, 2010; Misra & Panda Rajeev, 2017), but also on building a better connection with its stakeholders (e.g., customers) (Tanford & Malek, 2015; Wong et al., 2020). In other words, an institution's greening initiatives (i.e., green equity) render a powerful conduit in enabling the institution to attain a higher level of brand equity and relationship equity.

Take the lodging industry as an example. Eco-friendly hotel brands can improve customers' favorable impression towards them, which can ultimately arouse guests' brand preference and brand loyalty (Liu et al., 2014). Rosenbaum and Wong (2015) work on hotel greening programs point to the importance of eco-friendly practices, such as reducing energy consumption and using recyclable water, in the lodging industry as a means to enhance a hotel brand's image and quality, leading to a greater desire to repatronize the property through a stronger relational bond. The underlying logic may resonate with the premise of social identity theory (Löhndorf & Diamantopoulos, 2014), which posits a situation in which a social actor's greening goal is activated by a service provider's green initiatives to better categorize himself/herself as a part of (or in association with) the focal brand. We believe that destinations that embark on such green marketing initiatives would also benefit by improving destination brand equity and relationship equity. Accordingly, the following hypotheses were proposed:

**Hypothesis 2**: Green equity is positively related to brand equity.

**Hypothesis 3**: Green equity is positively related to relationship equity.


### The relationship between green equity and destination loyalty intention

Destination loyalty intention is defined as the propensity of a tourist to revisit a place (Oppermann, 2000); it is an important indicator to measure the relationship between tourists and a tourism locale (Yuksel et al., 2010). The relationship between greening and loyalty behaviors has been acknowledged in prior studies (e.g., Yusof et al., 2016). The body of literature has revealed that green practices are an effective way to enhance satisfaction and loyalty, and that they can serve as a main driving force for profitability and revenue growth of the providers (Gao et al., 2016; Kassinis & Soteriou, 2015). Taking the lodging business as an example again, it was identified that guests are more likely to patronize hotels with environment-friendly facilities, which further leads to increased propensity to staying in corresponding hotels in future trips (Han & Yoon, 2015). In sum, guests who are more eager to stay in eco-friendly hotels are more willing to pay extra and to provide positive word of mouth (Han et al., 2015). This phenomenon is also prevalent in other tourism products (Tolkes & Butzmann, 2018; Wong et al., 2015). Taken together, there is ample evidence to suggest a linkage between green practices of a tourism product or destination and revisit intention of tourists (Gao et al., 2016; Kassinis & Soteriou, 2015), despite most prior efforts germane to product-based experience. Here, we believe that a similar relationship is warranted in the context of a destination because it subsumes a wide array of tourism products; hence, the following hypothesis was proposed:

**Hypothesis 4**: Green equity is positively related to destination loyalty intention.

### The relationship between tourist equity drivers and destination loyalty intention

Loyalty intention is demonstrated by maintaining existing tourists through strengthening their tie with the hosting destination (Mao & Zhang, 2014). In the increasingly competitive market, efforts are made to recognize the role of customer equity as a major predictor of favorable touristic behaviors. Although prior research on the linkage between customer equity drivers and loyalty are primarily germane to the marketing discipline (Ou et al., 2017), recent tourism/hospitality literature has broadened its scope to cover travel-related phenomena. In particular, the linkage has been investigated primarily in tourism product related services such as restaurants (Hyun, 2009), casinos (Wong, 2013), retail stores (Yoon & Oh, 2016), and hotels (Lee & Park, 2019) with emphasis on customers' perceived loyalty to the particular service providers.

Yet, the tenet of the present research rests on tourist equity that conceptualizes value, brand, and relationship equity as a means to engender greater desire to revisit the host destination. A destination's marketing programs often serve as a conduit in building value propositions, developing favorable destination brand image and preference, and increasing relationship with tourists (including loyalty and affinity marketing programs) (Murphy et al., 2000; Ritchie & Geoffrey, 2010). That said, tourists perceptions of destination attributes such as functional benefits (i.e., travel value) (Bajs, 2015; Gallarza & Saura, 2006), symbolic benefits (i.e., destination brand equity) (Nam et al., 2011; San Martín et al., 2019), and relational benefits (Lam & Wong, 2020) have been noted in the body of literature to influence tourists' loyalty intentions and behaviors by enticing them and fulfilling their needs. Accordingly, the following hypotheses were proposed:

**Hypothesis 5**: Tourist equity (value equity, brand equity, and relationship equity) is positively related to destination loyalty intention.

### The moderating role of source market environmental performance

Environmental performance can be seen as individual's appraisal of the effect of environment-friendly initiatives. Performance of an ecosystem is crucial in tourism marketing, as there are growing environmental concerns that threaten humanity and the travel industry. Tourists consider environmental performance not only for their place of origin but also tourism destinations, especially when evaluating and selecting their place of choice (Mensah, 2006). Generally, tourists tend to seek eco-friendly places that can reconcile their ecological worries and malaise, while instilling their confidence in compliance with green measures (e.g., waste treatment measures, energy efficiency, usage of renewable energy sources, greenhouse gas emissions) (Cingoski & Petrevska, 2018; Yusof et al., 2016). ; In this study, environmental performance, such as energy efficiency and carbon emissions, render environmental cues embedded within the source market. Energy efficiency is reflected by introducing new energy initiatives to reduce operating costs, and by creating eco-friendly facilities to protect the environment (Cingoski & Petrevska, 2018). Energy-efficient practices achieve environmental protection by reducing harmful emissions (e.g., carbon dioxide, methane, nitrous oxide, etc.) that provoke global warming and climate change (Cingoski & Petrevska, 2018). In particular, carbon emissions may be even more alarming in recent years, as they have been acknowledged as a major factor that contributes to global warming. It was predicted that transport-related $CO_2$ emissions caused by tourism activity will increase 25% from 1,597 million tones in 2016 to 1,998 million tones by 2030. Moreover, tourism-related transport emissions represented 22% of all transport emissions in 2016 and will maintain a similar level in 2030 (UNWTO, 2019).

In general, a clean and well-preserved environment is considered as a main precondition for residents' well-being and livability (Wang et al., 2020). Thus, source markets that act on preserving the environment with endeavors to improve energy efficiency and lower carbon emission measures eventually transform into normative forces that "can trigger the activation of unconscious goals… [so] that even goal-directed behavior often takes place outside conscious awareness and… goals can be automatically activated by a multitude of environmental cues" (Dijksterhuis et al., 2005, p. 198). In other words, cues available from the source market's eco-achievement and greening initiatives may be consciously or unconsciously perceived, which intrigues people to pursue greening goals even when they travel to a foreign place (Wong et al., 2020). In this study, we focused on theorizing the influence of environmental cues as unconscious and mindless stimuli, because we operationalized environmental performance (i.e., regional energy efficiency and carbon emission) at the macro level that often exists without being noticed, rather than at the individual level where a tourist could consciously perceive the stimuli.

Drawing from the above assertions, with special emphasis on subliminal priming and persuasion, tourists originating from relatively environmentally friendly source markets may unconsciously place greater importance on green practices (e.g., more energy efficiency and lower carbon emissions) (Parkinson & Haggard, 2014). Such an embedded motive for, and placing of importance on, place greening should reinforce the ecological efforts enacted by the tourism destination. Here subliminal priming renders as a motivational force that activates tourists' innate goals (Cooper & Cooper, 2002) to green and to protect the mother Earth (Wang et al., 2020). In other words, there should be an interaction of both the host market's green equity and the source market's environmental performance that could jointly affect tourists' perceptions of the value, brand, relationship, and loyalty merits of the destination.

This logic is congruent with empirical results from the literature on subliminal influence that shows that "the behavioral effects of a subliminal prime seem to be contingent on a preexisting motivation to engage in these behaviors" (Albarracin & Vargas, 2010, p. 402). Hence, we argue that being green could better promote the value of a place, enhance the place's brand image and quality, and facilitate a stronger bond between the tourist and the destination, especially for tourists who are subliminally motivated by an embedded sphere of a healthy environment from their own habitat. In other words, drawing on subliminal persuasion and priming literature, we could say that the environmental performance of the source market should not directly affect a tourist's perceptions and behaviors; rather, its impact should be triggered based upon the tourist's encounter of a preexisting stimulus (i.e., greening of a destination) (Albarracin & Vargas, 2010; Wenke et al., 2010). However, the moderating effect of environmental performance could work both ways. On one hand, poor environmental performance of the source market could render a latent subliminal motive that impels tourists to seek greener places, perhaps for escapism and relaxation (Xu & Chan, 2010). On the other hand, superior performance of the source market's environment could reflect a subliminal motive commensurate with a normative force that intrigues the tourist to seek similar destination environmental appeals (Han & Hwang, 2015). Accordingly, the following hypotheses were postulated:

**Hypothesis 6**: Source market environmental performance moderates the relationship between green equity and tourist equity (value, brand, and relation equities).

**Hypothesis 7**: Source market environmental performance moderates the relationship between green equity and destination loyalty.

## Methods

### Sample

Data collected for this study came from two sources. The first data source was obtained from the literature (Wang et al., 2019; Zhao et al., 2019) in respect to regional environmental performance of the source market (more details are presented below). The second data source was based on a survey that was conducted on May, 2019 with inbound tourists visiting Macau, China. The destination was selected because the government has made an earnest endeavor to promote sustainable tourism with efforts in recycling paper, metal, and plastic; reducing and reusing waste water; and using renewable energy through building solar panels; as well as reducing energy consumptions through LED and other energy-reduction instruments (Wang et al., 2018; Watkinson, 2017). These initiatives were prevalent in the city and manifested through flyers and green facilities that are vividly available in major tourism attractions. Hence, respondents of the study were intercepted at major attractions. They were greeted with the objective of the study and informed about the green initiatives that the city has undergone. A person-administered approach was employed to assist tourists to complete the survey questionnaire. This approach helped to address questions raised by the respondents if they lacked knowledge about the measures and the city's greening program. To improve representation of the study, a systematic sampling method was used with a skip interview of three. Respondents who were unclear about the green initiatives of the city were excluded from the interview. The questionnaire was first developed in English and then back-translated to Traditional and Simplified Chinese with assistance from three bilinguals. A pilot test with eight subjects was conducted in order to improve the readability and clarity of the instrument.

The survey was completed by a total of 428 respondents, which reflects a response rate of 53.5%. Of these tourists, 46.4% were from mainland China, and the rest from other regions such as Hong Kong (13.5%), Taiwan (8.4%), Singapore (4.9%), Europe (9.1%), and more; 50.7% were males; 50.6% received a bachelor's degree of high education, while 25.8% received a high school graduate degree; 39.0% were between the age of 21 and 30, 31.8% were between the age of 31 and 40. On average, they stayed in the city for three days. These demographic characteristics were generally in line with the profile of inbound tourists in the city (Wong et al., 2019).

## Measures

As mentioned above, two sources of data were combined in order to perform the data analytics required. The first data sources were imported from the literature on China's regional *energy efficiency* (Zhao et al., 2019) and *carbon emission performance* (measures emission efficiency) (Wang et al., 2019) as proxy for environmental performance at the source market (i.e., province). These metrics were selected based on our preliminary investigation of the available environmental measures in China from English academic journals. According to Zhao et al. (2019), energy efficiency was assessed based on adjusted values obtained from exterior environmental factors on provincial energy efficiencies to eliminate the influences of exterior environmental factors; hence it provides a reliable measure of "real" energy performance of a region. Accordingly, higher values reflect better energy performance and efficiency in terms of energy consumption per capita of a particular locale. Wang et al. (2019) measure of carbon emission performance was used to assess regional efficiency of carbon emissions with higher values reflecting better efficiency in carbon emissions per capita within a region.

The second data source contains individual-level variables of interest, including green equity, value equity, brand equity, relationship equity, destination loyalty and demographic characteristics. Each item of the multi-item scale was evaluated using a 7-point Likert anchor ranging from 1 (strongly disagree) to 7 (strongly agree). *Green equity* was a four-item measure adopted from Rosenbaum and Wong (2015). The scale measures a destination's green initiatives from a tourist perspective with respect to four areas: energy saving, water reuse, pollution reduction, and recycling. Cornbach's alpha (α), used to assess consistency of the scale, was adequate with α = .93.

*Value equity* was adopted from Rust et al. (2004) and Rosenbaum and Wong (2010) with respect to three aspects of value proposition: convenience, price, and quality, of the destination.

**Table 1.** Means, standard deviations, and correlation matrix.

| | Mean | S.D. | AVE | 1 | 2 | 3 | 4 | 5 | 6 |
|---|---|---|---|---|---|---|---|---|---|
| 1. Value equity | 5.32 | 1.05 | .64 | .87 | | | | | |
| 2. Green equity | 6.29 | .90 | .76 | .25*** | .93 | | | | |
| 3. Brand equity | 4.75 | 1.17 | .61 | .26*** | .19*** | .81 | | | |
| 4. Relationship equity | 4.81 | 1.03 | .57 | .31*** | .17*** | .57*** | .70 | | |
| 5. Destination loyalty | 5.49 | 1.09 | .53 | .23*** | .24*** | .57*** | .52*** | .71 | |
| 6. Energy efficiency[a] | .92 | .12 | – | -.01 | -.01 | -.07 | -.12† | -.01 | – |
| 7. Carbon emission performance[a] | .91 | .12 | – | .02 | -.13† | -.11 | -.12† | -.04 | .34*** |

†Note: p < .10,
*p < .05,
**p < .01,
***p < .001.
[+]Data are disaggregated at the individual level.
Values presented at the diagonal are *Cronbach's alphas*.
AVE = average variance extracted.

The 10-item scale was fairly reliable with $\alpha = .87$. *Brand equity* was a three-item scale adopted from Rust et al. (2004). The scale assesses tourists' awareness, image, and quality of a destination brand; and it exercises adequate reliability with $\alpha = .81$. *Relationship equity* was a three-item scale adopted from Vogel et al. (2008). The scale assessed tourists' tendency to know and to stay connected with the destination; it's reliability is $\alpha = .70$. *Destination loyalty intention* assesses tourist propensity to express positive word of mouth and revisit the destination. It was evaluated using a two-item scale adopted from the literature (Wong et al., 2019). The scale is consistent, with $\alpha = .71$. Overall, all five scales of interest exhibited adequate scale validity, with factor loadings significant at the .001 level ($t \geq 12.37$), average variance extract (AVE) $\geq .53$, square root of each AVE $>$ correlations of the corresponding construct, $\alpha \geq .70$, and composite reliability $\geq .70$. The measurement model of the five multi-item scale produced adequate model fit: $\chi^2_{(177)} = 500.10$, comparative fit index (CFI) $= .97$, root mean square error of approximation (RMSEA) $= .06$, and standardized root mean square residual (SRMR) $= .09$.

Common method variance (CMV) poses a threat to the integrity and reliability of the results. Following the recommendation from the literature (Podsakoff et al., 2003), we first utilized two different data sources to mitigate the issue. Then we diagnosed CMV based on the marker variable method by partialling out a theoretically unrelated marker variable: satisfaction with cultural attraction, which is four-item scale adopted from Hui et al. (2007). Inclusion of the variable does not affect the results, suggesting that CMV is not a limitation. Multicollinearity was assessed through the variance inflation factor (VIF), and results indicate that all VIFs are below 2.0, suggesting that multicollinearity does not affect the reliability of the results.

## Findings

Table 1 presents descriptive statistics and zero-order correlation of the scales of interest. In general, results from Pearson-correlation analysis suggest that regional environmental performance has a moderate negative relationship to green equity ($r_{carbon\ emission\ performance\ -\ green\ equity} = -.13$, $p < .10$) and relationship equity ($r_{carbon\ emission\ performance/energy\ efficiency\ -\ green\ equity} = -.12$, $p < .10$). We then tested the individual-level hypothesized relationships in a structural equation model using LISREL 8.80. Results from Model 1 show that green equity is a significant predictor for value equity ($b = .33$, $p < .001$), brand equity ($b = .27$, $p < .001$), and relationship equity ($b = .21$, $p < .001$). In turn, value, brand, and relationship equities ($b \geq .29$, $p < .001$) are significant predictors for destination loyalty, supporting Hypotheses 1, 2, 3, and 5 (see Table 2). However, results from Model 2 reveal that green equity does not have a direct relationship with loyalty intention, thus Hypothesis 4 is not supported. We further assessed partial mediation using the Sobel test with results showing that $Z \geq 3.69$ ($p < .001$), indicating that value, brand, and

**Table 2.** Results of structural equation modeling parameter estimates.

| | Model 1 | | | | Model 2 |
|---|---|---|---|---|---|
| Main Effect | Value Equity | Brand Equity | Relationship Equity | Destination Loyalty | Destination Loyalty |
| Green equity | .33*** | .27*** | .21*** | | −.03 |
| Value equity | | | | .42*** | .43*** |
| Brand equity | | | | .29*** | .29*** |
| Relationship equity | | | | .45*** | .45*** |
| $R^2$ | .16 | .06 | .07 | .69 | .69 |

***Note: p < .001.
Fit indices: CFI = .95, RMSEA = .08, SRMR = .09.

relationship equities significantly mediate the green equity – destination loyalty relationships. We validated the mediation through Hayes' PROCESS procedure with a boostrapping sample of 5,000. Results reveal significant indirect effect for value equity ($b = .06$, $p < .01$, CI = .02 to .10), brand equity ($b = .12$, $p < .001$, CI = .07 to .18), and relationship equity ($b = .10$, $p < .001$, CI = .04 to .16).

Next we examined the multilevel design of the proposed model using hierarchical linear modeling (HLM) through HLM 6.06. HLM assumes that the endogenous measures including value, brand, and relationship equities as well as destination loyalty vary among a higher level of constituents (e.g., regions). Hence, we first tested this assumption through analysis of variance (ANOVA) with source market as the independent variable and the four endogenous measures as the dependent variables. Results reveal significant differences among regions ($F_{(30, 397)} \geq 1.69$, $p < .05$). Next we performed a series of HLMs to (re)assess the proposed relationships by taking the regional variance into account. Also, due to data availability from regional-level data, and to offer a finer control of extraneous influence from different cultures and economies persisting among countries, only regional data from China were used for the macro-level environmental performance measures. After cleaning the data, 21 dyadic datasets (i.e., regional and individual data) were created.

Results from Models 3 and 4 present evidence from a mediation model leading from green equity to destination loyalty through value, brand, and relationship equities; while results from Models 5 and 6 present further evidence for the mediation model by taking account of the two moderating factors: energy efficiency (EE) and carbon emission performance (CEP) (see Table 3). Results from these models provide an internal validation for those presented in Models 1 and 2 discussed above. In particular, green equity is a significant predictor for value equity ($b = .43$, $p < .001$) and brand equity ($b = .33$, $p < .05$), but not for relationship equity ($b = .11$, n.s.). In turn, value equity ($b = .15$, $p < .05$), brand equity ($b = .36$, $p < .001$), and relationship equity ($b = .11$, $p < .10$) are significant predictors for loyalty, but green equity ($b = .15$, n.s.) is not. These results generally support Hypotheses 1, 2, and 5.

The moderation of regional EE and CEP was examined through modeling their impacts based on random intercept (i.e., cross-level direct effect) and random slope (i.e., cross-level moderating effect). Results from Models 5 and 6 indicate that EE and CEP do not exercise direct influence on the mediating and dependent measures except for the relationship between EE and relationship equity ($b = -1.20$, $p < .10$). The cross-level EE × green equity ($b = -1.69$, $p < .01$) and CEP × green equity ($b = -.75$, $p < .05$) interactions on value equity were significant, while the CEP × green equity ($b = 1.85$, $p < .05$) interaction on brand equity was also significant; partially supporting Hypothesis 6, but not Hypothesis 7.

To graphically depict the cross-level moderating effect, we followed Aiken and West (1991) simple slope approach by redefining the independent variable and the moderator into plus and minus one standard deviation from the mean and portraying the values in Excel. Figures 2 and 3 consistently illustrate that the effect of green equity on value equity is more salient for tourists from regions where environmental performance (i.e., EE and CEP) is low. Results may suggest that tourists residing in regions where environmental protection and hence, environmental

**Table 3.** Results of hierarchical linear modeling estimates.

| | Model 3 | | | Model 4 | Model 5 | | | Model 6 |
|---|---|---|---|---|---|---|---|---|
| Main Effect | Value Equity | Brand Equity | Relationship Equity | Destination Loyalty | Value Equity | Brand Equity | Relationship Equity | Destination Loyalty |
| Green equity | .23** | | | .13† | .43*** | | | .15 |
| Value equity | | .38*** | | .15* | | .33* | | .15* |
| Brand equity | | | .24† | .36*** | | | .11 | .36*** |
| Relationship equity | | | | .10 | | | | .11† |
| *Cross-level Effect* | | | | | | | | |
| Energy efficiency (EE) | | | | | .00 | -.40 | $-1.20^{\dagger}$ | -.15 |
| Carbon emission performance (CEP) | | | | | 1.39 | .58 | -.73 | 1.33 |
| EE × Green equity | | | | | $-1.69^{**}$ | .07 | .59 | -.07 |
| CEP × Green equity | | | | | $-.75^{**}$ | 1.85* | 1.44 | -.15 |
| $R^2$ | .03 | .06 | .02 | .40 | .06 | .07 | .02 | .40 |
| $\chi^2_{(18)}$ | 47.31** | 34.96* | 25.14(n.s.) | 46.41*** | 49.50*** | 38.18** | 20.69(n.s.) | 59.42*** |

$^{\dagger}$*Note:* p < .10,
*p < .05,
**p < .01,
***p < .001.

performance are still relatively poor may subliminally influence them to crave greener places, where initiatives toward better energy consumption and environmental protection are of primary interest. Hence, a destination's green endeavors could help tourists from these source markets to better realize the valuation of their trips. Figure 4 illustrates that the effect of green equity on brand equity is more salient for tourists residing in high carbon emission performance regions. The result may hint at a congruency relationship between the source market's and the destination market's greening initiatives in improving the living environment (i.e., by reducing carbon emission); hence, being green helps to further promote the destination's brand image and quality, especially for those who perceive that the destination's greening endeavors are in line with their home regions.

## Discussion

The present inquiry is motivated by the extant research gaps on sustainable tourism with respect to a need to better understand an intricate linkage between source markets' and destination markets' environmental initiatives. In this study, we developed a multilevel model of green-induced tourist equity to test individual-level direct effects leading from green equity to destination loyalty intention through value equity, brand equity, and relationship equity germane to a tourism locale. We then tested the cross-level moderating effect emanating from the source market's environmental performance to the proposed direct relationships. Findings reveal that value and brand equity mediate the green equity – destination loyalty relationship. Moreover, the green equity – value/brand equity relationships are moderated by source market environmental performance. We close the article by recapping theoretical and practical implications that can further open a broader discourse on destination greening.

### *Theoretical implications*

This study contributes to the literature from three primary perspectives. First, this study conceptualizes green-induced tourist equity based upon the customer equity literature (Hyun, 2009; Lee

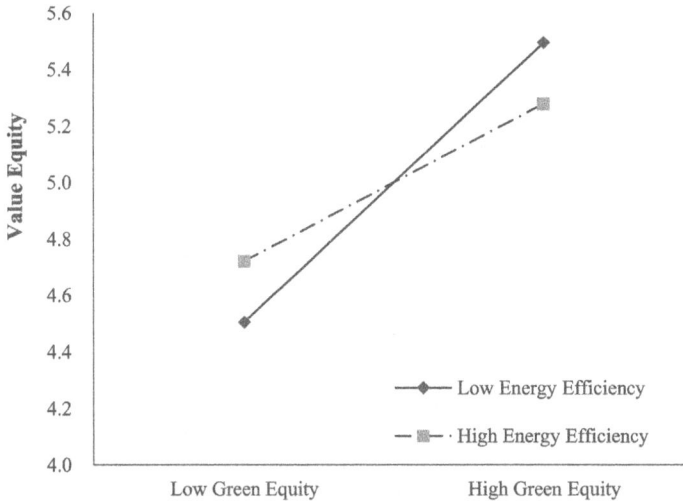

**Figure 2.** Cross-Level Green Equity × Energy Efficiency Interaction on Value Equity.

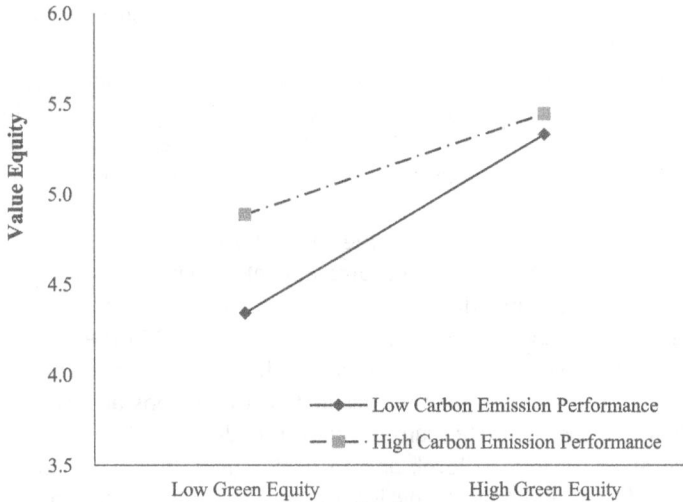

**Figure 3.** Cross-Level Green Equity × Carbon Emission Efficiency Interaction on Value Equity.

& Park, 2019; Rust et al., 2004) to describe tourist perceptions of a destination's greening efforts in delivering an elevated level of functional, symbolic, and relational benefits through value, brand, and relationship equities. On one hand, this new approach in assessing tourist equity helps foster a mediation impact of tourists' ecological concerns on their future behaviors. Such mediating effect of tourist equity is rather unique in the sustainable literature, as it illuminates that the linkage between greening and tourist behaviors is dependent upon a broad range of destination appeals that fall into three distinct pillars: value, brand, and relationship. More importantly, this study extends the empirical inquiry of customer equity to the sustainable tourism research domain, juxtaposing greening along with tourist equity in gauging the impact of destination ecological initiatives on tourists. Thus, the first contribution of this research lies in broadening the definition of customer/tourist equity from the firm level to the destination level. That is, we extend the concept – since it was germane merely as "determinant of the long-term value of the firm" as it "create[s] a significant competitive advantage" (Lemon et al., 2001, p. 21) in the marketing literature – to apply in the context of a territory. Accordingly, greening induced

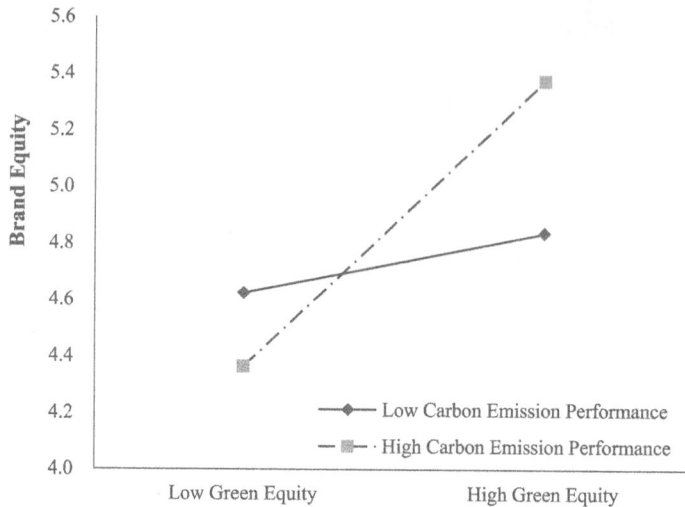

**Figure 4.** Cross-Level Green Equity × Carbon Emission Efficiency Interaction on Brand Equity.

equity goes beyond the profit-driven ideology to invigorate a more sustainable balance between tourism and the broader ecosystem.

Second, subliminal persuasion and priming are often used in communication and advertising research domains to articulate people's unconscious information processing, which gives rise to stimuli that inadvertently or mindlessly influence their attitudes, perceptions, and behaviors (Cooper & Cooper, 2002). In turn, people make unconscious choices that "are highly habitualized and based on attitudes that are automatically activated on the perception of a product [or situation]" (Dijksterhuis et al., 2005, p. 194). In the present study, we illustrate such a subliminal motivational force that is inherited from the source market's environmental conditions. As the literature predicts, this force does not directly impact people but it enacts upon other factors and situations to activate this true power. This discovery is rather novel in the tourism literature, as it opens a new avenue of research on how unconscious beliefs that are embedded in a place of origin could ultimately play a role in guiding destination evaluations and choices. In turn, applications of subliminal persuasion in the current study broaden the theoretical understanding of the role of attitude at an aggregate level, and they also help strengthen the linkage between source market innate forces and tourists' evaluations as they traverse through a web of foreign encounters. Given the plausible interaction between source market and host destination ecological initiatives, it opens new avenues for countries and territories to collaborate in the midst of tourism competitions. This coopetition logic (Fong et al., 2018) between origin and host markets paves the way for a new sustainable tourism paradigm that deserves great attention in future research.

Third, while a wide array of efforts have been acknowledged in the academic literature that are germane to the sustainability research stream in illustrating the role of greening on hotels (Han & Yoon, 2015; Tanford & Malek, 2015), events (Andersson et al., 2013; Wong et al., 2015), other tourism products (Yoon & Oh, 2016), and destinations (Fok & Law, 2018; Xu et al., 2020) from the tourist perspective, these initiatives are confined within the scope of individual-level analytics. This study contributes to the literature by bridging the micro and macro perspective of sustainable tourism and by integrating both individual-level and source market regional-level factors into a multilevel framework to better assess how source market environmental factors could ultimately influence tourists' greening perceptions and behavioral intentions during their excursions. The extension from mere micro-level or macro-level investigation to multilevel design is crucial to the social science inquiry tradition, as it takes into account the environmental situation that gives rise to tourist perceptions and behaviors. As such, it opens a new array of research opportunities not only to "reconcile the limitations of single-level analysis" (Wong, 2017, p. 809), but also to better gauge macro-level (i.e., source or destination market) "boundary conditions, contingencies, non-linearity, and contextual factors in order to gain a fuller understanding of the research problems at hand" (p. 812). In doing so, this study takes a systems thinking/theory tradition to offer a new approach in assessing the role of greening on destinations, taking account of individual tourist perceptions and innate ecological properties associated with his/her corresponding source market to portray an integrated model of how tourists' predispositions are indeed contingent upon the interaction of the source–host destination eco-forces.

### Managerial implications

Sustainable tourism has been an important issue since the 1990s, and it aims to better gauge the sustainability of a place balanced between three pillars: social, economical, and environmental. Yet, destination ecological concerns have lead to greater attention on green practices, especially in recent years, due to increased global awareness and severity of global warming, water shortage, natural disasters (e.g., flooding and extreme weather), and excessive waste, among others. As such, destination greening is gaining traction due to the ever growing environmental concerns among tourists and other stakeholders. As Satta et al. (2019, p. 265) note, "Implementing 'greener' and more sustainable strategies by tourism-based companies and destinations helps reflect the increasing environmental concerns influencing tourist behavior and consumption patterns." Satta et al.'s contention certainly is reflected in the present investigation, with findings pointing to a mediated relationship leading from green equity to destination loyalty intention through the three tourist equity drivers. In other words, destinations that are perceived to lift up their obligations to protect the environment through recycling and waste reduction programs, for example, are recognized as more ideal places for sojourns. Such places are also seen as possessing greater travel value and more favorable branding appeal; hence they represent better places for tourists to connect to.

The advent of tourist equity along with its three drivers could offer DMOs a key metric in gauging a destination's competitiveness and the attractiveness of its marketing programs. In fact, applications for tourist equity can certainly go beyond sustainable tourism to other areas of investigation. In sum, we believe that tourist equity can become a rather popular tool to evaluate the effectiveness of destination marketing and competitiveness from the tourist perspective. One area of application is tourist-area lifecycle (TALC), which takes growth and decline of tourist arrivals and expenditure as proxies for destination tourism development (Butler, 2009; McKercher & Wong, 2021). While these behavioral metrics are certainly important, they are nevertheless backward indicators of a place's development. Green equity and its outcomes (i.e., tourist equity) could serve as alternative indicators for TALC. Green equity is a perfect metaphor for a tourism-area ecological development, which helps a place to be environmentally sustainable. Tourist equity, on the other hand, focuses more on a wide array of functional, emotional, and symbolic benefits that offer social and economic sustainability to a destination. More importantly, these measures reflect forward indicators of TALC and hence, they entail trends for the future rather than from the past.

Green practices have largely permeated into our daily lives. Greening not only plays a salient role during our excursion in a foreign land, it is equally important to understand its impact on tourists at their common habitat: home. Although prior literature (e.g., Wong et al., 2020; Xu et al., 2020) has identified a spillover effect on pro-environmental behaviors and other greening perceptions, empirical evidence collated from the present study moves beyond the individual level to articulate a broader research context at the regional level. It showcases a phenomenon in which an origin market's environmental performance inadvertently influences tourist outbound greening-induced value and brand perceptions. In particular, the green-value equity relationship is more acute for tourists originating from low environmental performance regions, while the opposite is observed for the green-brand equity relationship. These findings hint at how destination greening could render as travel values especially for tourists whose source markets possess poor ecological quality.

That said, DMOs' promotion of functional benefits of their greening initiatives could be more effective in markets that have inferior environmental conditions (e.g., northern parts of China where pollution is rather severe due to heavy reliance on manufacturing), perhaps as a means of personal restoration. Such restoration potential is not only beneficial for the physical body, but it is also vital for tourists' mental well-being. On the other hand, DMOs' advertising appeals to place branding could be more effective in markets that have superior ecological practices, as a destination's green-induced branding appeals are more attractive for those who are also originating from regions that have superior environmental conditions. Thus, a destination can improve its attractiveness by joining forces with those regions to co-create more favorable reputation for environmental stewardship. Expanding the greening network in tourism certainly helps promote a better way of living and traveling and hence, more eco-conscious tourists.

### Limitations and future research directions

Findings of the study should be interpreted with the backdrop of their limitations. First, the study was conducted in China and hence, its generalizability is limited within a country-specific context. This limitation notwithstanding, the design of the study helps to control possible cultural and geo-political variations that may confound the results. Second, loyalty intention is a proxy for future revisit propensity, while actual revisit behavior may also be dependent on other factors that were not included in the present research. We encourage future research to assess the long-term effect of destination greening on tourist equity and other behavioral factors. This may require a longitudinal design with a continuous panel of cohorts and hence, it could also pose methodological challenges that inhibit execution of such a study.

## Disclosure statement

No potential conflict of interest was reported by the author(s).

## Funding

This research is supported by the National Natural Science Foundation of China (No. 72074230, 72004239), and Guangdong Provincial Department of Education (No. 2019GXJK052).

## ORCID

*IpKin Anthony Wong* iD http://orcid.org/0000-0003-4597-2228
*GuoQiong Ivanka Huang* iD https://orcid.org/0000-0001-9232-6822

## References

Aiken, L. S., & West, S. G. (1991). *Multiple regression: Testing and interpreting interactions*. Sage.

Albarracin, D., & Vargas, P. (2010). Attitudes and persuasion: From biology to social responses to persuasive intent. In S. T. Fiske, D. T. Gilbert & G. Lindzey (Eds.), *Handbook of social psychology* (5 ed., pp. 394–427). John Wiley & Sons.

Andersson, D. T., Jutbring, H., & Lundberg, E. (2013). When a music festival goes veggie: Communication and environmental impacts of an innovative food strategy. *International Journal of Event and Festival Management*, 4(3), 224–235.

Bajs, I. P. (2015). Tourist perceived value, relationship to satisfaction, and behavioral intentions: The example of the Croatian tourist destination Dubrovnik. *Journal of Travel Research*, 54(1), 122–134.

Bitner, M. J. (1992). Servicescapes: The impact of physical surroundings on customers and employees. *Journal of Marketing*, 56(2), 57–71. https://doi.org/10.1177/002224299205600205

Butler, R. (2009). Tourism in the future: Cycles, waves or wheels? Futures, 41(6), 346–352. https://doi.org/10.1016/j.futures.2008.11.002

Chen, Y.-S. (2010). The drivers of green brand equity: Green brand image, green satisfaction, and green trust. *Journal of Business Ethics*, 93(2), 307–319. https://doi.org/10.1007/s10551-009-0223-9

Chen, S. Y. (2016). Green helpfulness or fun? Influences of green perceived value on the green loyalty of users and non-users of public bikes. *Transport Policy*, 47, 149–159.

Cingoski, V., & Petrevska, B. (2018). Making hotels more energy efficient: the managerial perception. *Economic Research-Ekonomska Istraživanja*, 31(1), 87–101. https://doi.org/10.1080/1331677X.2017.1421994

Collins, A., & Cooper, C. (2017). Measuring and managing the environmental impact of festivals: The contribution of the Ecological Footprint. *Journal of Sustainable Tourism*, 25(1), 148–162. https://doi.org/10.1080/09669582.2016.1189922

Cooper, J., & Cooper, G. (2002). Subliminal motivation: A story revisited. *Journal of Applied Social Psychology*, 32(11), 2213–2227. https://doi.org/10.1111/j.1559-1816.2002.tb01860.x

Corte, V. D., & Aria, M. (2016). Coopetition and sustainable competitive advantage. The case of tourist destinations. *Tourism Management*, 54, 524–540.

Dijksterhuis, A., Smith, P. K., van Baaren, R. B., & Wigboldus, D. H. J. (2005). The unconscious consumer: Effects of environment on consumer behavior. *Journal of Consumer Psychology*, 15(3), 193–202. https://doi.org/10.1207/s15327663jcp1503_3

Erdogan, N., & Baris, E. (2007). Environmental protection programs and conservation practices of hotels in Ankara. *Tourism Management*, 28(2), 604–614. https://doi.org/10.1016/j.tourman.2006.07.003

Fok, K. W. K., & Law, W. W. Y. (2018). City re-imagined: Multi-stakeholder study on branding Hong Kong as a city of greenery. *Journal of Environmental Management*, 206, 1039–1051. https://doi.org/10.1016/j.jenvman.2017.11.045

Fong, V. H. I., Wong, I. A., & Hong, J. F. L. (2018). Developing institutional logics in the tourism industry through coopetition. *Tourism Management*, 66, 244–262. https://doi.org/10.1016/j.tourman.2017.12.005

Gallarza, M. G., & Saura, G. I. (2006). Value dimensions, perceived value, satisfaction and loyalty: An investigation of university students' travel behaviour. *Tourism Management*, 27(3), 437–452. https://doi.org/10.1016/j.tourman.2004.12.002

Gao, L., Melero-Polo, I., & Sese, F. J. (2020). Customer equity drivers, customer experience quality, and customer profitability in banking services: The moderating role of social influence. *Journal of Service Research*, 23(2), 174–193. https://doi.org/10.1177/1094670519856119

Gao, Y., & Mattila, A. S. (2014). Improving consumer satisfaction in green hotels: The roles of perceived warmth, perceived competence, and CSR motive. *International Journal of Hospitality Management*, 42, 20–31. https://doi.org/10.1016/j.ijhm.2014.06.003

Gao, Y., Mattila, A. S., & Lee, S. (2016). A meta-analysis of behavioral intentions for environment-friendly initiatives in hospitality research. *International Journal of Hospitality Management*, 54, 107–115. https://doi.org/10.1016/j.ijhm.2016.01.010

Gupta, A., Dash, S., & Mishra, A. (2019). All that glitters is not green: Creating trustworthy ecofriendly services at green hotels. *Tourism Management*, 70, 155–169. https://doi.org/10.1016/j.tourman.2018.08.015

Ham, S., & Han, H. (2013). Role of perceived fit with hotels' green practices in the formation of customer loyalty: Impact of environmental concerns. *Asia Pacific Journal of Tourism Research, 18*(7), 731–748. https://doi.org/10.1080/10941665.2012.695291

Han, H. (2015). Travelers' pro-environmental behavior in a green lodging context: Converging value-belief-norm theory and the theory of planned behavior. *Tourism Management, 47*, 164–177. https://doi.org/10.1016/j.tourman.2014.09.014

Han, H., & Hwang, J. (2015). Norm-based loyalty model (NLM): Investigating delegates' loyalty formation for environmentally responsible conventions. *International Journal of Hospitality Management, 46*, 1–14. https://doi.org/10.1016/j.ijhm.2015.01.002

Han, H., Hwang, J., Kim, J., & Jung, H. (2015). Guests' pro-environmental decision-making process: Broadening the norm activation framework in a lodging context. *International Journal of Hospitality Management, 47*, 96–107. https://doi.org/10.1016/j.ijhm.2015.03.013

Han, H., & Yoon, H. J. (2015). Hotel customers' environmentally responsible behavioral intention: Impact of key constructs on decision in green consumerism. *International Journal of Hospitality Management, 45*, 22–33. https://doi.org/10.1016/j.ijhm.2014.11.004

Han, H., Yu, J., Kim, H.-C., & Kim, W. (2018). Impact of social/personal norms and willingness to sacrifice on young vacationers' pro-environmental intentions for waste reduction and recycling. *Journal of Sustainable Tourism, 26*(12), 2117–2133. https://doi.org/10.1080/09669582.2018.1538229

Hirst, G., van Knippenberg, D., Chen, C.-H., & Sacramento, C. A. (2011). How does bureaucracy impact individual creativity? A cross-level investigation of team contextual influences on goal orientation-creativity relationships. *Academy of Management Journal, 54*(3), 624–641. https://doi.org/10.5465/amj.2011.61968124

Hui, T. K., Wan, D., & Ho, A. (2007). Tourists' satisfaction, recommendation and revisiting Singapore. *Tourism Management, 28*(4), 965–975. https://doi.org/10.1016/j.tourman.2006.08.008

Hyun, S. S. (2009). Creating a model of customer equity for chain restaurant brand formation. *International Journal of Hospitality Management, 28*(4), 529–539. https://doi.org/10.1016/j.ijhm.2009.02.006

Jarvis, N., Weeden, C., & Simcock, N. (2010). The benefits and challenges of sustainable tourism certification: A case study of the green tourism business scheme in the West of England. *Journal of Hospitality and Tourism Management, 17*(1), 83–93. https://doi.org/10.1375/jhtm.17.1.83

Ji, M., Wong, I. A., Eves, A., & Leong, A. M. W. (2018). A multilevel investigation of China's regional economic conditions on co-creation of dining experience and outcomes. *International Journal of Contemporary Hospitality Management, 30*(4), 2132–2152. https://doi.org/10.1108/IJCHM-08-2016-0474

Jones, P., Hillier, D., & Comfort, D. (2016). Sustainability in the hospitality industry. *International Journal of Contemporary Hospitality Management, 28*(1), 36–67. https://doi.org/10.1108/IJCHM-11-2014-0572

Karremans, J. C., Stroebe, W., & Claus, J. (2006). Beyond Vicary's fantasies: The impact of subliminal priming and brand choice. *Journal of Experimental Social Psychology, 42*(6), 792–798. https://doi.org/10.1016/j.jesp.2005.12.002

Kassinis, G. I., & Soteriou, A. C. (2015). Environmental and quality practices: using a video method to explore their relationship with customer satisfaction in the hotel industry. *Operations Management Research, 8*(3-4), 142–156. https://doi.org/10.1007/s12063-015-0105-5

Knezevic Cvelbar, L., Grün, B., & Dolnicar, S. (2020). "To clean or not to clean?" Reducing daily routine hotel room cleaning by letting tourists answer this question for themselves. *Journal of Travel Research*, 0047287519879779.

Koller, M., Floh, A., & Zauner, A. (2011). Further insights into perceived value and consumer loyalty: A 'Green' perspective. *Psychology & Marketing, 28*(12), 1154–1176.

Lam, I. K. V., & Wong, I. A. (2020). The role of relationship quality and loyalty program in tourism shopping: A multilevel investigation. *Journal of Travel & Tourism Marketing, 37*(1), 92–111.

Lee, B. Y., & Park, S. Y. (2019). The role of customer delight and customer equity for loyalty in upscale hotels. *Journal of Hospitality and Tourism Management, 39*, 175–184. https://doi.org/10.1016/j.jhtm.2019.04.003

Lemon, K. N., Rust, R. T., & Zeithaml, V. A. (2001). What drives customer equity? *Marketing Management, 10*(1), 20–25.

Liu, M., Wong, I. A., Shi, G., Chu, R., & Brock, J. (2014). The importance of corporate social responsibility (CSR) performance and perceived brand quality on customer-based brand preference. *Journal of Services Marketing, 28*(3), 181–194.

Löhndorf, B., & Diamantopoulos, A. (2014). Internal branding: Social identity and social exchange perspectives on turning employees into brand champions. *Journal of Service Research, 17*(3), 310–325. https://doi.org/10.1177/1094670514522098

Mao, I. Y., & Zhang, H. Q. (2014). Structural relationships among destination preference, satisfaction and loyalty in Chinese tourists to Australia. *International Journal of Tourism Research, 16*(2), 201–208. https://doi.org/10.1002/jtr.1919

Martínez García de Leaniz, P., Herrero Crespo, Á., & Gómez López, R. (2018). Customer responses to environmentally certified hotels: the moderating effect of environmental consciousness on the formation of behavioral intentions. *Journal of Sustainable Tourism, 26*(7), 1160–1177. https://doi.org/10.1080/09669582.2017.1349775

McKercher, B., & Wong, I. A. (2021). Do destinations have multiple lifecycles? *Tourism Management, 83*, 104232 https://doi.org/10.1016/j.tourman.2020.104232

Mehrabian, A., & Russell, J. A. (1974). *An approach to environmental psychology*. MIT Press.

Menguc, B., Auh, S., Katsikeas, C. S., & Jung, Y. S. (2016). When does (mis)fit in customer orientation matter for frontline employees' job satisfaction and performance? *Journal of Marketing, 80*(1), 65–83. https://doi.org/10.1509/jm.15.0327

Mensah, I. (2006). Environmental management practices among hotels in the greater Accra region. *International Journal of Hospitality Management, 25*(3), 414–431. https://doi.org/10.1016/j.ijhm.2005.02.003

Merli, R., Preziosi, M., Acampora, A., Lucchetti, M. C., & Ali, F. (2019). The impact of green practices in coastal tourism: An empirical investigation on an eco-labelled beach club. *International Journal of Hospitality Management, 77*, 471–482. https://doi.org/10.1016/j.ijhm.2018.08.011

Misra, S., & Panda Rajeev, K. (2017). Environmental consciousness and brand equity: An impact assessment using analytical hierarchy process (AHP). *Marketing Intelligence & Planning, 35*(1), 40–61.

Murphy, P., Pritchard, M. P., & Smith, B. (2000). The destination product and its impact on traveller perceptions. *Tourism Management, 21*(1), 43–52. https://doi.org/10.1016/S0261-5177(99)00080-1

Nam, J., Ekinci, Y., & Whyatt, G. (2011). Brand equity, brand loyalty and consumer satisfaction. *Annals of Tourism Research, 38*(3), 1009–1030. https://doi.org/10.1016/j.annals.2011.01.015

Oppermann, M. (2000). Tourism destination loyalty. *Journal of Travel Research, 39*(1), 78–84. https://doi.org/10.1177/004728750003900110

Ou, Y.-C., Verhoef, P., & Wiesel, T. (2017). The effects of customer equity drivers on loyalty across services industries and firms. *Journal of the Academy of Marketing Science, 45*(3), 336–356. https://doi.org/10.1007/s11747-016-0477-6

Palanica, A., Lyons, A., Cooper, M., Lee, A., & Fossat, Y. (2019). A comparison of nature and urban environments on creative thinking across different levels of reality. *Journal of Environmental Psychology, 63*, 44–51. https://doi.org/10.1016/j.jenvp.2019.04.006

Parkinson, J., & Haggard, P. (2014). Subliminal priming of intentional inhibition. *Cognition, 130*(2), 255–265. https://doi.org/10.1016/j.cognition.2013.11.005

Podsakoff, P. M., MacKenzie, S. B., Lee, J.-Y., & Podsakoff, N. P. (2003). Common method biases in behavioral research: A critical review of the literature and recommended remedies. *Journal of Applied Psychology, 88*(5), 879–903. https://doi.org/10.1037/0021-9010.88.5.879

Priporas, C.-V., Stylos, N., Rahimi, R., & Vedanthachari, L. N. (2017). Unraveling the diverse nature of service quality in a sharing economy: A social exchange theory perspective of Airbnb accommodation. *International Journal of Contemporary Hospitality Management, 29*(9), 2279–2301. https://doi.org/10.1108/IJCHM-08-2016-0420

Rahman, I., Park, J., & Chi, C. G-q. (2015). Consequences of "greenwashing": Consumers' reactions to hotels' green initiatives. *International Journal of Contemporary Hospitality Management, 27*(6), 1054–1081. https://doi.org/10.1108/IJCHM-04-2014-0202

Ritchie, J. R. B., & Geoffrey, I. C. (2010). A model of destination competitiveness/sustainability: Brazilian perspetives. *Rap — Rio De Janeiro, 44*(5), 1049–1066.

Rosenbaum, M. S., & Wong, I. A. (2010). Value equity in event planning: A case study of Macau. *Marketing Intelligence & Planning, 28*(4), 403–417.

Rosenbaum, M. S., & Wong, I. A. (2015). Green marketing programs as strategic initiatives in hospitality. *Journal of Services Marketing, 29*(2), 81–92. https://doi.org/10.1108/JSM-07-2013-0167

Rust, R. T., Lemon, K. N., & Zeithaml, V. A. (2004). Return on marketing: using customer equity to focus marketing strategy. *Journal of Marketing, 68*(1), 109–127. https://doi.org/10.1509/jmkg.68.1.109.24030

Rust, R. T., Zeithaml, V. A., & Lemon, K. N. (2000). *Driving customer equity: How customer lifetime value is reshaping corporate strategy*. Free Press.

San Martín, H., Herrero, A., & García de los Salmones, M. d M. (2019). An integrative model of destination brand equity and tourist satisfaction. *Current Issues in Tourism, 22*(16), 1992–2013. https://doi.org/10.1080/13683500.2018.1428286

Satta, G., Spinelli, R., & Parola, F. (2019). Is tourism going green? A literature review on green innovation for sustainable tourism. *Tourism Analysis, 24*(3), 265–280. https://doi.org/10.3727/108354219X15511864843803

Senge, P. S. (1994). *The fifth discipline: The art & practice of the learning organization*. Doubleday.

Tanford, S., & Malek, K. (2015). Segmentation of reward program members to increase customer loyalty: The role of attitudes towards green hotel practices. *Journal of Hospitality Marketing & Management, 24*(3), 314–343.

Teng, C.-C., Horng, J.-S., Hu, M.-L., Chien, L.-H., & Shen, Y.-C. (2012). Developing energy conservation and carbon reduction indicators for the hotel industry in Taiwan. *International Journal of Hospitality Management, 31*(1), 199–208. https://doi.org/10.1016/j.ijhm.2011.06.006

Tolkes, C., & Butzmann, E. (2018). Motivating pro-sustainable behavior: The potential of green events—A case-study from the Munich Streetlife Festival. *Sustainability, 10*(3731), 1–15.

Trvst. (2020). Importance of 4Rs – refuse, reduce, reuse, recycle. Retrieved May 20, 2020, from https://www.trvst.world/inspiration/importance-of-4rs-refuse-reduce-reuse-recycle/

UNWTO. (2019). Tourism's Carbon Emissions Measured in Landmark Report Launched at Cop25. Retrieved April 12, 2019, from https://www.unwto.org/news/tourisms-carbon-emissions-measured-in-landmark-report-launched-at-cop25

Vogel, V., Evanschitzky, H., & Ramaseshan, B. (2008). Customer equity drivers and future sales. Journal of Marketing, 72(6), 98–108. https://doi.org/10.1509/jmkg.72.6.098

Wang, S., Wang, H., Zhang, L., & Dang, J. (2019). Provincial carbon emissions efficiency and its influencing factors in China. Sustainability, 11, 1–21.

Wang, S., Wang, J., Li, J., & Yang, F. (2020). Do motivations contribute to local residents' engagement in pro-environmental behaviors? Resident-destination relationship and pro-environmental climate perspective. Journal of Sustainable Tourism, 28(6), 834–852. https://doi.org/10.1080/09669582.2019.1707215

Wang, X., Wu, N., Qiao, Y., & Song, Q. (2018). Assessment of energy-saving practices of the hospitality industry in Macau. Sustainability, 10(255), 1–15.

Watkinson, S. (2017). Greening Macau. Retrieved September 18, 2020, from https://macaucloser.com/en/content/greening-macau

Wenke, D., Fleming, S. M., & Haggard, P. (2010). Subliminal priming of actions influences sense of control over effects of action. Cognition, 115(1), 26–38. https://doi.org/10.1016/j.cognition.2009.10.016

Wolf, I. D., Ainsworth, G. B., & Crowley, J. (2017). Transformative travel as a sustainable market niche for protected areas: A new development, marketing and conservation model. Journal of Sustainable Tourism, 25(11), 1650–1673. https://doi.org/10.1080/09669582.2017.1302454

Wong, I. A. (2013). Exploring customer equity and the role of service experience in the casino service encounter. International Journal of Hospitality Management, 32, 91–101. https://doi.org/10.1016/j.ijhm.2012.04.007

Wong, I. A. (2017). Advancing tourism research through multilevel methods: Research problem and agenda. Current Issues in Tourism, 20(8), 809–824. https://doi.org/10.1080/13683500.2016.1186158

Wong, I. A., Wan, Y. K. P., Huang, G. I., & Qi, S. (2020). Green event directed pro-environmental behavior: An application of goal systems theory. Journal of Sustainable Tourism, 1–22. https://doi.org/10.1080/09669582.09662020.01770770

Wong, I. A., Wan, Y. K. P., & Qi, S. (2015). Green events, value perceptions, and the role of consumer involvement in festival design and performance. Journal of Sustainable Tourism, 23(2), 294–315. https://doi.org/10.1080/09669582.2014.953542

Wong, I. A., Xu, Y. H., Tan, X. S., & Wen, H. (2019). The boundary condition of travel satisfaction and the mediating role of destination image: The case of event tourism. Journal of Vacation Marketing, 25(2), 207–224. https://doi.org/10.1177/1356766718763691

Xu, F., Huang, L., & Whitmarsh, L. (2020). Home and away: Cross-contextual consistency in tourists' pro-environmental behavior. Journal of Sustainable Tourism, 28(10), 1443–1459. https://doi.org/10.1080/09669582.2020.1741596

Xu, J., & Chan, A. (2010). Service experience and package tours. Asia Pacific Journal of Tourism Research, 15(2), 177–194. https://doi.org/10.1080/10941661003629987

Xu, X., & Gursoy, D. (2015). Influence of sustainable hospitality supply chain management on customers' attitudes and behaviors. International Journal of Hospitality Management, 49, 105–116. https://doi.org/10.1016/j.ijhm.2015.06.003

Yoon, S., & Oh, J.-C. (2016). A cross-national validation of a new retail customer equity model. International Journal of Consumer Studies, 40(6), 652–664. https://doi.org/10.1111/ijcs.12289

Yuksel, A., Yuksel, F., & Bilim, Y. (2010). Destination attachment: Effects on customer satisfaction and cognitive, affective and conative loyalty. Tourism Management, 31(2), 274–284. https://doi.org/10.1016/j.tourman.2009.03.007

Yusof, N. A., Rahman, S., & Iranmanesh, M. (2016). The environmental practice of resorts and tourist loyalty: the role of environmental knowledge, concern, and behaviour. Anatolia, 27(2), 214–226. https://doi.org/10.1080/13032917.2015.1090463

Zhao, H., Guo, S., & Zhao, H. (2019). Provincial energy efficiency of China quantified by three-stage data envelopment analysis. Energy, 166, 96–107. https://doi.org/10.1016/j.energy.2018.10.063

# Hotels' sustainability practices and guests' familiarity, attitudes and behaviours

Hossein Olya ⓘ, Levent Altinay, Anna Farmaki ⓘ, Ainur Kenebayeva and Dogan Gursoy

**ABSTRACT**

This study investigates the effects of hotel's sustainability practices in three areas of sustainability and familiarity with those practices on hotel guest satisfaction and loyalty in the Kazakhstan hotel industry. Using a structural equation modelling (SEM) and fuzzy-set Qualitative Comparative Analysis (fsQCA), findings reveal that social and environmental dimensions play positive roles on guests' satisfaction and loyalty while the economic dimension and familiarity are not significantly related to guest loyalty, although they are likely to improve guest satisfaction. Furthermore, results of the analysis of the necessary conditions to achieve the expected model outcomes indicate that all three sustainability dimensions are necessary for sustainability efforts to have the most positive effect on guest satisfaction and loyalty. Results also indicate that although familiarity alone is insufficient, its combination with the social dimension increases guest satisfaction and loyalty. Findings provide theoretical and practical insights into sustainability practices in the hotel industry.

## Introduction

Arguably, sustainability is one of the most ideologically contested terms of recent times. Although there is no universally agreed definition of sustainability, it is generally regarded by intergovernmental organisations, policymakers and academics as the antidote to an array of problems facing societies, economies and the environment. While initial interpretations of sustainability were based on ecological principles, subsequent definitions offer a broader perspective by including social and economic dimensions alongside environmental goals in an effort to meet human needs in an equitable way (Jones et al., 2016). As the interest in sustainability grew over time, "sustainability has become one of business' most recent and urgent mandates" (Carroll & Buchholtz, 2014, p. 4), with companies being increasingly pressured to promote the "triple bottom line" approach (Elkington, 1997), where economic, social and environmental dimensions of sustainability are embedded into business strategies and practices (Han & Hyun, 2018b).

Although the hotel industry has been somewhat slow to adopt the sustainability paradigm (Jones et al., 2016; Trang et al., 2019), in recent years the concept has become "a defining issue" for hospitality companies (Deloitte, 2014, p. 41). Indeed, in light of the impacts of hotels' operations on local communities and the environment (de Grosbois, 2012), calls for hotels to align their practices to sustainability principles have intensified (Xu & Gursoy, 2015a). In fact, due to increased consumer awareness (Font & McCabe, 2017; Han et al., 2018), the sustainability practices of hotels have emerged as an important determinant influencing consumer attitudes and behaviours (Han et al., 2018; Law et al., 2016) including satisfaction, loyalty and decision-making process (Berezan et al., 2013; Han & Hyun, 2018a; Modica et al., 2020; Teng et al., 2012), and, thus, having a significant impact on financial performance (Garay & Font, 2012). As such, a significant body of literature that investigates consumer perceptions, attitudes and behaviours towards hotels' sustainable practices has begun to proliferate (e.g. Berezan et al., 2013; Chen, 2015; Kang et al., 2012; Ponnapureddy et al., 2017; Prud'homme & Raymond, 2013).

Nonetheless, a foray into extant literature reveals that the majority of past studies focussed solely on one dimension of sustainability such as the environmental or social aspect (e.g. Namkung & Jang, 2017; Martinez & del Bosque, 2013). Only a handful of studies have investigated the effects of all three sustainability dimensions within hotel settings (i.e. Modica et al., 2020; Xu & Gursoy, 2015b). This is surprising considering arguments that all three dimensions of sustainability (economic, social and environmental) need to be embedded in the strategies and practices of hotels (e.g. Farmaki, 2018). The hospitality industry is multi-faceted and complex in nature, comprising of several stakeholders whose performances and relations are interdependent (Farmaki, 2018); thus, a consideration of various stakeholders is needed for the successful implementation of sustainable practices in hotels (Modica et al., 2020) including employees, customers and hotel itself.

Therefore, this study examines the effects of hotel's sustainability practices in relation to employees, customers and hotel itself on guests' behaviours and attitudes and, specifically, on guest satisfaction and loyalty which have emerged as the most predominant attitudinal and behavioural aspects in the literature. In addition, this study considers the role of guests' familiarity with hotels' sustainability practices, which has been largely overlooked in past studies. Furthermore, this study draws insights from the Kazakhstan hotel industry; thus, responding to calls for more research on emerging economies (Goa, 2009) and within Asian contexts (Chen & Peng, 2016). Indeed, past studies on sustainability in hotel industry have mainly focussed on European (Modica et al., 2020) or American contexts (Xu & Gursoy, 2015b). Differences observed in previous studies between European and American consumers' attitudes and behaviours towards sustainability practices (Thompson, 2007) suggest that it is important to examine attitudes and behaviours of consumers from different geographical settings (Modica et al., 2020).

Sustainability and sustainable growth have been amongst the strategic priorities of all of the sectors of the economy in Kazakhstan, including tourism and hospitality (Marzhan, 2015). However, both the practices of the Kazakhstan hotels and pro-environmental behaviours of the Kazakh customers have been identified as areas of concern, recognised as not fully supportive of and integrated with the sustainability agenda of the country (Seilov, 2015). For example, independent hotels in major cities such as in Almaty and Nur-Sultan Astana, have only recently started to introduce the sustainable practices and customers awareness and familiarity of these practices stayed limited influencing their pro-environmental behaviours (Mussina & Bimerev, 2018). Thus, findings of this study are likely to make important theoretical and practical contributions to the advancement of existing knowledge on the influence of sustainability practices in the hotel sector on guest attitudes and behaviours.

This study evaluates both the sufficient and necessary factors that are needed to predict satisfaction and loyalty of hotel guests. Net effect of each predictor is investigated using SEM and sufficiency and necessity of the combination of predicators (i.e. causal recipes) is explored using fsQCA. Through identifying necessary conditions that are required to develop appropriate sustainability action plans by utilising a combination the three sustainability dimensions with familiarity, findings of this study will provide invaluable insight for hotel managers to further improve their guests' satisfaction and loyalty. Gannon et al. (2019, p. 245) noted that necessary conditions "are highly significant in terms of both theory and practice because without necessary predictors, the model outcome cannot occur and other predictors cannot play an alternative role in their absence". While previous research uses symmetrical analysis (e.g. SEM, regression, correlation) to identify sufficient factors affecting guest behaviours, there is a little knowledge of necessity of predictors of guests' satisfaction and loyalty. This is first empirical study that fills this research gap by investigating the conditions that are necessary to increase the satisfaction and loyalty of hotel guests through sustainability practices. Identification of the importance and necessity of the economic, social and environmental dimensions of sustainability as well as familiarity may help hotel managers properly allocate their scarce resources to meet the necessary conditions for attaining guest satisfaction and loyalty.

The rest of the paper is organised as follows. First, a review of the literature is provided to set the theoretical background of the study. Specifically, the importance of sustainability in hotels is explained before previous research is reviewed in terms of the impacts of hotel's sustainability practices on guests' behavioural and attitudinal responses. Then, the methodology adopted in this study is explained and justified before the results of the study are presented and discussed. Last, the implications, limitations and future research suggestions are drawn together as conclusions.

## Theoretical background

### Sustainability practices in hotel industry

Sustainability takes a long-term view of the future by considering ethical values and principles while endorsing responsible actions that incorporate environmental, societal and economic goals. The environmental dimension of sustainability aims at minimizing the negative environmental impacts brought about by hotel consumption through environmental monitoring (Vachon, 2007) and/or collaborative efforts including product redesign and greening of production processes (Tsai et al., 2011). The social dimension aims at enhancing the well-being of the employees, other suppliers, the guest and the local community at large (Gopalakrishnan et al., 2012) by providing appropriate training and a safe working environment (Vachon, 2007), fair trade (Schwartz et al., 2008) and establishing customer relations (Kleindorfer et al., 2009) and long-term partnerships (Farmaki, 2015) among others. Last, the economic dimension of sustainability involves the generation of profits and the growth of the company market share (Kassinis & Soteriou, 2009) while inflicting minimal social and environmental impacts.

It, thus, represents a balanced and holistic approach that recognises the role of all stakeholders and both present and future generations' entitlement to the use of resources (Font & McCabe, 2017). As such, sustainability has become an imperative goal (McDonagh & Prothero, 2014), intensifying the push for legislation from governmental agencies as well as the responsible actions by industries and individuals since a true sustainability requires companies to ensure long-term profitability, the social welfare of their stakeholders such as customers, suppliers and employees and the minimizing the negative impact of their activities on the environment.

In recent years, the concept of sustainability has gained currency within the hotel industry with businesses publicly committing to strategic corporate sustainability agendas (Jones et al., 2016). Nonetheless, questions have been raised over the motives driving the adoption of sustainable practices by hotels (Font & McCabe, 2017; Wymer & Polonsky, 2015). For example, it has been suggested that many of these sustainability programmes are adopted mainly because of the financial gains associated with cost-efficiency or attempts to enhance corporate image and build community relationships and employee loyalty through greenwashing activities (Font et al., 2012; Jones et al., 2016). Indeed, the profitability of companies has been found to improve as a result of sustainability practices adoption, which are assumed to positively influence consumer perceptions, attitudes and behaviours (Molina-Azorin et al., 2009). Considering the rise of consumer awareness of the importance of environmental-friendly and responsible consumption (Cronin et al., 2011; Sheth et al., 2011), an increasing number of hotels are adopting sustainability practices (Moise et al., 2018).

### Effects of sustainability practices on guest behaviour

A burgeoning number of studies have investigated the effects of sustainability on consumer behaviour (e.g. Chen & Tung, 2014; Kang et al., 2012; Lee et al., 2010; Olya et al., 2019; Verma et al., 2019), mainly focussing on "consumer satisfaction" and "loyalty", which are considered as the predominant attitudinal and behavioural aspects of sustainability. The overwhelming focus on these attitudinal and behavioural aspects is not surprising considering that consumer satisfaction and loyalty are clear indicators of a company's success in providing products and services efficiently.

### Satisfaction

Arguably, satisfaction emerges as the most widely investigated attitudinal trait by hospitality researchers (Cicerali et al., 2017). Defined as "a person's feeling of pleasure or disappointment which resulted from comparing a product's perceived performance or outcome against his/her expectations" (Kotler & Keller, 2006, p. 144), satisfaction has been examined in relation to a number of factors including company profits and market share, customer repeat purchase intention and positive word-of-mouth behaviour (Pizam & Ellis, 1999). While it represents an important construct in the mainstream management literature, in the hotel research domain satisfaction is of particular importance due to the specific characteristics of hotel products. Indeed, considering the perishable, tangible and intangible elements of hotel products, satisfaction emerges as the leading criterion for determining product and service quality (Hayes, 1998).

A large number of studies have also examined the effects of sustainability practices on guest satisfaction (e.g. Berezan et al., 2013; Lu & Stepchenkova, 2012; Slevitch et al., 2013; Yu et al., 2017), concluding that guest satisfaction is positively impacted by the sustainability actions of hotels. More specifically, satisfaction was found to increase when companies adopted environmental-friendly practices (Gao & Mattila, 2014; Yu et al., 2017). Likewise, customer satisfaction emerges as a mediator between environmental practices and company financial performance (Kassinis & Soteriou, 2009). Indeed, the financial performance of companies acts as an indication of their ability to offer high quality services, which is a determinant of consumer satisfaction (Sánchez-Fernández & Iniesta-Bonillo, 2009); hence, customer satisfaction may be an antecedent of company financial performance (Lo et al., 2015). As Sánchez-Franco et al. (2019) professed, satisfaction as an indicator of a service provider's performance efficacy is a key measure of a hotel's competitive advantage. Studies have also reported a positive relationship between satisfaction and social dimension of sustainability. For instance, high employee satisfaction was reported in companies where the management is concerned with the social welfare of their employees (Chi & Gursoy, 2009), which, in turn, leads to high customer satisfaction as satisfied employees are likely to perform better in their jobs (de Leaniz & Rodriguez, 2015). In the hotel industry, where there are a number of close interactions between employees and guests, the importance of employee satisfaction is particularly elevated.

## Loyalty

Loyalty emerges as an important construct in the hospitality literature since it determines future behavioural intentions and indicates guests' trust towards the company (Sipe & Testa, 2018). Defined as the strength of the relationship between one's relative attitudes and repeat purchase (Rather, 2018), customer loyalty reflects consumers' level of attachment to products and brands. According to Oliver (1999 p. 392) customer loyalty is "a deeply held commitment to rebuy or patronise preferred product or service consistently in the future". Thus, customer loyalty is central to success of any business (Toufaily et al., 2013) as it may help companies build long-term mutually beneficial relationships with their clients (Kandampully et al., 2015; Pan et al., 2012) and protect themselves against competitors as loyal customers resist switching (So et al., 2013). Indeed, loyal customers are easier and cheaper to serve than non-loyal customers (Tepeci, 1999), and loyal customers are likely to pay more for a company's offerings (Evanschitzky et al., 2012).

In light of the growing guest interest on sustainability practices within the hotel industry, the influence of environmental, social and economic sustainability practices on customer loyalty has been examined extensively, with findings indicating a positive relationship. For example, companies' environmental-friendly activities have been found to enhance customer loyalty (Chen, 2015; Han et al., 2019; Lee et al., 2010) and contribute to company reputation (Jang et al., 2015), thus, positively influencing customer loyalty. Moreover, the social dimension of sustainability, such as employee welfare schemes, has been found to contribute to brand loyalty (Chi & Gursoy, 2009) as it increases the attractiveness of a company. Likewise, the economic aspect of sustainable actions of companies seems to contribute to consumer loyalty as it improves perceived product quality (Shi et al., 2014) and strengthens company reputation (Pena et al., 2013). Interestingly, customer satisfaction and loyalty have been found to be interrelated (Kim et al., 2013). Specifically, guests' satisfaction was identified as an antecedent of their loyalty (Gursoy et al., 2014) with Loureiro and Kastenholz (2011) and Orel and Kara (2014) recognising the mediating effect of perceived quality between satisfaction and loyalty.

## The role of familiarity

The influence of consumer familiarity with a company or brand on behavioural aspects was considered within hospitality literature (e.g. Ha & Jang, 2010; Lin, 2013), albeit at a lesser extent. Defined as "the number of product-related experiences that have been accumulated by the consumer" (Alba & Hutchinson, 1987, p. 411), familiarity with a product or brand arises from past experiences of using a product. Even though the terms "knowledge" and "familiarity" are often used interchangeably, the latter is considered as an umbrella term that encompasses consumer experience, prior knowledge and strength of belief (Ha & Perks, 2005). Generally speaking, familiarity is of strategic importance to company managers as it directly affects customer behaviour (Türkel et al., 2016). On the one hand, it is influential on customer satisfaction as different levels of familiarity provide customers different frameworks of reference for evaluating products (Söderlund, 2002; Tam, 2008). Indeed, if consumers are unfamiliar with a brand, they may doubt its quality because they lack the information required to make an evaluation (Hoeffler & Keller, 2003). Hence, familiarity with a product reduces uncertainty in future purchase situations (Flavián et al., 2006). On the other hand, familiarity with a product increases the more frequently consumers use the product; therefore, it may be argued that familiarity with a company and its products is related to customer loyalty. According to Tepeci (1999) the more familiar a consumer is with a product/brand, the more likely he/she to purchase the product/brand.

Within the context of sustainability, familiarity with a company's economic, environmental or social practices has been previously studied. For instance, Bourke et al. (2020) argued that hotels promote their financial performances, earnings stability and share values in a way to manipulate the market. Hotels play this game to make their stocks more appealing to the stakeholders. Türkel et al. (2016) found that communication of environmental-friendly practices of familiar brands positively influences purchase intentions. Likewise, Perera and Chaminda (2013) suggested that brand familiarity plays a moderating role between a company's social responsibilities and product evaluation. On a similar note, Plewa et al. (2015) found that perceived familiarity with a company's sustainability actions may positively influence consumer perceptions of companies' sustainability image; thus, contributing to customer loyalty. Nonetheless, the role of familiarity of all three dimensions of sustainability on behavioural aspects has been largely overlooked, especially in relation to the hotel sector.

### Research aim

This study aims to examine the effects of the hotels' economic, social and environmental sustainability practices on guest satisfaction and loyalty, by considering the role of guest familiarity with hotels and their suppliers' sustainability practices. More specifically, this empirical study aims to address three research questions. First, what are the sufficient factors (i.e. economic, social and environmental dimensions and familiarity) in order to increase satisfaction and loyalty of hotel guests? Second, under which conditions does hotel guests' satisfaction and loyalty emerge in relation to the above-named predictors? Third, what factors are necessary to attain satisfaction and loyalty among hotel guests? To answer the first research question, a SEM analysis was conducted to investigate the net effects of predictors on expected outcomes (satisfaction and loyalty). In response to the second research question, fsQCA was conducted to explore causal recipes leading to the expected outcomes of guest satisfaction and loyalty. The third question was addressed using necessary condition analysis that identifies the necessary conditions to ensure guest satisfaction and loyalty.

The multi-faceted nature of the tourism industry (Farmaki, 2018) calls for a holistic consideration of all stakeholders involved in the supply and delivery of the hotel product. Generally speaking, the hotel sector requires collaboration across the supply chain for the effective delivery of the hotel product (Xu & Gursoy, 2015a). Given the fact that hotel sector is one where there is high interaction among key parties and, in particular, between employees and guests (Wells et al., 2016), it is necessary to consider key stakeholders in the examination of sustainability practices as they all share the responsibility for their implementation in hotels. To this end, this study considered guest perspectives on the economic, social and environmental sustainability practices in hotels in relation to in addition to the hotel itself, customers and employees whose interactions are important for the success of sustainability (Farmaki, 2018).

## Methodology

### Questionnaire design and measurement scales

A self-administered survey questionnaire was used to collect data. All constructs were measured utilising items adapted from previous studies (Casaló, Flavián, & Guinalíu, 2008; Lee & Kwon, 2011; Xu and Gursoy, 2015). Following Xu and Gursoy (2015), the survey instrument was designed to measure guest's satisfaction, loyalty, and attitudes to environmental, social and economic sustainability practices. As presented in Table 1, social dimension of sustainability practices was measured by 11 items, environmental dimension by four items and economic dimension by five items. All of the items used to measure environmental, social and economic sustainability practices were adopted from Xu and Gursoy's (2015) study. Environmental sustainability measures included items related to product design such as energy and water saving, use of solar power and communication to customers. Social sustainability measures included items related to work environment, employee safety and wellbeing, and customer orientation, quality and information dissemination. Economic sustainability measures included items related to occupancy and revenue growth, competitiveness and market share growth. Customer satisfaction was

**Table 1.** Descriptive statistics, normality, reliability and confirmatory factor analysis results.

| Scale item | λ | Mean | SD | Kurtosis | Skewness |
|---|---|---|---|---|---|
| *Social Dimension (α: 0.920 ; CR: 0.919; AVE: 0.512)* | | | | | |
| Create a safe and healthy work environment | 0.780** | 4.413 | 0.713 | 4.199 | −1.562 |
| Provide measures that ensure safe and healthy working conditions for all employees | 0.690** | 4.413 | 0.738 | 2.442 | −1.383 |
| Comply with labour legislation and employee contracts | 0.710** | 4.385 | 0.800 | 3.990 | −1.729 |
| Support all employees who want to pursue further education | 0.519** | 4.271 | 0.720 | 0.639 | −0.830 |
| Listen to employees' suggestions | 0.549** | 4.179 | 0.772 | 2.911 | −1.224 |
| Provide all employees with proper and fair wages that reward them for their work | 0.650** | 4.271 | 0.745 | 1.554 | −1.023 |
| Improve product quality and enhance added value | 0.735** | 4.358 | 0.691 | 2.282 | −1.116 |
| Be customer-oriented | 0.819** | 4.408 | 0.744 | 2.972 | −1.499 |
| Provide all customers with high quality services and products | 0.819** | 4.353 | 0.747 | 3.196 | −1.412 |
| Provide all customers with accurate and adequate information in making purchasing decisions. | 0.778** | 4.390 | 0.656 | −0.091 | −0.715 |
| Treat all customers fairly | 0.748** | 4.252 | 0.770 | 1.078 | −0.955 |
| *Environmental Dimension (α: 0.763; CR: 0.756; AVE: 0.509)* | | | | | |
| Implement an energy saving program | 0.654** | 3.766 | 0.951 | −0.613 | −0.257 |
| Use solar power instead of fuel | 0.707** | 3.945 | 0.839 | −0.136 | −0.459 |
| Use water-saving flush in bathrooms | 0.860** | 3.710 | 0.811 | −0.479 | −0.151 |
| Communicate the environmental policy to customers | 0.607** | 3.954 | 0.942 | 0.955 | −0.912 |
| *Economic Dimension (α: 0.793; CR: 0.796; AVE: 0.510)* | | | | | |
| High return on their assets | 0.701** | 3.881 | 0.751 | 0.632 | −0.520 |
| High net sales growth | 0.699** | 3.982 | 0.717 | 1.437 | −0.726 |
| High overall performance and success level | 0.802** | 4.078 | 0.795 | 1.768 | −1.024 |
| High competitive position | 0.700** | 4.124 | 0.735 | 1.428 | −0.829 |
| High occupation rate growth | 0.661** | 3.784 | 0.931 | 0.404 | −0.657 |
| *Familiarity (α:0.789 ; CR:0.795; AVE:0.569)* | | | | | |
| I am familiar with hotels and their suppliers' sustainability practices. | 0.636** | 3.216 | 0.988 | −0.765 | −0.330 |
| Compared to public, I am familiar with hotels and their suppliers' sustainability practices. | 0.903** | 3.248 | 0.983 | −0.844 | −0.194 |
| Compared to my friends and acquaintances, I am familiar with hotels and their suppliers' sustainability practices. | 0.699** | 3.344 | 0.965 | −0.512 | −0.368 |
| *Loyalty (α: 0.815; CR: 0.819; AVE:0.602 )* | | | | | |
| I will recommend this type of hotel to my friends, relatives or colleagues | 0.773** | 3.927 | 0.769 | 1.271 | −0.727 |
| I will spread positive recommendations of this type of hotel to others | 0.838** | 3.922 | 0.722 | 0.323 | −0.470 |
| I will stay at this type of hotel whenever possible | 0.713** | 3.922 | 0.703 | 0.394 | −0.449 |
| *Satisfaction (α:0.879 ; CR: 0.879; AVE: 0.596)* | | | | | |
| I will be very happy if I can stay at this type of hotel | 0.863** | 3.950 | 0.774 | 1.880 | −0.870 |
| I will be very satisfied if a hotel can provide such level of service | 0.860** | 4.000 | 0.684 | 1.204 | −0.520 |
| My choice to stay in the hotel will be a wise one | 0.713** | 4.119 | 0.617 | −0.419 | −0.079 |
| I think it would be the right thing to stay at this type hotel | 0.731** | 3.986 | 0.655 | −0.354 | −0.085 |
| I will be very happy if I can stay at this type of hotel | 0.672** | 4.014 | 0.681 | 0.247 | −0.369 |

*Note:* λ: Confirmatory factor loading value; α: Cronbach's Alpha CR: Composite Reliability; AVE: Average Variance Extracted; SD: standard deviation. Fit statistics: standardised root mean squared residual (SRMR): 0.061 (cut-off <0.08); Normed Fit Index (NFI): 0.934 (cut-off > 0.09).

measured by five items adopted from Caber et al. (2013) and Xu and Gursoy (2015). Customer loyalty was measured by three items that were adopted from Chi (2011) as presented in Table 1. Items for measuring familiarity were adapted from Casaló et al. (2008) and Lee and Kwon (2011). All items were measured on a 5-point Likert-type scale (1 = strongly disagree, 5 = strongly agree). Demographics of respondents including age, gender, education, and income levels were collected in the final section of questionnaire.

**Table 2.** Demographics of respondents.

| Gender | N | % | Age | N | % |
|---|---|---|---|---|---|
| Female | 145 | 66.51 | 17-27 years old | 78 | 35.78 |
| Male | 73 | 33.49 | 28-37 years old | 82 | 37.61 |
| Total | 218 | 100.00 | 38-47 years old | 30 | 13.76 |
| Marital Status | | | 48-57 years old | 20 | 9.17 |
| Single | 93 | 42.70 | Older than 57 years old | 8 | 3.67 |
| Married | 103 | 47.20 | Total | 218 | 100.00 |
| Divorced | 22 | 10.10 | Average Monthly Income ($) | | |
| Total | 218 | 100.00 | Below 1000 | 34 | 15.60 |
| Educational Level | | | 1001-2000 | 136 | 62.39 |
| Diploma and below | 24 | 11.00 | 2001-4000 | 27 | 12.39 |
| Some college degree | 10 | 4.60 | 40001-5000 | 16 | 7.34 |
| Bachelor | 122 | 56.00 | Above 5000 | 5 | 2.29 |
| Master | 58 | 26.60 | Total | 218 | 100.00 |
| PhD | 4 | 1.80 | | | |
| Total | 218 | 100.00 | | | |

A back-translation method was applied to prepare the questionnaire in the Russian and Kazakh languages. A pilot study was conducted by inviting 15 guests to complete the survey to assess the understandability of the questions, the time taken for completion and to identify any other issues. These guests were the loyal business customers of the hotels identified by the hotel and marketing managers of the properties. The outcome of the pilot study was satisfactory and no revisions were necessary.

## Data collection

The research access to the hotels and hotel guests was facilitated through the personal contacts of the local researcher. The research sample included Kazakh guests staying in four and five star independently owned hotels in Almaty. These hotels disclose their sustainable practices via their websites, brochures and flyers. Sustainable practices of the hotels cover wide range of areas including water usage; use of local producers as suppliers and waste recycling.

One of the researchers, fluent both in English and Kazakh languages, personally visited the hotels and run the survey with the guests by using face-to-face approach in order to boost up the response rate and also ensure that questions are clarified whenever it was needed. They were approached at the hotel lobby and invited to participate. Most of these guests were repeat guests visiting Almaty for business.

Using a convenience sampling method, 310 guests of hotels in Kazakhstan were directly approached from August 2017 to October 2017. A total of 220 completed survey questionnaires were returned (71.0% response rate). After removing the two incomplete questionnaires, 218 valid cases (70.3%) were retained for further analysis. The profile of the respondents is presented in Table 2.

## Data analysis

Before analysing the data to test the proposed relationships, the normality of data was tested and the Cronbach alpha score of each construct was estimated. Afterwards, a confirmatory factor analysis was conducted to assess the reliability, scale composition and the validity of the study measures. Construct validity (including convergent and divergent validity) and fit validity (standardised root mean squared residual-SRMR and Normed Fit Index- NFI) of the proposed model is evaluated (Taheri et al., 2019). Structural equation modelling (SEM) was then used to investigate the relationships between the constructs (social, economic, environmental and familiarity) and the effect on satisfaction and loyalty. As SEM is used to identify the net effect of dimensions on the outcome variables, a fuzzy-set Qualitative Comparative Analysis (fsQCA) was subsequently conducted to explore sufficient combinations of the dimensions called causal recipes, to predict the satisfaction and loyalty of hotel guests. fsQCA is set-theoretic approach that can generate new knowledge of sustainability of hotel management by exploring complex conditions leading to satisfaction and loyalty of hotel guests. fsQCA functions based on Boolean Algebra. It involves three stages of data calibration (transforming Likert scale data to fuzzy set data), truth tabulation analysis (computing all possible conditions leading to the expected outcomes) and counter-factual analysis (refining all possible conditions and selecting algorithms that consistently and sufficiently describe recipes for achieving the outcomes) (Olya et al., 2020). Two measures of "consistency" and "coverage" are used to refine algorithms calculated using fsQCA which are analogous to "correlation" and "determination coefficient", respectively (Olya & Mehran, 2017). Analysis of necessary condition was also performed to identify which dimensions are necessary to increase guests' satisfaction and loyalty. Unlike SEM and fsQCA that shows sufficient net and combination effects of predictors, Analysis of necessary condition reveals necessary predictor of the outcome (Gannon et al., 2019; Olya & Al-ansi, 2018; Olya & Han, 2020). This means that it is less likely to stimulate satisfaction and loyalty of guest in the absence of necessary predictors. Results of analysis of necessary condition helps hoteliers to focus on necessary predictors of satisfaction and loyalty of their guests.

## Results

### Measurement model testing

Results of the test of normality of data, descriptive statistics, reliability and confirmatory factor analysis are presented in Table 1. Results of Skewness and Kurtosis confirm that data were nor-mally distributed as Skewness and Kurtosis values fell within the recommend range of ±3 (Han et al., 2019). However, Kurtosis value for three items exceed the cut-off three. To address this issue, we used SEM because of its "ability to fit non-standard models, including flexible handling of longitudinal data, databases with autocorrelated error structures (time series analysis), and databases with non-normally distributed variables" (The Division of Statistics & Scientific Computation, 2012, p. 8). Means and standard division of the scale items were calculated and presented in Table 1. Results of Cronbach's alpha (α) and composite reliability (CR) scores were above recommended level of 0.70, providing satisfactory evidence for reliability of the measures. During the confirmatory factor analysis (CFA), one item from environment dimension (i.e. Develop an environmental policy) was excluded due to low factor loading. Other items significantly and sufficiently loaded to hypothesised factors ($> 0.5$, $p < 0.005$). Average Variance Extracted (AVE) was calculated to evaluate convergent validity of the constructs. Results of AVE showed that the constructs satisfied the recommended level of 0.50 (Khan et al., 2018).

Discriminant validity of the constructs were examined utilising Fornell-Larcker Criterion and Heterotrait-Monotrait Ratio (HTMT) (Table 3). As presented in Table 3, the square root of AVE of each construct was larger than the correlation of all pairs of the constructs, which satisfied the Fornell-Larcker Criterion requirements for discriminant validity. All HTMT values were less than 0.9, which provided further evidence of discriminate validity among study constructs. Results of standardised root mean squared residual (SRMR $< 0.08$) met the accepted criteria for the fit of the proposed model with the empirical data (Hair et al., 2017; Henseler et al., 2015).

### Structural model testing

Direct effects of sustainability dimensions (social, environmental and economic) and guest familiarity with sustainability practices on satisfaction and loyalty are illustrated in Figure 1. According to the results, satisfaction was significantly and positively influenced by familiarity

**Table 3.** Results of discriminate validity.

Fornell-Larcker Criterion

| Construct | 1 | 2 | 3 | 4 | 5 | 6 |
|---|---|---|---|---|---|---|
| 1. Economic Dimension | **0.666** | | | | | |
| 2. Environnemental Dimension | 0.457 | **0.668** | | | | |
| 3. Social Dimension | 0.469 | 0.492 | **0.715** | | | |
| 4. Familiarity | 0.166 | 0.108 | −0.016 | **0.755** | | |
| 5. Loyalty | 0.463 | 0.509 | 0.501 | 0.129 | **0.776** | |
| 6. Satisfaction | 0.492 | 0.436 | 0.428 | 0.153 | 0.704 | **0.772** |

Heterotrait-Monotrait Ratio (HTMT)

| Construct | 1 | 2 | 3 | 4 | 5 | 6 |
|---|---|---|---|---|---|---|
| 1. Economic Dimension | | | | | | |
| 2. Environnemental Dimension | 0.453 | | | | | |
| 3. Social Dimension | 0.464 | 0.466 | | | | |
| 4. Familiarity | 0.189 | 0.185 | 0.424 | | | |
| 5. Loyalty | 0.463 | 0.496 | 0.078 | 0.142 | | |
| 6. Satisfaction | 0.500 | 0.423 | 0.494 | 0.153 | 0.707 | |

Note: Diagonal bolded values are square root of AVE.

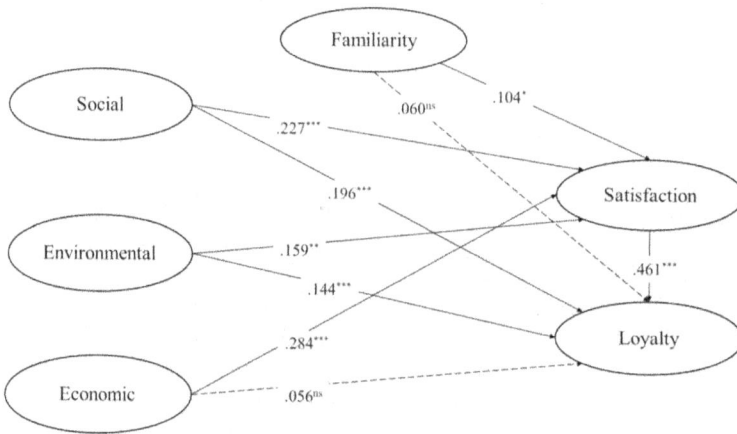

**Figure 1.** Results of regression analysis.
Note: non-significant effect indicated as dotted line. * $p < 0.05$, ** $p < 0.01$, *** $p < 0.001$.

($\beta = 0.104$, $p < 0.05$), the social dimension ($\beta = 0.227$, $p < 0.001$), the environmental dimension ($\beta = 0.159$, $p < 0.01$) and the economic dimension ($\beta = 0.284$, $p < 0.001$). Loyalty was also improved by the social ($\beta = 0.196$, $p < 0.001$) and environmental dimensions ($\beta = 0.144$, $p < 0.001$). There was no significant relationship between economic dimension and guest loyalty and none between familiarity and guest loyalty.

### Configurational model testing

In order to further understand the implications of the mix of those three dimensions of CSR, fsQCA was conducted to identify the causal recipes. The results of this configurational modelling are presented in Table 4. According to fsQCA results, there were three causal recipes that explained the conditions for achieving guests' satisfaction (coverage: 0.957, consistency: 0.933). Model 1 indicated that a combination of the environmental and social dimensions offered the causal conditions that predicted guests' satisfaction. Findings further indicated that a combination (Model 2) of the provision of social and economic dimensions was also a significant

**Table 4.** Results of configurational modelling.

| Causal model for predicting satisfaction | Raw coverage | Unique coverage | Consistency |
|---|---|---|---|
| M1: Environmental * Social | 0.884 | 0.023 | 0.948 |
| M2: Social * Economic | 0.911 | 0.039 | 0.949 |
| M3: Social * Familiarity | 0.682 | 0.009 | 0.980 |
| solution coverage: 0.957 | | | |
| solution consistency: 0.933 | | | |
| Causal model for predicting loyalty | | | |
| M1: Environmental * Social | 0.902 | 0.027 | 0.935 |
| M2: Social * Economic | 0.922 | 0.032 | 0.927 |
| M3: Social * Familiarity | 0.629 | 0.008 | 0.961 |
| solution coverage: 0.969 | | | |
| solution consistency: 0.913 | | | |

*Note*: M stands for model.

**Table 5.** Results of analysis of necessary condition.

| Antecedent necessary condition | Outcome: Satisfaction | | Outcome: Loyalty | |
|---|---|---|---|---|
| | Consistency | Coverage | Consistency | Coverage |
| Social | **0.974** | 0.902 | **0.983** | 0.880 |
| ~Social | 0.175 | 0.973 | 0.178 | 0.955 |
| Environmental | **0.892** | 0.939 | **0.909** | 0.924 |
| ~Environmental | 0.306 | 0.986 | 0.312 | 0.972 |
| Economic | **0.923** | 0.937 | **0.931** | 0.913 |
| ~Economic | 0.275 | 0.937 | 0.276 | 0.967 |
| Familiarity | 0.690 | 0.976 | 0.698 | 0.954 |
| ~Familiarity | 0.532 | 0.962 | 0.544 | 0.951 |
| Satisfaction | | | **0.965** | 0.932 |
| ~Satisfaction | | | 0.261 | 0.969 |

*Note*: necessary antecedent condition is highlighted in bold (consistency >0.9). ~ is outcome negation.

determinant of satisfaction. Results of Model 3 indicated that the social dimension and the familiarity of guests with sustainable practices presented a causal recipe, leading to satisfaction.

Similar to satisfaction, fsQCA identified three causal recipes that described the conditions that resulted in guests' loyalty (coverage: 0.969, consistency: 0.913) and in each case the social dimension was necessary. Results of analysis of necessary condition are provided in Table 5. Social, environmental and economic dimensions were necessary conditions to ensure guests' satisfaction and loyalty as the magnitude of consistency values was above or close to 0.90. Consistency above 0.85 is considered as an acceptable cut-off of necessary condition (Olya & Han, 2020). Familiarity was not necessary to improve satisfaction and loyalty of hotel guests. Satisfaction was a necessary factor to achieve guests' loyalty.

## Discussion and implications

This study aimed at examining the effects of the three dimensions of sustainability practices of hotels (economic, social and environmental) on hotel guest satisfaction and loyalty by considering the role of familiarity with supply chain's sustainability practices. A second aim of this study was to identify the conditions that are necessary for the achievement of guest satisfaction and loyalty. By being the first empirical study to examine the necessary conditions for guest satisfaction and loyalty, this study contributes to both theoretical and practical knowledge that may be of use to academics, industry practitioners and policymakers alike.

Overall, study findings confirm that there are a significant and positive relationships between the social, environmental and economic dimensions of sustainability and guest satisfaction, thus concurring with previous studies (Berezan et al., 2013; Modica et al., 2020; Xu & Gursoy, 2015b; Yu et al., 2017). This finding is not surprising considering the increasing awareness among hotel guests of the importance of sustainability (Cronin et al., 2011; Moise et al., 2018; Sheth et al.,

2011). As such, hotels need to incorporate sustainability practices in their business and marketing activities if they want to ensure guest satisfaction and firm competitiveness. Indeed, it was previously highlighted that customer satisfaction may contribute to company performance (Lo et al., 2015) and help a firm gain competitive advantage (Sánchez-Franco et al., 2019). With regard to the lesser examined variable of familiarity with sustainability practices, this study found that familiarity has a significant and positive effect on guest satisfaction. In other words, if guests are familiar with the hotels' sustainability practices, they will be more satisfied with the hotel. Therefore, study findings provide support to previous investigations asserting that familiarity may enhance customer satisfaction, as it reinforces product evaluation and purchase intention (Perera & Chaminda, 2013; Türkel et al., 2016). Within this context, it may be argued that hotels should adequately communicate their sustainability practices to their current and/or prospective guests.

Additionally, we identified positive relationships between the social and environmental dimensions of sustainability and guest loyalty but not between the economic dimension and guest loyalty. This finding suggest that guests are more likely to return to a hotel if the hotel is socially and environmentally responsible, however, the hotel's and its suppliers economic actions have no impact on their loyalty towards the hotel. This finding contradicts previous research on the influence of the sustainability dimensions on guest loyalty. For example, Modica et al. (2020) found the economic dimension of sustainability to directly influence hotel guest loyalty. Similarly, Xu and Gursoy (2015b) established a relationship between the economic and environmental dimensions and guest loyalty but not with the social dimension of sustainability. Evidently, hotels need to communicate their social and environmental practices to guests in order to enhance loyalty in Kazakhstan, as previously suggested (Chen, 2015; Lee et al., 2010). Indeed, such communication tactics may help improve the reputation of the company among its customers (Jang et al., 2015). In addition, as customer attitudes and behaviours towards the firm are informed by its approach to sustainability and by the cultural orientation of customers (Choi et al., 2016), it may be argued that the lack of a relationship between the economic dimension of sustainability and guest loyalty results from the cultural context within which the study was undertaken. Indeed, cultural differences were acknowledged as an antecedent to customer attitudes and behaviours towards sustainability (Miska et al., 2018).

Interestingly, in relation to familiarity, we found no effect of guest's familiarity with sustainability practices on their loyalty contrary to past studies that reported a significant relationship between these two variables (e.g. Plewa et al., 2015). However, results from fsQCA indicated that a combination of familiarity and social dimension increases guest loyalty. This is in line with Woodside (2014) and Olya and Altinay (2016) who argued that the role of each antecedent depends on the attribute (presence or absence) of other antecedents in a given recipe leading to the expected outcome.

This study contributes to the sustainability literature in the field of hospitality by exploring sufficient recipes (combination of three sustainability dimension along with familiarity) leading to guest satisfaction and loyalty. The fsQCA results suggest that for significant groups within the population investigated, the social dimension is of significance; thus, supporting arguments of previous studies, which suggest that the sustainable practices of a firm are closely linked to the social background of Asian cultures (Xu & Yang, 2010). Many studies have suggested that collectivist and long-term oriented societies, such the Kazakh society, place higher emphasis on sustainability practices that can benefit the society (Lee & Herold, 2016; McCarty & Shrum, 2001). Thus, collectivist and long-term oriented nature of Kazakh culture lead Kazakh consumers to place greater emphasis on social and environmental benefits of sustainability practices for their communities than the economic and financial indulgence, which result in support for businesses that initiate and implement social and environmental sustainable business practices. Considering the fact that hospitality companies have historically emphasised the environmental practices of sustainability (Chen & Peng, 2016) with the social dimension of sustainability receiving less attention in company sustainability reports (Bohdanowicz & Zientara, 2009), it is crucial that hospitality businesses located Kazakhstan not only focus on environmental sustainability practices but also place great emphasis on sustainability practices related to social dimension.

Study findings indicate that all three dimensions of sustainability are necessary conditions for hotel guest satisfaction. Therefore, results of this study provide significant insight to hotel managers in Central Asian contexts for highlighting the development and implementation of sustainability practices that encompass all three dimensions. It is also important that managers understand the interactions between social, economic and environmental dimensions of sustainability practices, which can significantly increase benefits of sustainability practices for the success of a company (Xu and Gursoy 2015). For example, hotels may provide training and advancement opportunities that can improve social welfare while motivating employees to provide higher quality services and, in turn, improve customer satisfaction and revenue. It is also critical that the marketing departments of hotels need to disseminate information on practices related to these three dimensions of sustainability and promote their implementation accordingly through print and online channels.

Moreover, this study found that all three dimensions of sustainability are necessary factors to achieve loyalty among hotel guests. Thus, hotel managers need to pay close attention to raising awareness of their sustainability practices among guests in order to enhance guest loyalty. Indeed, previous studies highlight the importance of sustainability on guest loyalty as such practices may improve company attractiveness (Chi & Gursoy, 2009), perceived product quality (Shi et al., 2014) and company reputation (Pena et al., 2013). Hotel managers should also be aware of the familiarity level of their guests in terms of sustainability practices, as its combination with social dimension may increase their loyalty.

## Conclusion

This empirical study examined the effects of the Kazakh hotels' sustainable practices, namely the economic, social and environmental dimensions, on guest satisfaction and loyalty. This study also investigated the role of familiarity with sustainability practices in predicting guest satisfaction and loyalty, which has been largely overlooked in extant literature. Furthermore, this study is the first to examine the conditions that are necessary for the achievement of guest satisfaction and loyalty. Thus, this study contributes to existing theoretical and practical knowledge on the implementation of sustainability within the hotel industry.

Overall, this study concludes that familiarity and all three dimensions of sustainability are *sufficient* to increase hotel guests' satisfaction. While social and environmental dimensions emerged as sufficient antecedents of loyalty, familiarity and the economic dimension alone are found to be *insufficient* to achieve guest loyalty. The fsQCA results indicated how a combination of these three dimensions of sustainability with guest familiarity of sustainable practices may influence satisfaction and loyalty. Specifically, a combination of the social dimension with each of the three antecedents (familiarity and environmental and economic dimensions of sustainability) form *recipes* to improve satisfaction and loyalty. These results show that although familiarity alone is insufficient, its combination with the social dimension may increase guest satisfaction and loyalty. Similarly, the economic dimension is insufficient to increase loyalty but along with the social dimension, it may encourage guests to revisit the hotel. In terms of the necessity of antecedents of satisfaction and loyalty, the social, environmental and economic dimensions are found to be *necessary* to achieve guest satisfaction and loyalty. Study findings indicate that familiarity is *unnecessary* to obtain the expected outcomes; however, satisfaction is a requirement in order to obtain loyalty entailing that without satisfaction, it is less likely for hotels to have loyal guests.

Findings of this study provide important practical insights with regard to sustainability in the hotel industry, particularly in Central Asian cultural contexts, which have insofar received very little scholarly attention. For instance, study findings indicate that hotel managers need to implement and communicate all three dimensions of sustainability practices to hotel guests as these dimensions contribute to the enhancement of guest satisfaction and loyalty. Guests with high levels of familiarity with sustainability practices of hotel are most likely to be satisfied if the hotel follows sustainability principles. Managers can, therefore, enhance guest loyalty by improving their familiarity along with knowledge on the social dimensions of sustainability practices, which appear to be greatly valued within Kazakhstan contexts (Xu & Yang, 2010).

In conclusion, this study addresses important gaps in the literature on hotel sustainability in terms of geographical scope and scale, as it considers Kazakhstan guests' perceptions of the sustainability of the hotel sector. To this end, this study yields significant theoretical and practical insights. Notwithstanding, the study is not free of limitations. Study findings are based on cross-sectional data collected from one Central Asian country, which may inhibit the generalisation of findings to the hotel industries in other Asian countries. This study examined the perceptions of hotel guests only. Therefore, future research may want to conduct experimental research to involve the views of other stakeholders across the entire hotel supply chain in other Asian cultural contexts as well. Specifically, future research may be performed to consider the views of hotel managers in bridging the gap between suppliers and guests regarding hotel sustainability practices. Finally, as the literature suggests, young, educated and female guests are most likely to be informed and engaged in sustainability practices (Muralidharan & Sheehan, 2018), future research may examine the influence of guest demographics on the perceptions of sustainability in hotel.

## Disclosure statement

No potential conflict of interest was reported by the author(s).

## ORCID

Hossein Olya (iD) http://orcid.org/0000-0002-0360-0744
Anna Farmaki (iD) http://orcid.org/0000-0002-9996-5632

## References

Alba, J. W., & Hutchinson, J. W. (1987). Dimensions of consumer expertise. *Journal of Consumer Research*, *13*(4), 411–454. https://doi.org/10.1086/209080

Berezan, O., Raab, C., Yoo, M., & Love, C. (2013). Sustainable hotel practices and nationality: The impact on guest satisfaction and guest intention to return. *International Journal of Hospitality Management, 34*, 227–233. https://doi.org/10.1016/j.ijhm.2013.03.010

Bohdanowicz, P., & Zientara, P. (2009). Hotel companies' contribution to improving the quality of life of local communities and the well-being of their employees. *Tourism and Hospitality Research, 9*(2), 147–158. https://doi.org/10.1057/thr.2008.46

Bourke, J. G., Izadi, J., & Olya, H. G. (2020). Failure of play on asset disposals and share buybacks: application of game theory in the international hotel market. *Tourism Management, 77*, 103984. https://doi.org/10.1016/j.tourman.2019.103984

Caber, M., Albayrak, T., & Loiacono, E. T. (2013). The Classification of Extranet Attributes in Terms of Their Asymmetric Influences on Overall User Satisfaction. *Journal of Travel Research, 52*(1), 106–116. https://doi.org/10.1177/0047287512451139

Carroll, A., & Buchholtz, A. (2014). *Business and Society: ethics, sustainability, and stakeholder management.* Nelson Education.

Chen, M. F., & Tung, P. J. (2014). Developing an extended theory of planned behavior model to predict consumers' intention to visit green hotels. *International Journal of Hospitality Management, 36*, 221–230. https://doi.org/10.1016/j.ijhm.2013.09.006

Chen, R. J. (2015). From sustainability to customer loyalty: A case of full-service hotels' guests. *Journal of Retailing and Consumer Services, 22*, 261–265. https://doi.org/10.1016/j.jretconser.2014.08.007

Chen, X., & Peng, Q. (2016). A content analysis of corporate social responsibility: Perspectives from China's top 30 hotel-management companies. *Hospitality & Society, 6*(2), 153–181. https://doi.org/10.1386/hosp.6.2.153_1

Chi, G-q. (2011). Destination Loyalty Formation and Travelers' Demographic Characteristics: A Multiple Group Analysis Approach. *Journal of Hospitality & Tourism Research, 35*(2), 191–212. https://doi.org/10.1177/1096348010382233

Chi, C. G., & Gursoy, D. (2009). Employee satisfaction, consumer satisfaction, and financial performance: an empirical examination. *International Journal of Hospitality Management, 28*(2), 245–253. https://doi.org/10.1016/j.ijhm.2008.08.003

Choi, J., Chang, Y. K., Li, Y. J., & Jang, M. G. (2016). Doing good in another neighborhood: Attributions of CSR motives depend on corporate nationality and cultural orientation. *Journal of International Marketing, 24*(4), 82–102. https://doi.org/10.1509/jim.15.0098

Cicerali, E. E., Kaya Cicerali, L., & Saldamlı, A. (2017). Linking psycho-environmental comfort factors to tourist satisfaction levels: Application of a psychology theory to tourism research. *Journal of Hospitality Marketing & Management, 26*(7), 717–734. https://doi.org/10.1080/19368623.2017.1296395

Cronin, J. J., Smith, J. S., Gleim, M. R., Martinez, J., & Ramirez, E. (2011). Green marketing strategies: an examination of stakeholders and the opportunities they present. *Journal of the Academy of Marketing Science, 39*(1), 158–174. https://doi.org/10.1007/s11747-010-0227-0

de Grosbois, D. (2012). Corporate social responsibility reporting by the global hotel industry: Commitment, initiatives and performance. *International Journal of Hospitality Management, 31*(3), 896–905. https://doi.org/10.1016/j.ijhm.2011.10.008

de Leaniz, P. M. G., & Rodriguez, I. R. D. B. (2015). Exploring the antecedents of hotel consumer loyalty: a social identity perspective. *Journal of Hospitality Marketing and Management, 24*(1), 1–23.

Deloitte (2014). Hospitality 2015: game changers or spectators. Available at: https://www2.deloitte.com/ie/en/pages/consumer-business/articles/hospitality-2015.html.

Elkington, J. (1997). *Cannibals with forks: The triple bottom line of twentieth century business.* Oxford.

Evanschitzky, H., Ramaseshan, B., Woisetschläger, D. M., Richelsen, V., Blut, M., & Backhaus, C. (2012). Consequences of customer loyalty to the loyalty program and to the company. *Journal of the Academy of Marketing Science, 40*(5), 625–638. https://doi.org/10.1007/s11747-011-0272-3

Farmaki, A. (2015). Regional network governance and sustainable tourism. *Tourism Geographies, 17*(3), 385–407. https://doi.org/10.1080/14616688.2015.1036915

Farmaki, A. (2018). Corporate social responsibility in hotels: a stakeholder approach. *International Journal of Contemporary Hospitality Management, 31*(6), 2297–2320. https://doi.org/10.1108/IJCHM-03-2018-0199

Flavián, C., Guinalíu, M., & Gurrea, R. (2006). The role played by perceived usability, satisfaction and consumer trust on website loyalty. *Information & Management, 43*(1), 1–14. https://doi.org/10.1016/j.im.2005.01.002

Font, X., & McCabe, S. (2017). Sustainability and marketing in tourism: its contexts, paradoxes, approaches, challenges and potential. *Journal of Sustainable Tourism, 25*(7), 869–883. https://doi.org/10.1080/09669582.2017.1301721

Font, X., Walmsley, A., Cogotti, S., McCombes, L., & Hausler, N. (2012). Corporate social responsibility: The disclosure-performance gap. *Tourism Management, 33*(6), 1544–1553. https://doi.org/10.1016/j.tourman.2012.02.012

Gannon, M., Taheri, B., & Olya, H. (2019). Festival quality, self-connection, and bragging. *Annals of Tourism Research, 76*, 239–252. https://doi.org/10.1016/j.annals.2019.04.014

Gao, Y., & Mattila, A. S. (2014). Improving consumer satisfaction in green hotels: the roles of perceived warmth, perceived competence and CSR motive. *International Journal of Hospitality Management, 42*, 20–31. https://doi.org/10.1016/j.ijhm.2014.06.003

Garay, L., & Font, X. (2012). Doing good to do well? Corporate social responsibility reasons, practices and impacts in small and medium accommodation enterprises. *International Journal of Hospitality Management, 31*(2), 329–337. https://doi.org/10.1016/j.ijhm.2011.04.013

Gopalakrishnan, K., Yusuf, Y. Y., Musa, A., Abubakar, T., & Ambursa, H. M. (2012). Sustainable supply chain management: A case study of British aerospace systems. *International Journal of Production Economics, 140*(1), 193–203. https://doi.org/10.1016/j.ijpe.2012.01.003

Gursoy, D., Chen, S. J., G., & Chi, C. (2014). Theoretical examination of destination loyalty formation. *International Journal of Contemporary Hospitality Management, 26*(5), 809–827. https://doi.org/10.1108/IJCHM-12-2013-0539

Ha, H. Y., & Perks, H. (2005). Effects of consumer perceptions of brand experience on the web: Brand familiarity, satisfaction and brand trust. *Journal of Consumer Behaviour, 4*(6), 438–452. https://doi.org/10.1002/cb.29

Ha, J., & Jang, S. S. (2010). Effects of service quality and food quality: The moderating role of atmospherics in an ethnic restaurant segment. *International Journal of Hospitality Management, 29*(3), 520–529. https://doi.org/10.1016/j.ijhm.2009.12.005

Hair, J. F., Hult, G. T. M., Ringle, C. M., & Sarstedt, M. (2017). *A primer on partial least squares structural equation modeling (PLS-SEM)* (2nd ed.). Sage.

Han, H., & Hyun, S. (2018a). Eliciting customer green decisions related to water saving at hotels: Impact of customer characteristics. *Journal of Sustainable Tourism, 26*(8), 1437–1452. https://doi.org/10.1080/09669582.2018.1458857

Han, H., & Hyun, S. (2018b). What influences water conservation and towel reuse practices of hotel guests?. *Tourism Management, 64*, 87–97. https://doi.org/10.1016/j.tourman.2017.08.005

Han, H., Lee, J., Trang, H., & Kim, W. (2018). Water conservation and waste reduction management for increasing guest loyalty and green hotel practices. *International Journal of Hospitality Management, 75*, 58–66. https://doi.org/10.1016/j.ijhm.2018.03.012

Han, H., Yu, J., Lee, J., & Kim, W. (2019). Impact of hotels' sustainability practices on guest attitudinal loyalty: Application of loyalty chain stages theory. *Journal of Hospitality Marketing & Management, 28*(8), 905–925. https://doi.org/10.1080/19368623.2019.1570896

Hayes, B. E. (1998). *Measuring customer satisfaction: Survey design, use, and statistical analysis methods.* ASQ Quality Press.

Henseler, J., Ringle, C. M., & Sarstedt, M. (2015). A New Criterion for Assessing Discriminant Validity in Variance-based Structural Equation Modeling. *Journal of the Academy of Marketing Science, 43*(1), 115–135. https://doi.org/10.1007/s11747-014-0403-8

Hoeffler, S., & Keller, K. L. (2003). The marketing advantages of strong brands. *Journal of Brand Management, 10*(6), 421–445. https://doi.org/10.1057/palgrave.bm.2540139

Jang, Y. J., Kim, W. G., & Lee, H. Y. (2015). Coffee shop consumers' emotional attachment and loyalty to green stores: the moderating role of green consciousness. *International Journal of Hospitality Management, 44*, 146–156. https://doi.org/10.1016/j.ijhm.2014.10.001

Jones, P., Hillier, D., & Comfort, D. (2016). Sustainability in the hospitality industry: Some personal reflections on corporate challenges and research agendas. *International Journal of Contemporary Hospitality Management, 28*(1), 36–67. https://doi.org/10.1108/IJCHM-11-2014-0572

Kandampully, J., Zhang, T., & Bilgihan, A. (2015). Customer loyalty: a review and future directions with a special focus on the hospitality industry. *International Journal of Contemporary Hospitality Management, 27*(3), 379–414. https://doi.org/10.1108/IJCHM-03-2014-0151

Kang, K. H., Stein, L., Heo, C. Y., & Lee, S. (2012). Consumers' willingness to pay for green initiatives of the hotel industry. *International Journal of Hospitality Management, 31*(2), 564–572. https://doi.org/10.1016/j.ijhm.2011.08.001

Kassinis, G. I., & Soteriou, A. C. (2009). Greening the service profit chain: the impact of environmental management practices. *Production and Operations Management, 12*(3), 386–403. https://doi.org/10.1111/j.1937-5956.2003.tb00210.x

Khan, H. u R., Ali, M., Olya, H. G. T., Zulqarnain, M., & Khan, Z. R. (2018). Transformational leadership, corporate social responsibility, organizational innovation, and organizational performance: Symmetrical and asymmetrical analytical approaches. *Corporate Social Responsibility and Environmental Management, 25*(6), 1270–1283. https://doi.org/10.1002/csr.1637

Kim, H., Cha, J., Singh, A. J., & Knutson, B. (2013). A longitudinal investigation to test the validity of the American consumer satisfaction model in the U.S. hotel industry. *International Journal of Hospitality Management, 35*, 193–202. https://doi.org/10.1016/j.ijhm.2013.05.004

Kleindorfer, P. R., Singhal, K., & Wassenhove, L. N. V. (2009). Sustainable operations management. *Production and Operations Management, 14*(4), 482–492. https://doi.org/10.1111/j.1937-5956.2005.tb00235.x

Kotler, P., & Keller, K. (2006). *Marketing management.* Prentice Hall.

Law, A., De Lacy, T., Lipman, G., & Jiang, M. (2016). Transitioning to a green economy: the case of tourism in Bali. *Journal* of Cleaner Production, *111*, 295–305. https://doi.org/10.1016/j.jclepro.2014.12.070

Lee, J. S., Hsu, (Jane, ), L.-T., Han, H., & Kim, Y. (2010). Understanding how consumers view green hotels: how a hotel's green image can influence behavioural intentions. *Journal of Sustainable Tourism, 18*(7), 901–914. https://doi.org/10.1080/09669581003777747

Lee, K. H., & Herold, D. M. (2016). Cultural relevance in corporate sustainability management: a comparison between Korea and Japan. *Asian Journal of Sustainability and Social Responsibility, 1*(1), 1–21. https://doi.org/10.1186/s41180-016-0003-2

Lee, Y., & Kwon, O. (2011). Intimacy, familiarity and continuance intention: An extended expectation–confirmation model in web-based services. *Electronic Commerce Research and Applications, 10*(3), 342–357. https://doi.org/10.1016/j.elerap.2010.11.005

Lin, Y. C. (2013). Evaluation of co-branded hotels in the Taiwanese market: the role of brand familiarity and brand fit. *International Journal of Contemporary Hospitality Management, 25*(3), 346–364. https://doi.org/10.1108/09596111311311017

Lo, A., Wu, C., & Tsai, H. (2015). The impact of service quality on positive consumption emotions in resort and hotel spa experiences. *Journal of Hospitality Marketing & Management, 24*(2), 155–179. https://doi.org/10.1080/19368623.2014.885872

Loureiro, S. M. C., & Kastenholz, E. (2011). Corporate reputation, satisfaction, delight, and loyalty towards rural lodging units in Portugal. *International Journal of Hospitality Management, 30*(3), 575–583. https://doi.org/10.1016/j.ijhm.2010.10.007

Lu, W., & Stepchenkova, S. (2012). Ecotourism experiences reported online: Classification of satisfaction attributes. *Tourism Management, 33*(3), 702–712. https://doi.org/10.1016/j.tourman.2011.08.003

Martinez, P., & del Bosque, I. R. D. (2013). CSR and consumer loyalty: the roles of trust, consumer identification with the company and satisfaction. *International Journal of Hospitality Management, 35*, 89–99.

Marzhan, T. (2015). Social, environmental and economic sustainability of Kazakhstan: a long-term perspective. *Journal of Central Asian Survey, 34*(4), 456–483.

McCarty, J. A., & Shrum, L. J. (2001). The influence of individualism, collectivism, and locus of control on environmental beliefs and behavior. *Journal of Public Policy & Marketing, 20*(1), 93–104. https://doi.org/10.1509/jppm.20.1.93.17291

McDonagh, P., & Prothero, A. (2014). Sustainability marketing research: Past, present and future. *Journal of Marketing Management, 30*(11-12), 1186–1219. https://doi.org/10.1080/0267257X.2014.943263

Miska, C., Szőcs, I., & Schiffinger, M. (2018). Culture's effects on corporate sustainability practices: A multi-domain and multi-level view. *Journal of World Business, 53*(2), 263–279. https://doi.org/10.1016/j.jwb.2017.12.001

Modica, P. D., Altinay, L., Farmaki, A., Gursoy, D., & Zenga, M. (2020). Consumer perceptions towards sustainable supply chain practices in the hospitality industry. *Current Issues in Tourism, 23*(3), 358–375. https://doi.org/10.1080/13683500.2018.1526258

Moise, M. S., Gil-Saura, I., & Ruiz-Molina, M. E. (2018). Effects of green practices on guest satisfaction and loyalty. *European Journal of Tourism Research, 20*(20), 92–104.

Molina-Azorin, J. F., Claver-Cortes, E., Lopez-Gamero, M. D., & Tari, J. T. (2009). Green management and financial performance: a literature review. *Management Decisions, 47*(7), 1080–1110.

Muralidharan, S., & Sheehan, K. (2018). The Role of Guilt in Influencing Sustainable Pro-Environmental Behaviors among Shoppers: Differences in Response by Gender to Messaging about England's Plastic-Bag Levy. *Journal of Advertising Research, 58*(3), 349–362. https://doi.org/10.2501/JAR-2017-029

Mussina, K. P., & Bimerev, R. T. (2018). Analysis of Modern Hospitality Industry in Kazakhstan. *International Innovation Research, 15*, 164–167.

Namkung, Y., & Jang, S. (2017). Are consumers willing to pay more for green practices at restaurants? *Journal of Hospitality & Tourism Research, 41*(3), 329–356. https://doi.org/10.1177/1096348014525632

Oliver, R. L. (1999). Whence consumer loyalty?. *Journal of Marketing, 63*(4_suppl1), 33–44. https://doi.org/10.1177/00222429990634s105

Olya, H. G., & Altinay, L. (2016). Asymmetric modeling of intention to purchase tourism weather insurance and loyalty. *Journal of Business Research, 69*(8), 2791–2800. https://doi.org/10.1016/j.jbusres.2015.11.015

Olya, H. G., & Han, H. (2020). Antecedents of space traveler behavioral intention. *Journal of Travel Research, 59*(3), 528–544. https://doi.org/10.1177/0047287519841714

Olya, H. G., & Mehran, J. (2017). Modelling tourism expenditure using complexity theory. *Journal of Business Research, 75*, 147–158. https://doi.org/10.1016/j.jbusres.2017.02.015

Olya, H. G., Bagheri, P., & Tümer, M. (2019). Decoding behavioural responses of green hotel guests: A deeper insight into the application of the theory of planned behaviour. *International Journal of Contemporary Hospitality Management, 31*(6), 2509–2525. https://doi.org/10.1108/IJCHM-05-2018-0374

Olya, H., Jung, T. H., Dieck, M. C. T., & Ryu, K. (2020). Engaging visitors of science festivals using augmented reality: asymmetrical modelling. *International Journal of Contemporary Hospitality Management, 32*(2), 769–796. https://doi.org/10.1108/IJCHM-10-2018-0820

Orel, F. D., & Kara, A. (2014). Supermarket self-checkout service quality, customer satisfaction, and loyalty: Empirical evidence from an emerging market. *Journal of Retailing and Consumer Services, 21*(2), 118–129. https://doi.org/10.1016/j.jretconser.2013.07.002

Pan, Y., Sheng, S., & Xie, F. T. (2012). Antecedents of customer loyalty: An empirical synthesis and re-examination. *Journal of Retailing and Consumer Services, 19*(1), 150–158. https://doi.org/10.1016/j.jretconser.2011.11.004

Pena, A. I. P., Jamilena, D. M. F., & Molina, M. A. R. (2013). Antecedents of loyalty toward rural hospitality enterprises: the moderating effect of the consumer's previous experience. *International Journal of Hospitality Management, 34*, 127–137.

Perera, L. C. R., & Chaminda, J. W. D. (2013). Corporate social responsibility and product evaluation: The moderating role of brand familiarity. *Corporate Social Responsibility and Environmental Management, 20*(4), 245–256. https://doi.org/10.1002/csr.1297

Pizam, A., & Ellis, T. (1999). Customer satisfaction and its measurement in hospitality enterprises. *International Journal of Contemporary Hospitality Management, 11*(7), 326–339. https://doi.org/10.1108/09596119910293231

Plewa, C., Conduit, J., Quester, P., & Johnson, C. (2015). The Impact of corporate Volunteering on CSR Image: A Consumer Perspective. *Journal of Business Ethics, 127*(3), 643–659. https://doi.org/10.1007/s10551-014-2066-2

Ponnapureddy, S., Priskin, J., Ohnmacht, T., Vinzenz, F., & Wirth, W. (2017). The influence of trust perceptions on German tourists' intention to book a sustainable hotel: A new approach to analysing marketing information. *Journal of Sustainable Tourism, 25*(7), 970–988. https://doi.org/10.1080/09669582.2016.1270953

Prud'homme, B., & Raymond, L. (2013). Sustainable development practices in the hospitality industry: An empirical study of their impact on customer satisfaction and intentions. *International Journal of Hospitality Management*, *34*, 116–126.

Rather, R. A. (2018). Investigating the impact of customer brand identification on hospitality brand loyalty: A social identity perspective. *Journal of Hospitality Marketing & Management*, *27*(5), 487–513. https://doi.org/10.1080/19368623.2018.1404539

Sánchez-Fernández, R., & Iniesta-Bonillo, M. Á. (2009). Efficiency and quality as economic dimensions of perceived value: Conceptualization, measurement, and effect on satisfaction. *Journal of Retailing and Consumer Services*, *16*(6), 425–433. https://doi.org/10.1016/j.jretconser.2009.06.003

Sánchez-Franco, M. J., Navarro-García, A., & Rondán-Cataluña, F. J. (2019). A naive Bayes strategy for classifying customer satisfaction: A study based on online reviews of hospitality services. *Journal of Business Research*, *101*, 499–506. https://doi.org/10.1016/j.jbusres.2018.12.051

Schwartz, K., Tapper, R., & Font, X. (2008). A sustainable supply chain management framework for tour operator. *Journal of Sustainable Tourism*, *16*(3), 298–314. https://doi.org/10.2167/jost785.0

Seilov, G. A. (2015). Does the adoption of customer and competitor orientations make small hospitality businesses more entrepreneurial? *International Journal of Contemporary Hospitality Management*, *27*(1), 71–86. https://doi.org/10.1108/IJCHM-12-2013-0547

Sheth, J. N., Sethia, N. K., & Srinivas, S. (2011). Mindful consumption: a customer-centric approach to sustainability. *Journal of the Academy of Marketing Science*, *39*(1), 21–39. https://doi.org/10.1007/s11747-010-0216-3

Shi, Y., Prentice, C., & He, W. (2014). Linking service quality, consumer satisfaction and loyalty in casinos, does membership matter?. *International Journal of Hospitality Management*, *40*, 81–91. https://doi.org/10.1016/j.ijhm.2014.03.013

Sipe, L. J., & Testa, M. R. (2018). From satisfied to memorable: An empirical study of service and experience dimensions on guest outcomes in the hospitality industry. *Journal of Hospitality Marketing & Management*, *27*(2), 178–195. https://doi.org/10.1080/19368623.2017.1306820

Slevitch, L., Mathe, K., Karpova, E., & Scott-Halsell, S. (2013). "Green" attributes and customer satisfaction: Optimization of resource allocation and performance. *International Journal of Contemporary Hospitality Management*, *25*(6), 802–822. https://doi.org/10.1108/IJCHM-07-2012-0111

So, K. K. F., King, C., Sparks, B. A., & Wang, Y. (2013). The influence of customer brand identification on hotel brand evaluation and loyalty development. *International Journal of Hospitality Management*, *34*, 31–41. https://doi.org/10.1016/j.ijhm.2013.02.002

Söderlund, M. (2002). Customer familiarity and its effects on satisfaction and behavioral intentions. *Psychology & Marketing*, *19*(10), 861–879. https://doi.org/10.1002/mar.10041

Taheri, B., Olya, H., Ali, F., & Gannon, M. J. (2019). Understanding the influence of airport servicescape on traveler dissatisfaction and misbehavior. *Journal of Travel Research*, https://doi.org/10.1177/0047287519877257

Tam, J. L. (2008). Brand familiarity: its effects on satisfaction evaluations. *Journal of Services Marketing*, *22*(1), 3–12. https://doi.org/10.1108/08876040810851914

Teng, C. C., Horng, J. S., Hu, M. L. M., Chien, L. H., & Shen, Y. C. (2012). Developing energy conservation and carbon reduction indicators for the hotel industry in Taiwan. *International Journal of Hospitality Management*, *31*(1), 199–208. https://doi.org/10.1016/j.ijhm.2011.06.006

Tepeci, M. (1999). Increasing brand loyalty in the hospitality industry. *International Journal of Contemporary Hospitality Management*, *11*(5), 223–230. https://doi.org/10.1108/09596119910272757

The Division of Statistics and Scientific Computation (2012). Structural Equation Modeling Using AMOS. The University of Texas at Austin. https://stat.utexas.edu/images/SSC/Site/AMOS_Tutorial.pdf.

Thompson, A. (2007). Survey: Europeans More 'Green' than Americans. *Live Science*. Available at: http://www.livescience.com/4695-survey-europeans-green-americans.html.

Toufaily, E., Ricard, L., & Perrien, J. (2013). Customer loyalty to a commercial website: Descriptive meta-analysis of the empirical literature and proposal of an integrative model. *Journal of Business Research*, *66*(9), 1436–1447. https://doi.org/10.1016/j.jbusres.2012.05.011

Trang, H., Lee, J., & Han, H. (2019). How do green attributes elicit guest pro-environmental behaviors? The case of green hotels in Vietnam. *Journal of Travel & Tourism Marketing*, *36*(1), 14–28. https://doi.org/10.1080/10548408.2018.1486782

Tsai, W. H., Chen, H. C., Liu, J. Y., Chen, S. P., & Shen, Y. S. (2011). Using activity-based costing to evaluate capital investments for green manufacturing systems. *International Journal of Production Research*, *49*(24), 7275–7292. https://doi.org/10.1080/00207543.2010.537389

Türkel, S., Uzunoğlu, E., Kaplan, M. D., & Vural, B. A. (2016). A Strategic Approach to CSR Communication: Examining the Impact of Brand familiarity on Consumer Responses. *Corporate Social Responsibility and Environmental Management*, *23*(4), 228–242. https://doi.org/10.1002/csr.1373

Vachon, S. (2007). Green supply chain practices and the selection of environmental technologies. *International Journal of Production Research*, *45*(18-19), 4357–4379. https://doi.org/10.1080/00207540701440303

Verma, V. K., Chandra, B., & Kumar, S. (2019). Values and ascribed responsibility to predict consumers' attitude and concern towards green hotel visit intention. *Journal of Business Research*, *96*, 206–216. https://doi.org/10.1016/j.jbusres.2018.11.021

Wells, V. K., Smith, D. G., Taheri, B., Manika, D., & McCowlen, C. (2016). An exploration of CSR development in heritage tourism. *Annals of Tourism Research*, *58*, 1–17. https://doi.org/10.1016/j.annals.2016.01.007

Woodside, A. G. (2014). Embrace• perform• model: Complexity theory, contrarian case analysis, and multiple realities. *Journal of Business Research*, *67*(12), 2495–2503. https://doi.org/10.1016/j.jbusres.2014.07.006

Wymer, W., & Polonsky, M. J. (2015). The limitations and potentialities of green marketing. *Journal of Nonprofit & Public Sector Marketing*, *27*(3), 239–262. https://doi.org/10.1080/10495142.2015.1053341

Xu, X., & Gursoy, D. (2015). Influence of sustainable hospitality supply chain management on customers' attitudes and behaviors. *International Journal of Hospitality Management*, *49*, 105–116. https://doi.org/10.1016/j.ijhm.2015.06.003

Xu, S., & Yang, R. (2010). Indigenous characteristics of Chinese corporate social responsibility conceptual paradigm. *Journal of Business Ethics*, *93*(2), 321–333. https://doi.org/10.1007/s10551-009-0224-8

Xu, X., & Gursoy, D. (2015a). A conceptual framework of sustainable hospitality supply chain management. *Journal of Hospitality Marketing & Management*, *24*(3), 229–259. https://doi.org/10.1080/19368623.2014.909691

Xu, X., & Gursoy, D. (2015b). Influence of sustainable hospitality supply chain management on customers' attitudes and behaviors. *International Journal of Hospitality Management*, *49*, 105–116. https://doi.org/10.1016/j.ijhm.2015.06.003

Yu, Y., Li, X., & Jai, T. (2017). The impact of green experience on consumer satisfaction: evidence from TripAdvisor. *International Journal of Contemporary Hospitality Management*, *29*(5), 1340–1361. https://doi.org/10.1108/IJCHM-07-2015-0371

# The anchoring effect of aviation green tax for sustainable tourism, based on the nudge theory

Haeok Liz Kim and Sunghyup Sean Hyun

**ABSTRACT**

Anchors functions as a reference point for human behavior to adjust the boundaries of a plausible range of values for a question, assuming that a given anchor is greater than the boundary value for a range of plausible answers. This study tests the anchoring effect on travelers who may need to pay aviation green tax. The analyses results reveal that the anchor had a significant effect on approximately 65% of the total of 333 participants (excluding those who entered the same amount before and after (N = 216). This study contributes to the literature on sustainable tourism by investing the factors that induce voluntary change in consumer behavior. The practical implications are to improve transparency and to minimize tax resistance by positively influencing converting travelers' perceptions of aviation green taxes. Further, in countries where aviation green tax is not introduced, these results could serve as a guide to encourage travelers' voluntary participation to pay aviation green taxes. As such, tourists' voluntary participation will contribute to sustainable tourism in the long-term.

## Introduction

According to the distribution of emissions by the tourism industry as reported in 2008 by the World Tourism Organization (UNWTO), air transport is expected to increase to 53% by 2035. The tourism industry is increasingly and excessively dependent on air transport (Gössling et al., 2007). In Europe, 11% of traveler movements were by air, but air travel occupied 46% of traveler transport-related emission (UNWTO, 2008). Especially, international aviation significantly exacerbates climate change through fossil fuel consumption and greenhouse gas emissions (Becken, 2007). The aviation industry currently accounts for 2% of all artificial $CO_2$ emissions. If nothing changes, this ratio is expected to increase to 22% by 2050 (CNN, Dec. 17, 2019). Previous studies reported that about 17% of global tourism by air transport accounts for approximately 40% ($CO_2$ only) to 75% (maximum estimate) of radiative forcing, including the non-$CO_2$ impact on climate change (Nawijn & Peeters, 2010; Sausen et al., 2005). The aviation industry comprises the largest share (40%) of total $CO_2$ emissions, followed by car transport (32%) (Scott et al., 2010). The aviation industry's significant impact on climate change is causing increasing concern on environment sustainability. Its exacerbation of global warming, measured by radiative forcing, is much greater due to the effects of non-carbon emissions and cruise altitudes (Lee et al., 2009; Scott et al., 2010).

As a significant part of the modern transportation systems, airlines have a great impact on the development of modern society (Li et al., 2016). In recent years, in concern of the $CO_2$ emissions by airplanes, Swedish environmental activist Greta Thunberg initiated the environmental movement called "flight shame" (flygskam in Swedish); the movement has spread to Europe (*Business Insider*, Aug. 24, 2019). To counter climate change problems caused by tourism, European countries have introduced aviation green taxes (McKercher et al., 2010).

Environmental tax reform (ETR), also known as ecological tax reform, environmental fiscal reform, green tax reform, green tax shifting, or green tax swap (Bosquet, 2000; European Commission, 1997b; Goulder, 1995; Hamond et al., 1997; OECD, 1997; von Weizsacker & Jesinghaus, 1992), is a significant development in the fields of environmental policy and public fiscal reform (Bosquet, 2000). The target of green tax is to raise awareness on environmental policy as well as achieve public fiscal reform.

Many tourism companies have also introduced carbon-offset programs. For example, the United Airlines introduced its Eco-Skies® CarbonChoice carbon offset sponsorship program that calculates each traveler's carbon footprint and invests in projects that go beyond offset to reduce carbon and return it to its destination (United, Carbon Choice carbon offset program; UNWTO, 2017). Moreover, European countries such as France and the Netherlands are imposing environmental aviation tax on passengers using aircrafts. Furthermore, the KLM Royal Dutch Airlines launched a sustainability campaign in 2019 called "Fly Responsibly." The campaign encourages travelers to pack lighter to diminish weight carried by aircrafts so that they burn less fuel, and it offers passengers an easy way to pay offsets for $CO_2$ emissions generated by their travels (*The Washington Post*, Jul. 9, 2019).

The mandatory aviation green tax limits consumers' freedom of choice and is criticized not only by tourists traveling to or from Europe but also by the International Air Transport Association (IATA). The IATA points out the negative aspects of implementing aviation green taxes, such as passenger travel costs, airline revenues, and government revenues (Lange et al., 2018). This study attempts to analyze the anchoring effect of existing aviation tax to travelers' willingness to pay aviation tax, based on nudge theory. As an aspect of anchoring, labeling aims to contribute to voluntary environmental protection behaviors by raising environmental awareness of consumers.

The objective of this study is to discuss the legitimacy of imposing aviation green taxes on passengers in countries that recognize the importance of the environment and climate change. In addition, it examines the anchoring effect of the nudge theory that affects the decisions of consumers who are voluntarily willing to pay environmental tax.

## Theoretical background and hypotheses

### The nudge theory and the anchoring effect

Human behavior is complicated. In the book *Nudge*, "nudge" is defined as "any aspect of the choice architecture that alters people's behavior in a predictable way without forbidding any option or significantly changing their economic incentives" (Thaler & Sunstein, 2009: 7). People often behave in ways that can be predicted by economic theories (Kosters & Van der Heijden, 2015; Thaler & Sunstein, 2009). Thus, "nudge" deviates behavior from those projected by standard economic forecasts (Hansen & Jespersen, 2013). Thaler and Sunstein insist that nudges imply only minimal costs; therefore, they are not troublesome to libertarians (Hausman & Welch, 2010).

Currently, influential behavioral scientists including Daniel Kahneman (2011), Roberto Cialdini (2009), and Dan Ariely (2008) have proven that people behave in ways with the knowledge that it may not benefit their own interests (bounded will power), or they do so, because they deem benevolent behavior to be more fair than selfish behavior (bounded self-interest) (Kosters & Van der Heijden, 2015). To describe the anchoring effect, Tversky and Kahneman (1974) proposed a mechanism by which anchors act as a starting point for adjustment (Strack & Mussweiler, 1997; Tversky & Kahneman, 1974). Strack and Mussweiler (1997) argued that "Anchor values serve as the reference point for people to adjust the boundary of the range of plausible values for the question, presuming that the given anchor is more extreme than the boundary value for the range of plausible answers."

According to *Nudge*, the anchoring effect involves an anchor serving as a nudge, adjustments in the way one thinks is appropriate, and the numbers one knows. Bias occurs, because anchoring is typically insufficient. Anchors affect how people think about their lives. For example, a college student was asked two questions: (a) "How happy are you?" and (b) "How often do you date?" When asked in the order of a then b, the relationship between the two questions was very weak. However, when the order of the questions was switched (first b then a), the correlation increased by .62. The students referred to dating heuristics (how happy they are) in their answers (Thaler & Sunstein, 2009; Strack et al., 1988).

In his book *Predictably Irrational*, Dan Ariely (2008) provides an "arbitrary coherence" of examples of the impact of anchors on purchasing decisions. Arbitrary coherence means that if the price is set as "arbitrary," it not only ends up being low at the time when it is settled but also later on. A total of 55 students from the MIT Sloan School of Business were asked to write down the last two digits of their social security number and bid on items including wine. Analysis results revealed that the last two digits of one's social security number served as an anchor (Ariely, 2008).

Lange et al. (2018) explored a new paradigm measurement tool under controlled laboratory conditions to determine whether people actually practice pro-environmental behavior tasks (PEBTs). The PEBT paradigm is the theoretical basis for the prediction of pro-environmental behavior, and each variable that constitutes the PEBTs significantly influences the other variables (Lange et al., 2018). The anchoring effect and the effect of choice architecture on consumer choice can be explained using the tip culture of the United States. Most restaurants in the United States provide tip choices (15%, 18%, or 20%) on receipts, which act as an another for customer choice. Furthermore, Nelson et al. (2019) studied the donation activity of tourists who have been asked to donate to coastal conservation to solve land and maritime problems. Their analysis results show that the majority of tourists do not think the island will be environmentally sustainable under the practices of the current management and that they have the willingness to pay (WTP) an eco-fee to fund conservation efforts on the island (Nelson et al., 2019).

### Aviation green tax

Environmental taxes are an important tool for governments to adjust the relative prices of goods and services (OECD, 2017). While airport expansions may be justified by the need to support economic prosperity, the transport sector, including aviation, must also cover the incurred environmental costs (Eddington, 2006). The imposition of aviation green taxes affects a broad section of the economy including passengers, governments/revenue authorities, the broader tourism sector, and airlines. Emissions due to growth in the aviation industry will increase the proportion of carbon budgets that it accounts for Anderson et al. (2007) and Ryley et al. (2010). For this reason, governments should encourage the development and use of sustainable aviation fuels, new technologies to decrease aviation industry's carbon emissions and imposing inefficient aviation green taxes.

The "price" elasticity of demand is generally high in air travel (i.e., sensitive to price changes). Thus, levying additional forms of taxes on air travel prices will have a negative impact on the overall demand in air travel (IATA, 2018). From an academic point of view, a complication in the designing of aviation green tax is that there is little research about passenger's willingness to pay (WTP) for mandatory taxes (Sonnenschein & Smedby, 2019). Also, customers generally do not trust that the government will spend the revenue generated from environmental taxes on environmental action. A survey that asked consumers, "Do you trust governments to spend money from environment taxes specifically on environmental protection programs?" demonstrated that consumers were uniformly skeptical in many key European Union markets (*Travel Weekly*, Jun. 20, 2019).

Government interventions to reduce CO2 emissions caused by air transport, through measures such as tax or emissions trading schemes with strict global or sectoral limits that affect air travel costs, can limit tourists in choosing travel destinations (Nawijn & Peeters, 2010). Therefore, limiting air travel will certainly reduce the emission of greenhouse gases; however, it will also affect some of the 17% of all holiday travel by air transport, as there might not be a reasonable alternative transportation mechanism to travel to some destinations (Nawijn & Peeters, 2010). These tax aversion studies have been carried out only in the United States, and the results could be smaller in Northern Europe where taxes are less stigmatized (Hardisty et al., 2019).

Sonnenschein and Smedby (2019) studied the valuation of WTP for mitigating air travel emissions to increase its policy relevance for a mandatory aviation green tax.

### *Measurement factor: the anchoring effect of aviation green tax*

One of the variables that affect environmental behavior is pro-environmental self-identity (PESI). Pro-environmental behavior emphasizes the diverse factors that affect environmental behavior (Whitmarsh & O'Neill, 2010), and self-identity is differentiating oneself from others or social groups in terms of one's beliefs, values, and behaviors (Christensen, Rothberger, Wood, & Matz, 2004; Whitmarsh & O'Neill, 2010). Self-identity is an important predictor of behavior related to pro-environmental actions (Fekadu & Kraft, 2001; Sparks et al., 1995; Sparks & Shepherd, 1992; Terry et al., 1999). Self-identity has a significant influence on the predictor of behavior as well as TPB variables (action, subjective norm and perceived behavioral control) related to eco-friendly behavior (Fekadu & Kraft, 2001; Sparks et al., 1995; Sparks & Shepherd, 1992; Terry et al., 1999; Whitmarsh & O'Neill, 2010). The TPB model has proven strong predictive power for many pro-environmental behaviors (Carfora et al., 2017; Gatersleben, Murtagh, & Abrahamse, 2014; Sparks, Hinds, Curnock, & Pavey, 2014; Whitmarsh & O'Neill, 2010). Therefore, based on these previous studies, PESI determined that it is closely related to TPB variables (GPV, ACC and CCRP) and proposed hypotheses H1, H2 and H3.

The second variable affecting environmental behavior is green perceived value (GPV). Wu et al. (2018) examined the green switching intentions of airline customers by studying the relationship among experiential quality dimensions, GPV, green experiential satisfaction, and green corporate image and reputation regarding green experiential loyalty. Their findings show that GPV is the most significant prerequisite for a green experience, followed by enjoyment quality, green corporate reputation, and physical environmental quality. Green corporate image and interaction quality were found to be the least important (Wu et al., 2018).

The third variable is action for climate change (ACC). Self-identity seeks to establish consistency between actions and attitudes (Christensen, Rothberger, Wood, & Matz, 2004), inducing specific intentions (Carfora et al., 2017). Becken (2004) studied how tourism experts and tourists perceive climate change and carbon offsetting schemes. Respondents were asked if they would be willing to participate in tree-planting to reduce greenhouse gas emissions and whether they considered "climate change to be an issue for tourism." Only the segment of green tourists responded that they think that tourism has a significant effect on climate change and that planting a tree has a positive effect on the environment (Becken, 2004). Ward et al. (2011) study revealed that customers were more willing to pay for an ENERGY STAR certified refrigerator than others. Further, the survey highlighted that participants' perceived need for action for climate change and the environment were motivated by their concern for the environment.

The fourth variable is climate change risk perception (CCRP) introduced by van der Linden (2015) who examined four factors related to CCRP (socio-demographic, socio-cultural, cognitive, and experiential). According to the author, societal and personal risk were highly related to each other in converting public risk perceptions into personal risk perception (Xie et al., 2019). Environmental risk perception demonstrated individuals' perception of the influence of human behavior on the environment and nature surrounding themselves (Han & Hyun, 2017; Kollmuss & Agyeman, 2002). Environmental perception is often used interchangeably with the term adverse outcome of valued objects (Han, 2015; Han & Hyun, 2017). Han and Hyun (2017) found that one's cognitive level of environmental perception derives his/her expected feeling of guilt, which is a key aspect of negative anticipated affect.

The fifth variable is attitude toward green policy (AGP). According to the TPB, the predictor of behavior is person's intention to involve in the behavior; This intention is determined by perceived behavioral control, subjective norms, and attitudes toward perceived behavior (Ajzen, 1991; Han, 2015; Han et al., 2010). Attitude is a person's overall evaluation of participating in concrete behavior (Ajzen, 1991).

Chen and Peng (2012) found that visitors with sufficient environmental knowledge about green hotel products, build intention to visit if they are convinced that they are physically capable and if they have positive attitude toward green hotel stay (Han, 2015). Also, Ru et al. (2019) studied attitudes toward fine particulate matter (PM 2.5) reduction behavior based on young people's negative and positive evaluations of PM 2.5 and found that personal moral norms and behavioral control significantly affect PM 2.5 reduction intentions (Ru et al., 2019).

The sixth and last variable is willingness to pay (WTP). An environmental behavior intention refers to the likelihood that a customer revisits and pays for an environmental product and service (González-Rodríguez et al., 2020). In this context, WTP a price is related to the amount of money a customer would be willing to pay for a product or service's perceived value of environmental benefit (Han et al., 2009; Lee et al., 2010, González-Rodríguez et al., 2020). The anchoring affects consumers' willingness to accept and pay. The results of the study suggested that circumstantial differences between selling and buying decisions affect systematic differences to anchoring effects, the impact of anchoring on judgments of WTA and WTP are essentially the same (Simonson & Drolet, 2004). Environmental concern has been demonstrated to be positively related to respondents' actual WTP for a product with positive environmental externalities (Krarup & Russell, 2005; Ward et al., 2011). Also, Olya et al. (2019) studied tourists' willingness to purchase climate insurance and loyalty based on an expectation-disconfirmation model. As a result, climate insurance in the tourism industry plays key role as a practical implication for addressing uncertain spatial and temporal climatic conditions.

## Moderating effects of materialism and hedonism

Human value and lifestyles relate concretely to freedom, equality, altruism, security and hedonism (Arlt et al., 2011). Schwartz (1992) suggested personal values were determined using the simple version of the Schwarz Value Survey (SVS). He identified that the 10 motivationally distinct 'universal' values of power, hedonism, achievement, stimulation, universalism, self-direction, benevolence, tradition, security and conformity can be divided into the two dimensions of openness to change versus conservation and self-enhancement and self-transcendence versus to investigate into public skepticism about uncertain climate change (Poortinga et al., 2011; Schwartz, 1992).

People have a range of disparate conflicting values such as materialism and personal ambition, those who identify with self-enhancement value. The core self-transcendence and self-enhancement are significant role in personal value research regarding public engagement with climate change (Corner et al., 2014).

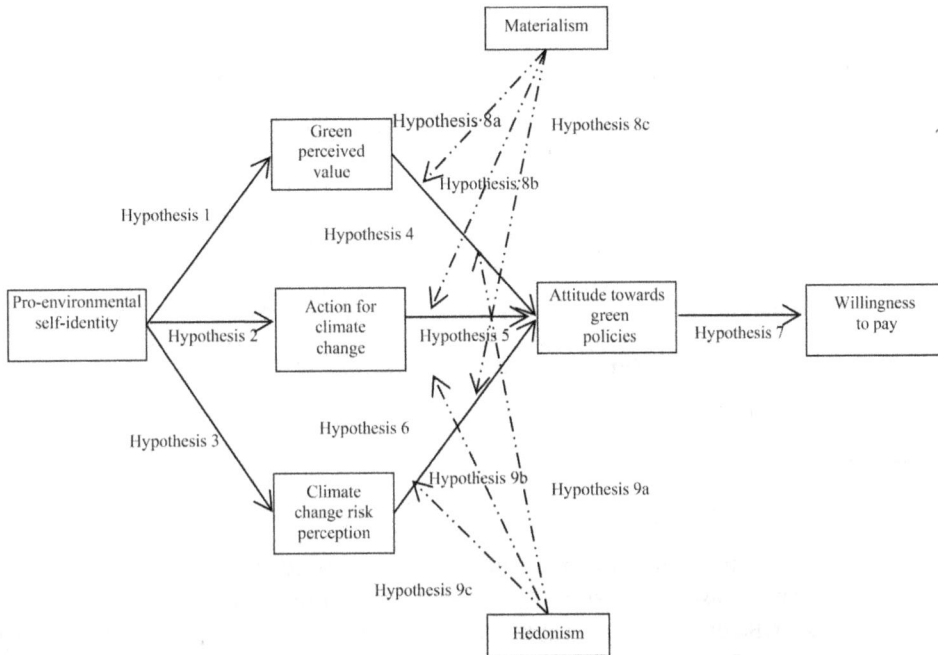

**Figure 1.** Research model. Note: Two identical models (models for high and low materialism and hedonism groups) are proposed.

Our study adapted the six items from the materialism scale devised by Richins and Dawson (1992) as moderating effects. Materialism has been identified as the significance individuals place on property as a personal value (Felix & Almaguer, 2019; Richins & Dawson, 1992). Further, Richins (1994) defined the relationship between individual values and material possessions, concluding that individuals find various meanings in their possessions regarding their personal values. Materialism has been characterized as a value that guides conduct in a variety of consumption contexts (Richins & Dawson, 1992). Felix and Almaguer (2019) found that the happiness side of materialism is positively related to purchase intentions for pro-environmental products. Banerjee and McKeage (1994) found a negative correlation between environmentalism and materialism, the relationship of two constructs were as opposite indication of personal orientation toward consumption.

Our study also adapted the five items of hedonism from Bozkurt et al. (2008) study as moderating effects. Hedonism is an individual's tendency to pursue psychological benefits such as emotional happiness as well as intangible, experiential, and multi-sensory benefits (Ahtola, 1985; Holbrook & Hirschman, 1982; Hyun & Kang, 2014). Hedonists seek fun and immediate pleasure (Holbrook & Hirschman, 1982; Hyun & Kang, 2014), and they have strong behavior intentions compared to that of non-hedonists when obtaining pleasure from consumption (Hyun & Kang, 2014). Based on the literature review, this study proposes a conceptual model (Figure 1).

**H1:** PESI has a positive impact on GPV.

**H2**: PESI has a positive impact on ACC.

**H3**: PESI has a positive impact on CCRP.

**H4**: GPV has a positive impact on attitude toward green policies (AGP).

**H6**: CCRP has a positive impact on AGP.

**H7**: AGP has a positive impact on WTP.

**H8a**: Materialism has a moderating effect on the relationship between GPV and AGP.

**H8b**: Materialism has a moderating effect on the relationship between ACC and AGP.

**H8c**: Materialism has a moderating effect on the relationship between CCRP and AGP.

**H9a**: Hedonism has a moderating effect on the relationship between GPV and AGP.

**H9b**: Hedonism has a moderating effect on the relationship between ACC and AGP.

**H9c**: Hedonism has a moderating effect on the relationship between CCRP and AGP.

## Methodology

### *Measurement instruments*

This research aims to study the environmental issues associated with aviation, and attitudes toward existing aviation taxes and use of additional revenue from aviation taxes, based on the research methods of Bloor (2001). The questionnaire applied in our research gathered demographic characteristics of participants comprising personal information, age, gender, and income, as well as information on participants' environmental awareness, attitudes, and willingness to pay aviation green taxes. The comprised six theoretical variables in the proposed conceptual model: PESI (Whitmarsh & O'Neill, 2010), GPV (Wu et al., 2018), ACC (Becken, 2004), CCRP (van der Linden, 2015; Xie et al., 2019), AGP (Ru et al., 2019), and WTP as endogenous variable (Han et al., 2009; González-Rodríguez et al., 2020; Lee et al., 2010). Moderating variables were measured using two items, hedonism (Arlt et al., 2011; Corner et al., 2014; Poortinga et al., 2011; Schwartz, 1992) and materialism (Banerjee & McKeage, 1994; Felix & Almaguer, 2019; Richins, 1994; Richins & Dawson, 1992).

We used a total of 20 measurement items based on previous research. The questionnaire items were based on a five-point Likert scale ranging from "Not at all" (1) to "Very much" (5).

Furthermore, we used the software programs, SPSS 23 and AMOS 24, to analyze the collected sample data. First, we conducted a confirmatory factor analysis (CFA) to estimate a measurement model. Second, we used structural equation modeling (SEM) and a modeling comparison to evaluate the hypothesized and proposed model relationships among the constructs. This two-step process and the moderating impact for test metric invariance were based on the suggestions by Anderson and Gerbing (1988).

To investigate the anchoring effect, we asked 333 respondents about their willingness to pay aviation green taxes: "How much would you pay if you had to pay an aviation environmental tax when using an aircraft?" The base amount for aviation green tax was presented as about €25 (approximately $28), based on Sweden and Germany's (destination of less than 6,000 km) aviation tax. This study assumes that travelers with high PESI (Whitmarsh & O'Neill, 2010) would be willing to pay green tax and that pro-environmental identity has a significant impact on environmental protection behavior intentions. It can also be assumed that the higher the pro-environmental identity, the more environmentally friendly behavior intentions and attitudes travelers will have.

### *Data collection and sample characteristics*

This study examines travelers' perceptions and behavioral intentions regarding sustainable tourism and the environment, through questions relating to pro-environmental identity. Travelers

**Table 1.** Correlation, AVE, composite reliability, and mean, SD.

| Demographic profiles of participants | | Frequency (N = 333) | Percent (%) |
|---|---|---|---|
| Gender | Male | 165 | 49.5 |
| | Female | 168 | 50.5 |
| Marital status | Single | 120 | 36.0 |
| | Married | 213 | 64.0 |
| Age | 20-30 | 65 | 19.5 |
| | 30-40 | 64 | 19.2 |
| | 30-50 | 68 | 20.4 |
| | 50-60 | 68 | 20.4 |
| | 60 years and older | 68 | 20.4 |
| Education level | High school degree | 51 | 15.3 |
| | College degree | 35 | 10.5 |
| | Bachelor's degree | 192 | 57.7 |
| | Graduate degree or above | 55 | 16.5 |
| Monthly income | Under $2,000 | 18 | 5.4 |
| | Under $2,000 ∼ over $3,000 | 62 | 18.6 |
| | Under $3,000 ∼ over $4,000 | 56 | 16.8 |
| | Under $4,000 ∼ over $5,000 | 55 | 16.5 |
| | Over $5,000 | 143 | 42.6 |
| Total | | 333 | 100 |

were asked to respond to two questionnaires, one with a brief description that functions as a label (anchoring effect) of environmental protectionism and one without it. We conducted a comparison analysis of two groups of travelers to find out how labeling affects their willingness to pay aviation green taxes as well as their behavioral intentions of environmental protection. Using a combination of qualitative and quantitative research methods, the study aims to analyze travelers' intent to act as influenced by the anchoring effect of the nudge theory.

To collect the data, we used the survey system of an online research firm in South Korea. The online marketing research firm sent the questionnaire invitation via e-mail to travelers who were randomly selected from the firm's database. This study was conducted by distributing online surveys to 400 potential participants including those who had traveled within the past year from February 7, 2020 to March 27, 2020. As a result, 333 (83.25%) valid questionnaires were analyzed excluding sixty-seven (16.75) unreliable responses.

## Results

### Demographic profiles

We evaluated the respondents' characteristics and information. After excluding 30 unreliable responses, we analyzed the final sample (N = 333). The survey was conducted over a two-week period from March 13, 2020 to March 27, 2020. Table 1 shows the profile of survey respondents.

Among the 333 travelers, 165 (49.5%) participants were male and 168 (50.5%) were female. The respondents' age ranged from 20 to 70. The most recent airline use was within the last year. Most respondents were highly educated. Most participants (192, 57.7%) had a bachelor's degree and 55 (16.5%) had a graduate degree or above. Among all participants, 51 (15.3%) had only a high school diploma and 35 (10.5%) had a college degree. Most participants (143, 42.6%) reported a monthly income of over $5,000. The monthly income of 62 (18.6%) participants were between $2,000 and $2,999; that of 56 participants (16.8%) between $3,000 and $3,999; that of 55 participants (16.5%) between $4,000 and $4,999; and that of 18 participants (5.4%) were below $2,000.

### Anchoring effects

Table 2 shows the comparison results of the respondents' WTP before and after the anchoring amount (approximately $28) is presented. In our analysis, we excluded the responses of 15

**Table 2.** The results regarding anchoring effect.

| Total (N = 333) | Survey respondents | Average of difference before anchoring (A) | Average of difference after anchoring (B) | Average of difference before and after anchoring (B-A) |
|---|---|---|---|---|
| Excluding respondents not willing to pay | 318 | USD 21.34 | USD 15.27 | USD −6.07 |
| Excluding respondents with same amount before and after anchoring | 216 | USD 21.99 | USD 13.09 | USD −8.89 |

respondents who did not intend to pay environmental taxes. The average amount of the difference between the responses and the anchor amount decreased from $21.4 to $15.3 (by $6.1). In addition, we excluded the responses of 117 who did not change their responses after the anchor was presented; the average amount of the difference for the remaining 216 respondents decreased from $22 to $13.1 (by $8.9). The results show that the average amount of the difference from the anchor decreased especially for the respondents who were willing to pay aviation taxes. These results support our research hypothesis that the anchoring effect is significant.

### Confirmatory factor analysis (CFA)

We conducted an assessment of the measurement model; the CFA with a method of maximum likelihood estimation indicated that the goodness-of-fit (GFI) statistics for the measurement model is acceptable ($\chi^2 = 345.543$, $df = 174$, $\chi^2/df = 1.986$, $p < .001$, GFI = .910, CFI = .955, IFI = .956, TLI = .946, RMSEA = .054). Details are shown in Table 3. All variables were significant in terms of their relevant latent construct ($p < .01$). The composite reliability decreased to between .80 and .95 and exceeded the criterion of .70 for the strong internal consistency suggested by Anderson and Gerbing (1988). Then, we calculated the average variance extracted (AVE) for convergent validity assessment. The values decreased to between .514 and .864, all AVE exceeding the criterion of .500 suggested by Fornell and Larcker (1981) and Hair et al. (2017). Furthermore, as shown in Table 4, the correlations among variables were all lower than the AVE values. Therefore, the measures of convergent and discriminant validity were evident.

### Structural equation modeling (SEM)

We used SEM to estimate the proposed model. The model fit statistics revealed that the data is an adequate fit to the values ($\chi^2 = 445.093$, $df = 163$, $\chi^2/df = 2.731$, $p < .001$, GFI = .879, CFI = .915, IFI = .916, TLI = .901, RMSEA = .072). SEM results and the hypotheses testing results are shown in Figure 2 and Table 5 in detail. As presented in Figure 2, the results showed that the variables (PESI, GPV, ACC, and CCRP) and the factors (AGP and WTP) are positively significant. The standardized coefficient values were .724 for PESI, .848 for GPV, .782 for ACC, .652 for CCRP, .351 for AGP), and .351 for WTP.

As shown in Table 5 and Figure 2, we evaluated the hypothesized constructs according to the proposed theoretical framework. First, the results indicate that PESI has a significant influence on positive GPV ($\beta = .724$, p < .05), ACC ($\beta = .848$, p < .05), and CCRP ($\beta = .778$, p < .01). As such, Hypotheses 1, 2, and 3 were supported. Second, the results showed that GPV ($\beta = .170$, p > .05) and CCRP ($\beta = .184$, p > .05) did not have a significant effect on AGP. However, ACC ($\beta = .406$, p < .05) has a significant effect on AGP. As such, Hypothesis 5 was supported but Hypotheses 4 and 6 were not. Third, the results revealed that AGP is a significant function of WTP ($\beta = .617$, p < .05), thus supporting Hypothesis 7.

**Table 3.** Confirmatory factor analysis: Items and loadings.

| Construct and scale item | Standardized loading |
|---|---|
| *Pro-environmental self-identity* | |
| I think of myself as an environment-friendly consumer. | 0.592 |
| I think of myself as someone who is very concerned about environmental issues. | 0.936 |
| I would be embarrassed to be seen as having an environmentally friendly lifestyle | 0.873 |
| *Green Perceived Value* | |
| This airline's environmental functions (e.g., eco-friendly dishes, weight lightening seats, eco-friendly aircrafts) provide very good value. | 0.682 |
| This airline's environmental performance meets my expectations. | 0.782 |
| I use this airline, because it addresses environmental concerns more than the other airlines do. | 0.859 |
| *Action for climate change* | |
| The production of electricity from renewable sources such as solar, wind, and biomass is an effective way to combat global climate change. | 0.651 |
| We need more government regulations to force people to protect the environment. | 0.705 |
| There is no urgent need to take measures to prevent global climate change today. | 0.879 |
| Global climate change will have a noticeably negative impact on the environment in which my family and I live. | 0.687 |
| *Climate change risk perception* | |
| In your judgment, how likely are you, sometime during your life, to experience serious threats to your health or overall well-being as a result of climate change? | 0.783 |
| How often do you worry about the potentially negative consequences of climate change? | 0.763 |
| In your judgment, how likely do you think it is that climate change will have very harmful, long-term impacts on our society? | 0.749 |
| How serious would you rate current impacts of climate change worldwide? | 0.816 |
| *Attitude towards green policies* | |
| I am positive about environmental policies. | 0.724 |
| I am generally satisfied with environmental policies. | 0.554 |
| I think environmental policies are worth it. | 0.742 |
| *Willingness to pay* | |
| I am willing to pay the green aviation tax to protect the environment. | 0.918 |
| I am willing to pay the green aviation tax to protect future generations. | 0.950 |
| I am willing to pay the green aviation tax for sustainable tourism. | 0.898 |
| *Materialism* | |
| I admire people who own expensive homes, cars, and clothes. | |
| Some of the most important achievements in life include acquiring material possessions. | |
| I place much importance on the amount of material objects people own as a sign of success. | |
| The things I own say a lot about how well I am doing in life. | |
| I like to own things that impress people. | |
| I pay much attention to the material objects that other people own. | |
| *Hedonism* | |
| I think people should live their lives in accordance with their feelings and desires. | |
| People cannot take their money with them when they die, so we should live for today. | |
| People should always pursue pleasure in their lives. | |
| People should always live for the present moment. | |
| People should not sacrifice the pleasure of the present for the possibility of something better in the future. | |

Note: All factor loadings are significant at $p < .001$.

## *Structural invariance model*

Based on previous studies, we tested for the moderating effects of materialism and hedonism (H8a, H8b, H8c, H9a, H9b and H9c). We divided the total of 333 samples into "low" and "high" materialism and "low" and "high" hedonism based on analysis results of the K-means cluster. Those with low materialism comprised 120 cases whereas those with high materialism comprised 110 cases. The assessment results of the baseline model comprising these high and low groups for materialism and hedonism are displayed in Table 6 and Figure 2. We conducted an empirical comparison between baseline and nested models by using a difference test of Chi-square for the degree of freedom (Anderson & Gerbing, 1988). The moderating score results show that the

**Table 4.** Correlation, AVE, composite reliability, and mean, SD.

| | PESI | GPV | ACC | CCRP | AGP | WTP |
|---|---|---|---|---|---|---|
| PESI | – | .496[b] | 0.485 | 0.526 | 0.441 | 0.412 |
| GPV | .246[a] | – | 0.720 | 0.559 | 0.458 | 0.490 |
| ACC | 0.235 | 0.518 | – | 0.631 | 0.463 | 0.410 |
| CCRP | 0.277 | 0.312 | 0.398 | – | 0.473 | 0.455 |
| AGP | 0.194 | 0.210 | 0.214 | 0.224 | – | 0.541 |
| WTP | 0.170 | 0.240 | 0.168 | 0.207 | 0.293 | – |
| Mean | 3.613 | 4.033 | 4.173 | 4.040 | 3.6046 | 3.430 |
| SD | .660 | .531 | .473 | .572 | .630 | .899 |
| CR | .904 | .922 | .842 | .924 | .800 | .912 |
| AVE | .765 | .799 | .514 | .688 | .575 | .864 |

1PESI: pro-environmental self-identity, GPV: green perceived value, ACC: action for climate change, CCRP: climate change risk perception, AGP: attitude towards green policies, WTP: willingness to pay.
2Goodness-of-fit statistics for the measurement model: $\chi^2 = 445.093$, $df = 163$, $\chi^2/df = 2.731$, $p < .001$, GFI = .879, CFI = .915, IFI = .916, TLI = .901.
aCorrelations between variables are below the diagonal.
bSquared correlations between variables are above the diagonal.

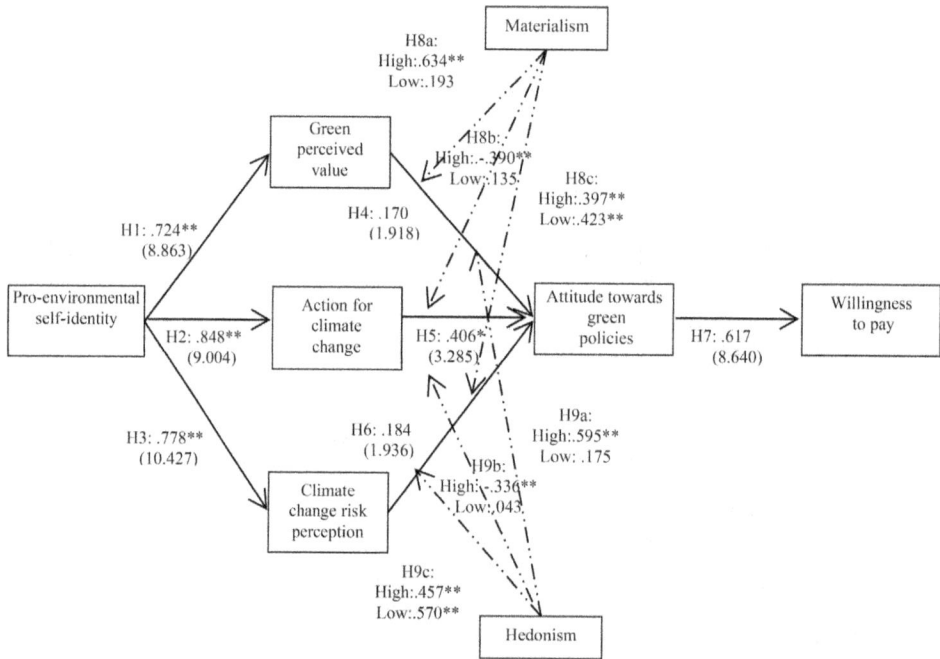

**Figure 2.** Results of the structural equation modeling. Notes. 1. Two identical models (models for high and low materialism and hedonism groups) are proposed. 2. Goodness-of-fit statistics for the structural model: $\chi^2 = 445.093$, df = 163, $\chi^2$ /df = 2.731, p < .001, GFI = .879, CFI = .915, IFI = .916, TLI = .901, RMSEA = .072 (*p < .05, **p < .01).

effects of GPV, ACC, and CCRP on AGP are significantly different between high and low groups ($\Delta\chi^2$ [1] = .001, p < .05).

We then calculated the moderating effect of materialism on the relationship between GPV and AGP (H8a), ACC and AGP (H8b), and CCRP and AGP (H8c). The chi-square difference for Hypothesis 8a between the constrained and unconstrained models was found to be significant ($\Delta\chi^2$ (1) =4.777, p <.05). The chi-square difference for Hypothesis 8 b between the constrained and unconstrained models was also found to be significant ($\Delta\chi^2$ (1) =4.941, p <.05). However, the chi-square difference for Hypothesis 8c between the constrained and unconstrained models was not found to be significant ($\Delta\chi^2$ (1) =0.05, p > .05). Thus, the moderating effect of materialism in Hypothesis 8a and 8 b was supported while in Hypothesis 8c, it was not.

**Table 5.** Coefficient, t-value, $R^2$ ($n = 333$).

| Hypothesis | Linkage | Standardized estimates | t-value | Hypothesis |
|---|---|---|---|---|
| Hypothesis 1 | PESI →GPV | .724 | 8.863** | Supported |
| Hypothesis 2 | PESI → ACC | .848 | 9.004** | Supported |
| Hypothesis 3 | PESI → CCRP | .778 | 10.427** | Supported |
| Hypothesis 4 | GPV → AGP | .170 | 1.918 | Not supported |
| Hypothesis 5 | ACC →AGP | .406 | 3.285** | Supported |
| Hypothesis 6 | CCRP → AGP | .184 | 1.936 | Not supported |
| Hypothesis 7 | AGP → WTP | .617 | 8.640** | Supported |
| Hypothesis 8a | Moderating effect of MTL on the relationship between GPV and AGP | | | |
| Hypothesis 8b | Moderating effect of MTL on the relationship between ACC and AGP | | | |
| Hypothesis 8c | Moderating effect of MTL on the relationship between AGP and AGP | | | |
| Hypothesis 9a | Moderating effect of HDN on the relationship between GPV and AGP | | | |
| Hypothesis 9b | Moderating effect of HDN on the relationship between ACC and AGP | | | |
| Hypothesis 9c | Moderating effect of HDN on the relationship between AGP and AGP | | | |

1PESI: pro-environmental self-identity, GPV: green perceived value, ACC: action for climate change, CCRP: climate change risk perception, AGP: attitude towards green policies, WTP: willingness to pay, MTL: materialism, HDN: hedonism.
2Goodness-of-fit statistics for the structural model: $\chi^2 = 445.093$, $df = 163$, $\chi^2/df = 2.731$, $p < .001$, GFI $= .879$, CFI $= .915$, IFI $= .916$, TLI $= .901$, RMSEA $= .072$.
*$p < .05$, **$p < .01$.

**Table 6.** Baseline model and nested model assessment, chi-square test.

| Paths | High MTL group (n = 110) | | Low MTL group (n = 120) | | Baseline model (freely estimated) | Nested model (equally restricted) |
|---|---|---|---|---|---|---|
| | β | t-value | β | t-value | | |
| GPV lueGP | .749 | 5.182** | .237 | 1.822 | $\chi^2$ (452) = 1116.325 | $\chi^2$ (453) = 1121.102 [a] |
| ACC ) = 1 | -.432 | −4.130** | .177 | 1.457 | $\chi^2$ (452) = 1116.325 | $\chi^2$ (453) = 1121.266 [b] |
| CCRP = 11 | .330 | 5.582** | .330 | 5.582** | $\chi^2$ (452) = 1116.325 | $\chi^2$ (453) = 1116.375 [c] |

| Paths | High HDN group (n = 110) | | Low HDN group (n = 106) | | Baseline model (freely estimated) | Nested model (equally restricted) |
|---|---|---|---|---|---|---|
| | β | t-value | β | t-value | | |
| GPV → AGP | .737 | 4.982** | .206 | 1.647 | $\chi^2$ (452) = 1024.774 | $\chi^2$ (453) = 1030.209 [b] |
| ACC → AGP | -.433 | −3.437** | .048 | .422** | $\chi^2$ (452) = 1024.774 | $\chi^2$ (453) = 1028.295 [b] |
| CCRP → AGP | .425 | 4.764** | .503 | 4.312 | $\chi^2$ (452) = 1024.774 | $\chi^2$ (453) = 1024.926 [c] |

Chi-square difference test:
[a] $\Delta\chi^2$ (1) = 4.777, p < .05
[b] $\Delta\chi^2$ (1) = 4.941, p < .05
[c] $\Delta\chi^2$ (1) = 0.05, p > .05
[a] $\Delta\chi^2$ (1) = 5.435, p < .05
[b] $\Delta\chi^2$ (1) = 3.521, p > .05
[c] $\Delta\chi^2$ (1) = 0.152, p > .05

Hypothesis testing results:
H8a: supported
H8b: supported
H8c: not supported
H9a: supported
H9b: not supported
H9c: not supported

1PESI: pro-environmental self-identity, GPV: green perceived value, ACC: action for climate change, CCRP: climate change risk perception, AGP: attitude towards green policies, WTP: willingness to pay, MTL: materialism, HDN: hedonism (*$p < .05$, **$p < .01$).

Then, we calculated hedonism's moderating effect on the relationship between GPV and AGP (H9a), ACC and AGP (H9b), and CCRP and AGP (H9c). The chi-square difference for Hypothesis 9a between the unconstrained and constrained models was found to be significant ($\Delta\chi^2$ (1) =5.435, $p < .05$). The chi-square difference for Hypothesis 9b between the unconstrained and constrained models was not found to be significant ($\Delta\chi^2$ (1) =3.521, p > .05). The chi-square difference for Hypothesis 9c between the unconstrained and constrained models was also found to be not significant ($\Delta\chi^2$ (1) = 0.152, p > .05). Thus, the moderating effect of hedonism in Hypothesis 9a was supported but not in Hypotheses 9b and 9c.

## Discussions and implications

This research aimed to study the anchoring effect of the nudge theory on changing the behaviors and perceptions of tourists. Although it has recently emerged as a social issue, there is little research on aviation green taxes in the sustainable tourism sector. This study aimed to test the effect of the anchor on travelers who may need to pay aviation green taxes. The results revealed that the anchor had a significant effect on about 65% of all participants (N = 333), excluding those who entered the same amount before and after (N = 216).

The results of the analysis show that PESI (environment-friendly consumer, concerned with environmental issues, environmentally friendly lifestyle) has a significant effect on environmental business (GPV), supporting Hypothesis 1. The results also showed that PESI has a significant impact on ACC (an effective way to compound global climate change, measures to prevent global climate change), supporting Hypothesis 2. However, the analysis results did not support Hypothesis 3 as PESI did not display a significant impact on CCRP. In addition, Hypothesis 4 that assumed GPV to have a significant effect on AGP and Hypothesis 6 that assumed CCRP to have a significant effect on AGP were not supported. On the other hand, Hypothesis 5 that assumed ACC to have a significant impact on AGP and Hypothesis 7 that assumed AGP to have a significant impact on WTP were supported.

According to the results of the validation test for moderating effect, respondents with a strong tendency toward materialism (owning expensive material possessions) demonstrated a significant effect on the path that GPV impacts ACC and did not display a statistically significant impact on the path that CCRP impacts AGP. Respondents with high hedonism (people who live in accordance with their feelings, desires, the present moment, and pleasure) displayed a significant correlation on the path that GPV impacts AGP. In other words, hedonists have GPVs such as an environmental business and performance as well as a positive AGP. On the other hand, GPV did not have a significant impact on the correlation between ACC and AGP.

This study contributes to the literature of sustainable tourism by examining the factors that induce voluntary change in consumer behavior. In theoretical contributions, hospitality research has been mainly focused on marketing and human resource management. However, the nudge theory was applied to the aviation industry, and the economic approach was focused on consumer behavior. Therefore, this study has significance in that it attempted to converge studies of hospitality and economics.

The practical implications of this study are to improve transparency and to minimize tax resistance by positively nudging travelers' perceptions of aviation green taxes. First, workable implication to increase willingness to pay green tax for the aviation industry is to share the burden environmental financial support and to raise environmental issues with passengers. Second, in countries where aviation green tax is not introduced, the purpose is to decrease any tax resistance that traveler may have. Through this research, government can present examples of aviation green tax of other countries, as well as emphasize the importance of climate change to people and assert the justification of aviation green tax to people. Third, the IATA will have to actively review how to increase airline revenues while reducing passenger's travel costs, and not to focus on the negative aspects of aviation green taxes, taking social responsibility for the environment. To conclude, tourists' voluntary participation in these matters will contribute to long-term sustainable tourism.

The limitations of this study are as follows. First, the data of socio-demographic variables (age, gender, occupation, or income) were limited to the population of South Korean. This limit measuring the anchoring effect because the research results cannot be generalized as it has not been studied in various populations. Therefore, future study should verify the proposed conceptual model by considering various countries. Second, the data was collected via online survey by an online marketing researcher company. Therefore, future research should require to verify the proposed conceptual model using data collected by face-to-face and mail survey.

## Funding

This work was supported by the National Research Foundation of Korea(NRF) grant funded by the Korea government(*MSIT) (No.2018R1A5A7059549). *Ministry of Science and ICT.

## References

Ahtola, O. T. (1985). Hedonic and utilitarian aspects of consumer behavior: An attitudinal perspective. *Advances in Consumer Research, 12*(1).

Anderson, K., Bows, A., Footitt, A. (2007). Aviation in a low carbon EU. *A research report by the Tyndall Centre, University of Manchester. Friends of the Earth.* Retrieved November 21, 2019, from http://www.foe.co.uk/resource/reports/aviation_tyndall_07_main.pdf

Anderson, J. C., & Gerbing, D. W. (1988). Structural equation modeling in practice: A review and recommended two-step approach. *Psychological Bulletin, 103*(3), 411–423. https://doi.org/10.1037/0033-2909.103.3.411

Ariely, D. (2008). *Predictably irrational: The hidden forces that shape our decisions.* HarperCollins.

Arlt, D., Hoppe, I., & Wolling, J. (2011). Climate change and media usage: Effects on problem awareness and behavioural intentions. *International Communication Gazette, 73*(1-2), 45–63. https://doi.org/10.1177/1748048510386741

Ajzen, I. (1991). The theory of planned behavior. *Organizational Behavior and Human Decision Processes, 50*(2), 179–211. https://doi.org/10.1016/0749-5978(91)90020-T

Banerjee, B., & McKeage, K. (1994). How green is my value: Exploring the relationship between environmentalism and materialism. *ACR North American Advances, 21*(1), 147–152.

Becken, S. (2004). How tourists and tourism experts perceive climate change and carbon-offsetting schemes. *Journal of Sustainable Tourism, 12*(4), 332–345. https://doi.org/10.1080/09669580408667241

Becken, S. (2007). Tourists' perception of international air travel's impact on the global climate and potential climate change policies. *Journal of Sustainable Tourism, 15*(4), 351–368. https://doi.org/10.2167/jost710.0

Bloor, M. (Ed.) (2001). *Focus groups in social research.* Sage.

Bosquet, B. (2000). Environmental tax reform: Does it work? A survey of the empirical evidence. *Ecological Economics, 34*(1), 19–32.c. https://doi.org/10.1016/S0921-8009(00)00173-7

Bozkurt, V., Bayram, N., Furnham, A., & Dawes, G. (2008). The Protestant work ethic and hedonism among Kyrgyz, Turkish, and Australian students. *Journal for General Social Issues, 4*(5), 749–769.

Carfora, V., Caso, D., Sparks, P., & Conner, M. (2017). Moderating effects of pro-environmental self-identity on pro-environmental intentions and behaviour: A multi-behaviour study. *Journal of Environmental Psychology, 53,* 92–99. https://doi.org/10.1016/j.jenvp.2017.07.001

Christensen, P. N., Rothgerber, H., Wood, W., & Matz, D. C. (2004). Social norms and identity relevance: A motivational approach to normative behavior. *Personality & Social Psychology Bulletin, 30*(10), 1295–1309. https://doi.org/10.1177/0146167204264480

Cialdini, R. B. (2009). We have to break up. *Perspectives on Psychological Science, 4*(1), 5–6. https://doi.org/10.1111/j.1745-6924.2009.01091.x

Corner, A., Markowitz, E., & Pidgeon, N. (2014). Public engagement with climate change: The role of human values. *Wiley Interdisciplinary Reviews: Climate Change, 5*(3), 411–422.

Eddington, R. (2006). The Eddington Transport Study. Main Report: Transport's Role in Sustaining the UK's Productivity and Competitiveness.

European Commission. (1997b). Tax provisions with a potential impact on environmental protection. Office for Official Publications of the European Communities, Luxembourg.

Fekadu, Z., & Kraft, P. (2001). Self-identity in planned behavior perspective: Past behavior and its moderating effects on self-identity-intention relations. *Social Behavior and Personality: An International Journal, 29*(7), 671–685. https://doi.org/10.2224/sbp.2001.29.7.671

Felix, R., & Almaguer, J. (2019). Nourish what you own: Psychological ownership, materialism and pro-environmental behavioral intentions. *Journal of Consumer Marketing, 36*(1), 82–91. https://doi.org/10.1108/JCM-10-2017-2417

Fornell, C., & Larcker, D. F. (1981). Structural equation models with unobservable variables and measurement error: Algebra and statistics. *Journal of Marketing Research, 18*(3), 382–388. https://doi.org/10.1177/002224378101800313

Hamond, M. J., DeCanio, S. J., Duxbury, P., Sanstad, A. H., & Stinson, C. H. (1997). Tax waste, not work. *Challenge, 40*(6), 53–62. https://doi.org/10.1080/05775132.1997.11472003

Han, H. (2015). Travelers' pro-environmental behavior in a green lodging context: Converging value-belief-norm theory and the theory of planned behavior. *Tourism Management, 47*, 164–177. https://doi.org/10.1016/j.tourman.2014.09.014

Han, H., Hsu, L. T. J., & Lee, J. S. (2009). Empirical investigation of the roles of attitudes toward green behaviors, overall image, gender, and age in hotel customers' eco-friendly decision-making process. *International Journal of Hospitality Management, 28*(4), 519–528. https://doi.org/10.1016/j.ijhm.2009.02.004

Han, H., Hsu, L. T. J., & Sheu, C. (2010). Application of the theory of planned behavior to green hotel choice: Testing the effect of environmental friendly activities. *Tourism Management, 31*(3), 325–334. https://doi.org/10.1016/j.tourman.2009.03.013

Han, H., & Hyun, S. S. (2017). Fostering customers' pro-environmental behavior at a museum. *Journal of Sustainable Tourism, 25*(9), 1240–1256. https://doi.org/10.1080/09669582.2016.1259318

Hair, J. F., Jr., Sarstedt, M., Ringle, C. M., & Gudergan, S. P. (2017). *Advanced issues in partial least squares structural equation modeling.* Sage.

Hansen, P. G., & Jespersen, A. M. (2013). Nudge and the manipulation of choice: A framework for the responsible use of the nudge approach to behavior change in public policy. *European Journal of Risk Regulation, 4*(1), 3–28. https://doi.org/10.1017/S1867299X00002762

Hardisty, D. J., Beall, A. T., Lubowski, R., Petsonk, A., & Romero-Canyas, R. (2019). A carbon price by another name may seem sweeter: Consumers prefer upstream offsets to downstream taxes. *Journal of Environmental Psychology, 66*, 101342. https://doi.org/10.1016/j.jenvp.2019.101342

Hausman, D. M., & Welch, B. (2010). Debate: To nudge or not to nudge. *Journal of Political Philosophy, 18*(1), 123–136. https://doi.org/10.1111/j.1467-9760.2009.00351.x

Holbrook, M. B., & Hirschman, E. C. (1982). The experiential aspects of consumption: Consumer fantasies, feelings, and fun. *Journal of Consumer Research, 9*(2), 132–140. https://doi.org/10.1086/208906

Hyun, S. S., & Kang, J. (2014). A better investment in luxury restaurants: Environmental or non-environmental cues? *International Journal of Hospitality Management, 39*, 57–70. https://doi.org/10.1016/j.ijhm.2014.02.003

IATA. (2018). *Green Taxes.* Retrieved May 1, 2019, from https://www.iata.org/contentassets/1be02a5889fb439c902-f654737e89fbe/environmental_tax_pdf.pdf

González-Rodríguez, M. R., Díaz-Fernández, M. C., & Font, X. (2020). Factors influencing willingness of customers of environmentally friendly hotels to pay a price premium. *International Journal of Contemporary Hospitality Management, 32*(1), 60–80. https://doi.org/10.1108/IJCHM-02-2019-0147

Goulder, L. H. (1995). Effects of carbon taxes in an economy with prior tax distortions: An intertemporal general equilibrium analysis. *Journal of Environmental Economics and Management, 29*(3), 271–297. https://doi.org/10.1006/jeem.1995.1047

Gössling, S., Broderick, J., Upham, P., Ceron, J. P., Dubois, G., Peeters, P., & Strasdas, W. (2007). Voluntary carbon offsetting schemes for aviation: Efficiency, credibility, and sustainable tourism. *Journal of Sustainable Tourism, 15*(3), 223–248. https://doi.org/10.2167/jost758.0

Kahneman, D. (2011). *Thinking, fast and slow.* Macmillan.

Kosters, M., & Van der Heijden, J. (2015). From mechanism to virtue: Evaluating nudge theory. *Evaluation, 21*(3), 276–291. https://doi.org/10.1177/1356389015590218

Lange, F., Steinke, A., & Dewitte, S. (2018). The pro-environmental behavior task: A laboratory measure of actual pro-environmental behavior. *Journal of Environmental Psychology, 56*, 46–54. https://doi.org/10.1016/j.jenvp.2018.02.007

Lee, D. S., Fahey, D. W., Forster, P. M., Newton, P. J., Wit, R. C. N., Lim, L. L., Owen, B., & Sausen, R. (2009). Aviation and global climate change in the 21st century. *Atmospheric Environment, 43*(22), 3520–3537. https://doi.org/10.1016/j.atmosenv.2009.04.024

Li, Y., Wang, Y. Z., & Cui, Q. (2016). Has airline efficiency affected by the inclusion of aviation into European Union Emission Trading Scheme? Evidences from 22 airlines during 2008-2012. *Energy, 96*, 8–22. https://doi.org/10.1016/j.energy.2015.12.039

McKercher, B., Prideaux, B., Cheung, C., & Law, R. (2010). Achieving voluntary reductions in the carbon footprint of tourism and climate change. *Journal of Sustainable Tourism, 18*(3), 297–317. https://doi.org/10.1080/09669580903395022

Nawijn, J., & Peeters, P. M. (2010). Travelling "green": Is tourists' happiness at stake? *Current Issues in Tourism, 13*(4), 381–392. https://doi.org/10.1080/13683500903215016

Nelson, K. M., Partelow, S., & Schlüter, A. (2019). Nudging tourists to donate for conservation: Experimental evidence on soliciting voluntary contributions for coastal management. *Journal of Environmental Management, 237*, 30–43. https://doi.org/10.1016/j.jenvman.2019.02.003

OECD. (1997). *Environmental taxes and green tax reform.* Retrieved May 21, 2018, from https://www.oecd.org/sd-roundtable/papersandpublications/39372634.pdf

OECD. (2017). *Environmental tax*. Retrieved November 21, 2018, from https://data.oecd.org/envpolicy/environmental-tax.htm

Olya, H. G., Alipour, H., Peyravi, B., & Dalir, S. (2019). Tourism climate insurance: Implications and prospects. *Asia Pacific Journal of Tourism Research, 24*(4), 269–280. https://doi.org/10.1080/10941665.2018.1564338

Poortinga, W., Spence, A., Whitmarsh, L., Capstick, S., & Pidgeon, N. F. (2011). Uncertain climate: An investigation into public scepticism about anthropogenic climate change. *Global Environmental Change, 21*(3), 1015–1024. https://doi.org/10.1016/j.gloenvcha.2011.03.001

Richins, M. L., & Dawson, S. (1992). A consumer values orientation for materialism and its measurement: Scale development and validation. *Journal of Consumer Research, 19*(3), 303–316. https://doi.org/10.1086/209304

Ru, X., Qin, H., & Wang, S. (2019). Young people's behavior intentions towards reducing PM 2.5 in China: Extending the theory of planned behavior. *Resources, Conservation, and Recycling, 141*, 99–108. https://doi.org/10.1016/j.resconrec.2018.10.019

Ryley, T., Davison, L., Bristow, A., & Pridmore, A. (2010). Public engagement on aviation taxes in the United Kingdom. *International Journal of Sustainable Transportation, 4*(2), 112–128. https://doi.org/10.1080/15568310802471735

Sausen, R., Isaksen, I., Grewe, V., Hauglustaine, D., Lee, D. S., Myhre, G., Köhler, M. O., Pitari, G., Schumann, U., Stordal, F., & Zerefos, C. (2005). Aviation radiative forcing in 2000: An update on IPCC (1999). *Meteorologische Zeitschrift, 14*(4), 555–561. https://doi.org/10.1127/0941-2948/2005/0049

Schwartz, S. (1992). Universals in the content and structure of values: Theoretical advances and empirical tests in 20 countries. *Advances in Experimental Social Psychology, 25*, 1–65.

Scott, D., Peeters, P., & Gössling, S. (2010). Can tourism deliver its "aspirational" greenhouse gas emission reduction targets? *Journal of Sustainable Tourism, 18*(3), 393–408. https://doi.org/10.1080/09669581003653542

Simonson, I., & Drolet, A. (2004). Anchoring effects on consumers' willingness-to-pay and willingness-to-accept. *Journal of Consumer Research, 31*(3), 681–690. https://doi.org/10.1086/425103

Sonnenschein, J., & Smedby, N. (2019). Designing air ticket taxes for climate change mitigation: Insights from a Swedish valuation study. *Climate Policy, 19*(5), 651–663. https://doi.org/10.1080/14693062.2018.1547678

Sparks, P., & Shepherd, R. (1992). Self-identity and the theory of planned behavior: Assessing the role of identification with "green consumerism". *Social Psychology Quarterly, 55*(4), 388–399. https://doi.org/10.2307/2786955

Sparks, P., Shepherd, R., & Frewer, L. J. (1995). Assessing and structuring attitudes toward the use of gene technology in food production: The role of perceived ethical obligation. *Basic and Applied Social Psychology, 16*(3), 267–285. https://doi.org/10.1207/s15324834basp1603_1

Strack, F., & Mussweiler, T. (1997). Explaining the enigmatic anchoring effect: Mechanisms of selective accessibility. *Journal of Personality and Social Psychology, 73*(3), 437–446. https://doi.org/10.1037/0022-3514.73.3.437

Terry, D. J., Hogg, M. A., & White, K. M. (1999). The theory of planned behavior: Self-identity, social identity and group norms. *British Journal of Social Psychology, 38*(3), 225–244. https://doi.org/10.1348/014466699164149

Thaler, R. H., & Sunstein, C. R. (2009). *Nudge: Improving decisions about health, wealth, and happiness*. Penguin.

Tversky, A., & Kahneman, D. (1974). Judgment under uncertainty: Heuristics and biases. *Science, 185*(4157), 1124–1131. https://doi.org/10.1126/science.185.4157.1124

Xie, B., Brewer, M. B., Hayes, B. K., McDonald, R. I., & Newell, B. R. (2019). Predicting climate change risk perception and willingness to act. *Journal of Environmental Psychology, 65*, 101331. https://doi.org/10.1016/j.jenvp.2019.101331

van der Linden, S. (2015). The social-psychological determinants of climate change risk perceptions: Towards a comprehensive model. *Journal of Environmental Psychology, 41*, 112–124. https://doi.org/10.1016/j.jenvp.2014.11.012

von Weizsacker, E. V., & Jesinghaus, J. (1992). Ecological tax reform: A policy proposal for sustainable development (No. INVES-ET P01 W436e). CEPAL, Santiago (Chile).

Ward, D. O., Clark, C. D., Jensen, K. L., Yen, S. T., & Russell, C. S. (2011). Factors influencing willingness-to-pay for the ENERGY STAR® label. *Energy Policy, 39*(3), 1450–1458. https://doi.org/10.1016/j.enpol.2010.12.017

Whitmarsh, L., & O'Neill, S. (2010). Green identity, green living? The role of pro-environmental self-identity in determining consistency across diverse pro-environmental behaviors. *Journal of Environmental Psychology, 30*(3), 305–314. https://doi.org/10.1016/j.jenvp.2010.01.003

Wu, H. C., Cheng, C. C., & Ai, C. H. (2018). An empirical analysis of green switching intentions in the airline industry. *Journal of Environmental Planning and Management, 61*(8), 1438–1468. https://doi.org/10.1080/09640568.2017.1352495

UNWTO. (2008). *Climate change and tourism – responding to global challenges, Carbon Choice carbon offset program*. Retrieved May 21, 2018, from https://www.united.com/ual/en/us/fly/company/globalcitizenship/environment/carbon-offset-program.html

World Tourism Organization (UNWTO). (2017). *20 reasons sustainable tourism counts for development*. Retrieved May 17, 2019, from http://documents.worldbank.org/curated/en/558121506324624240/20-reasons-sustainable-tourism-counts-for-development

# Application of internal environmental locus of control to the context of eco-friendly drone food delivery services

Jinsoo Hwang, Jin-soo Lee, Jinkyung Jenny Kim and Muhammad Safdar Sial

**ABSTRACT**

This study was designed to apply the concept of internal environmental locus of control (INELOC) to the context of eco-friendly drone food delivery services. In particular, this study examined how four subdimensions of INELOC, namely, green consumers, activists, advocates, and recyclers, affect anticipated emotions, such as positive and negative anticipated emotions. In addition, the effects of positive and negative anticipated emotions on intention to use were proposed. A research model with 10 hypotheses was developed and tested on the basis of theoretical relationships using 324 samples collected in South Korea. Data analysis results indicated that all four subdimensions of INELOC significantly affect positive anticipated emotions. Meanwhile, only green consumers and advocates significantly affect negative anticipated emotions. Lastly, significant relationships were found between anticipated emotions and intention to use.

## Introduction

The role of drones in the Fourth Industrial Revolution has elicited attention because of the high utilization of drones in various industries, such as agriculture, broadcasting, distribution, military, and fire detection (Bamburry, 2015; CBS News, 2018). The expanding application of drones is based on advancements in their capabilities that involve widespread accessibility, high-quality imaging and live streaming, and respectable speed (CBS News, 2018; Shavarani et al., 2018). Thus, drones are regarded as a disruptive innovation that is reshaping the way business is conducted.

The food service industry is not exempted from the significance of drones. In the food service industry, drones are used for food delivery, and because they bypass traffic congestion, drone food delivery services can deliver quickly to customers (Hwang et al., 2019). According to Hwang and Kim (2019, p. 872), drone food delivery services refer to "a service that uses drones to deliver food to a place where the customer wants it." Furthermore, numerous experiments have been conducted to test drones in food delivery services, and their superiority in many aspects, such as cost, efficiency, and safety, has been confirmed (Doole et al., 2018; Park et al., 2018). Moreover, many recent studies have shown that drone-based delivery services play an important role in protecting the environment because in contrast with current delivery methods, such as cars and motorcycles, drones are powered by electricity (Hwang & Kim, 2019). Figliozzi (2017) proved that drone-based delivery is crucial for reducing energy consumption and carbon dioxide ($CO_2$) emissions. Stolaroff et al. (2018) argued that drone-based delivery helps decrease greenhouse gas (GHG) emissions. Although drone food delivery services facilitate environmental protection, only a few studies have focused on this topic.

The concept of internal environmental locus of control (INELOC) is stemming from internal locus of control, which refers to the degree that people believe they have control over the outcome of events (Rotter, 1966). People with a strong internal locus of control consider that the results are attributed to their efforts and capabilities. For example, when individuals receive the great performance appraisal, their own actions take the credit for the results rather than other external factors such as market demands or competitions. Mirroring the concept of internal locus of control, INELOC refers to individuals' beliefs that influence their environmental outcomes (McCarty & Shrum, 2001); it includes the following four subdimensions: green consumers, activists, advocates, and recyclers (Cleveland et al., 2012). The group of people who possess a sense of INELOC devote themselves to creating eco-friendly environment by conducting or participating related activities. That is, people with high INELOC levels tend to exhibit responsible environmental behavior than those with low INELOC levels (Hines et al., 1987). Cleveland et al. (2012) asserted that INELOC captures a consumer's multifaceted attitudes pertaining to his/her personal ability to affect environmental outcomes and individual responsibility toward sustainability. In addition, these authors argued that an individual's environment-friendly creeds are not translated into pro-environmental deeds and cannot address the needs of INELOC in predicting consumer pro-environmental behavior/intention. For this reason, identifying the potential INELOC of consumers is critical for predicting whether they use eco-friendly services.

An individual's behavioral intention is affected by the emotion that arises during actual experience and the positive or negative anticipation of such emotions (Baumgartner et al., 2008; Ha, 2018). Various customer behavior models in the environmental context have been built on existing theories, such as theory of planned behavior (TPB) and the norm activation model (NAM), which are derived from social psychology. Thus, as pointed out by Kals and Maes (2002), these studies frequently exclude the role of emotions. Thereafter, emotions and forward-looking emotions, namely, anticipated emotions, have been adopted in studies that explicate consumer behavior in diverse settings (Ahn & Kwon, 2020; Hwang et al., 2019; Rezvani et al., 2017; Xie et al., 2015). For example, Perugini and Bagozzi (2001) asserted that an individual's prospect-based emotions are a critical predictor of customer intentions to use. To the best of our knowledge, however, anticipated emotions have not been considered in explaining the associations between INELOC and customer ecological behavioral intention in drone food delivery services.

Recognizing a range of environmental challenges, people at present have increased their attention toward current patterns of food consumption (Koenig-Lewis et al., 2014; Siddiqui et al., 2018). In this respect, drones are part of the progression of refining ways of food delivery in the age of technological innovation, and the role of eco-friendly drones is inevitable in the current marketplace. Accordingly, studies have been performed in the context of drone-based delivery (Bamburry, 2015; Doole et al., 2018; Hwang & Kim, 2019; Stolaroff et al., 2018). However, these studies have focused on the benefits of using drone food delivery services to consumers. By contrast, research on the eco-friendly role of such services is highly insufficient. Distinct from prior research, the current study examines potential consumer propensity to eco-friendly drone food delivery services using INELOC because understanding consumer propensity is significant for developing an effective marketing strategy (Hwang & Hyun, 2017).

Accordingly, the current study aims to fill the research gaps by being the first to empirically examine INELOC and its outcomes in the context of eco-friendly drone food delivery services for the first time. In particular, the objectives of this study are to explore (1) the effects of the four subdimensions of INELOC, namely, green consumers, activists, advocates, and recyclers, on antici-pated emotions, including positive and negative emotions; and (2) the causal relationships between the two anticipated emotions on behavioral intention. Likewise, the proposed frame-work of the present research links the multi-dimensional INELOC, which is an essential personal trait, with anticipated emotions in understanding customer pro-environmental behavioral inten-tion. More specifically, it illustrates the direct impact of multifaceted INELOC on positive and negative anticipated emotions which, in turn, influence intention to use drone food delivery serv-ices. This approach is relatively infrequent in the current literature and it is thereby unique and differentiated from prior studies. The findings of this study are accordingly expected to provide empirical evidence for explaining the intricate associations among INELOC, anticipated emotions, and intention to use eco-friendly drone food delivery services which have not been discovered to date. These findings have meaningful implications for theory and practice.

## Literature review

### *Drone food delivery services*

Food delivery services are changing the landscape of food consumption. The growing demand for food delivery services has been related to different factors, such as an increased number of single households and dual-income families, changing lifestyles, new generations, and new eat-ing patterns (Forbes, 2018; UBS, 2018). The use of doorstep food delivery has been further accel-erated by technological progress, including a platform-to-consumer business model (Ray et al., 2019). That is, the advancement of digital technology is one notable impetus for developing food delivery services by enabling customers to purchase online through applications or web-sites to bring food to their doorstep. Food delivery services are considered a fast-growing sector, and food delivery sales can rise annually by more than 20%, resulting in estimated sales of US\$365 billion worldwide by 2030 (UBS, 2018).

Drones are currently considered innovative delivery tools because of their many advantages, such as cost, time, and effort. Doole et al. (2018) conducted an experiment in the field of fast-food delivery and confirmed that drone food delivery can cut the unit cost of traditional food delivery services in half. Thus, drone-based delivery is regarded as a promising solution to increasing traffic on the ground and a system for minimizing total costs (Shavarani et al., 2018). Moreover, drone food delivery services have been proven to play a significant role in reducing environmental footprint by minimizing $CO_2$ emissions and global warming (Goodchild & Toy, 2018; Koiwanit, 2018). Park et al. (2018) assessed the environmental impact of drones versus cur-rent delivery modes and demonstrated the substantial contributions of a drone delivery system. Similarly, the eco-friendly roles of drone food delivery services were comprehensively described by Kim and Hwang (2020); that is, drones can play an initiating and central role in the advance-ment of sustainability in the context of food delivery services.

Considering the high potential of drones, numerous companies have entered the drone food delivery service market. Wing, a subsidiary of Google's parent company, Alphabet Inc., was granted permission to fly drones to deliver food in Canberra, Australia (Reuters, 2019). An Uber Eats delivery drone design was revealed at the Forbes Under 30 Summit in Detroit, USA, and the company announced plans to test meal delivery using this new design in San Diego, USA in summer 2020 (Forbes, 2019). Recently, Manna, an Irish drone startup company, and its partner-ships with food ordering company Flipdish and Cubic Telecom were addressed in CES 2020 (CNET, 2019). The deal will connect drones through a 5 G network, allowing customers to track their order and know exactly when the drone will "drop" it at their door. In particular, the drones' use of an electric power system with zero carbon emissions was highlighted. A system for drone food delivery services is due to be installed in Europe.

Health, Safety, Environment, and Quality (HSEQ) is a well-known discipline that organizations are encouraged to or must comply with to ensure that their activities will not cause any harm (Rahimi, 1995). It involves developing safe and eco-friendly processes and building a systematic approach to abide by environmental regulations from the health and environment perspectives. HSEQ is a management trend in modern society that will certainly be accentuated further. Moreover, HSEQ is no longer limited to a certain industry but is also applied to the food service industry. For example, Siddiqui et al. (2018) introduced examples of innovative packaging of fruits and vegetables that contribute to HSEQ. In this respect, eco-friendly drone food delivery services must be exemplary in HSEQ as a disruptive innovation in the field of food delivery services.

## Explanation of each construct

### INELOC

The environment-friendly behavior/intention of individuals has been extensively studied, and many scholars have conducted research that focused on existing theories, such as theory of reasoned action, TPB, NAM, and value-belief-norm (VBN) theory. For example, Fielding et al. (2008) incorporated TPB to examine an individual's behavioral intention in contributing to sustainability. Their results confirmed that people with a more favorable attitude toward and a stronger sense of environmental activism exhibit a greater intention to engage in eco-friendly behavior. Kim and Hwang (2020) focused on the pro-environmental role of drone-based food delivery services and incorporated NAM and TPB into explaining the formation of consumer behavioral intention toward drone food delivery services. Their analysis results based on 401 samples provided empirical evidence that each construct of NAM (i.e., problem awareness, ascribed responsibility, and personal norm) and TPB (i.e., attitude, subjective norm, and perceived behavioral control) is a strong indicator of individual behavioral intention. Meanwhile, the premise of VBN theory is that personal values contribute to building beliefs, and consequently, personal norm; this theory has been proven to be powerful in predicting an individual's environmentally responsible behavior/intention (Han & Hwang, 2017; Van Riper & Kyle, 2014). A stream of these studies has validated the key constructs stemming from current theories in explaining consumer pro-environmental behavioral intention. However, additional attempts have been made to extend existing theories by adopting other variables and providing new insights of prominent variables, such as personality traits, in explaining consumer eco-friendly behavior/intention with improved predicting power.

Studies have investigated the roles of self-efficacy, self-esteem, and environmental consciousness level on the basis of several concepts related to individual psychological mechanisms (Kornilaki et al., 2019; Martínez García de Leaniz et al., 2018). In this regard, Bradley and Sparks (2002) recognized locus of control as the most enduring and predictive construct among various personality traits identified in the social science literature. Locus of control is described as the degree to which individuals believe or perceive that they can affect outcomes through their behavior; it is categorized into internal and external loci of control (Rotter, 1966). The former refers to people who hold control and perceive that outcomes depend on their input; the latter describes people who believe that they are powerless and that outcomes are beyond their control (Cleveland et al., 2012; Lefcourt, 1991; Rotter, 1966; Yang & Weber, 2019). In line with the notion of internal locus of control, INELOC is defined as "a construct that captures individuals' multifaceted attitudes pertaining to personal responsibility toward and ability to affect environmental outcomes" (Cleveland et al., 2012, p. 293). Similarly, INELOC has been proposed as an important variable for building consumer intention toward pro-environmental behavior (Aguilar et al., 2008; McCarty & Shrum, 2001). Despite studies adopting Rotter's locus of control (1966) in the field of environmental behavior, Guagnano (1995) argued that locus of control should be regarded as multifaceted because subtleties will not be articulated in the environmental context. Accordingly, Cleveland et al. (2012) addressed the necessity for multidimensional characteristics of INELOC and presented four underlying dimensions: green consumers, activists, advocates, and recyclers.

Green consumers refer to a group of individuals who live by a personal ethic and have complete confidence in their ability to make a difference toward sustainable development by using pro-environmental products or services and boycotting environmentally unfriendly companies. By contrast, Dono et al. (2010) described activists as people who are committed to public actions to improve the environmental quality of a policy/system and influence the broader population. Similarly, participating in demonstrations and providing financial support are frequently exemplified as their active involvement in sustainability (Fielding et al., 2008). Compared with activists, advocates engage in environmental movements that are less public and require lower commitment (Larson et al., 2015). Hence, advocates typically persuade families, friends, or colleagues to participate in environmentally responsible activities. Lastly, recyclers refer to people with a routine recycling behavior. The effort and time or inconvenience of recycling is part of individual volition; thus, recyclers are recognized as citizens who have a relatively simple and affordable environmental commitment (Iyer & Kashyap, 2007). This multidimensional INELOC, which distinguishes customers' propensity, has been adopted in the ecological context; its superior predicting power was articulated by Cleveland et al. (2012) and Colebrook-Claude (2019). Cleveland et al. (2012) argued that a unidimensional conceptualization of INELOC fails to capture the fine details in the context of sustainability. By contrast, multidimensional INELOC provides a more comprehensive understanding because it illustrates individual expressions of control over the environment.

### Anticipated emotions

Emotions are the affective responses of individuals to the perception of a certain situation (Clore et al., 1987). Emotions exert a significant impact and account for differences in individual sustainable behavior (Kals & Maes, 2002). Numerous studies on emotions have regarded emotions as informative cues that induce customer behavior/intention. For example, Xie et al. (2015) studied corporate green actions and investigated the mediating role of moral emotions on individual characteristics in consumer responses toward pro-environmental behavior. Their results showed significant associations among varied individual characteristics, moral emotions, and consumer responses. In particular, negative moral emotions (i.e., contempt, anger, and disgust) generate negative responses, such as negative word of mouth and complaints, among consumers. By contrast, positive emotions lead to positive responses, such as positive word of mouth and investment, among consumers. Nonetheless, this stream of research fails to assess the relevance of anticipation of the affective outcome of future decisions. In addition, the concept of prospect-based emotions, which refers to an individual's anticipated post-behavioral affective responses, was proposed.

Considering that drone food delivery services are not yet fully commercialized at present, the current study adopted the notion of anticipated emotions. Mellers and McGraw (2001) conceptualized anticipated emotions as guides to consumer decision-making. These authors presented decision affect theory and claimed that individuals anticipate emotions of future outcomes when they make decisions, which, in turn, influences the choice of options. That is, people generally predict the emotional consequences of their future decisions prior to making decisions, and these anticipated emotions influence their decision-making. Similarly, Baumgartner et al. (2008) asserted that anticipated emotions will encourage the formation of an individual's behavioral intention to engage in a certain behavior. Furthermore, numerous scholars have proposed anticipated emotions that consist of positive anticipated emotions (e.g., delighted, excited, happy, and proud) and negative anticipated emotions (e.g., depressed, disappointed, guilty, and uncomfortable) (Ahn & Kwon, 2020; Han et al., 2018; Onwezen et al., 2013; Perugini & Bagozzi, 2001). In this regard, Bagozzi et al. (2016) explained that people's decisions are generally influenced by the pursuit of positive emotions or the avoidance of negative emotions.

Following this line of research, the role of anticipated emotions in eco-friendly behavioral intention and innovative product adoption has been supported in diverse settings (Baumgartner et al., 2008; Han & Hwang, 2017; Han et al., 2018; Hwang et al., 2019; Piçarra & Giger, 2018). Onwezen et al. (2013) conducted analyses to examine the function of anticipated pride and guilt in environment-friendly behavior within the NAM model and an integrated NAM–TPB framework. The results based on 617 Dutch respondents indicated the mediating role of anticipated pride and guilt on the link between personal norm and behavior. Han et al. (2017) explored the progression of customers' eco-friendly behavior and developed a value–belief–emotion–norm framework based on the VBN model. They found that perceived ability exerts a significant impact on reducing the threat to an anticipated feeling of pride; they also confirmed that emotional states play an important role in generating intention. Rezvani et al. (2017) examined consumer adoption of sustainable products, and their analysis results demonstrated the direct influence of anticipated emotions on consumer adoption of pro-environment products. Similarly, anticipated emotions have been found to play a significant role in technology adoption. For example, Piçarra and Giger (2018) adopted a goal-directed model to assess the level of customer intention to work with robots. Their results confirmed the significance of anticipated emotions in the formation of an intention to work with robots.

## Intentions to use

Behavioral intention refers to the likelihood of individual engagement with a specific behavior (Chua et al., 2019; Moon & Han, 2019; Oliver, 1997); it has been measured using various subdimensions, such as intention to use, word-of-mouth intention, and willingness to pay a higher price (Han et al., 2018; Hwang et al., 2019; Kim & Hwang, 2020; Zeithaml et al., 1996). Among the proposed underlying dimensions of behavioral intention, intention to use has been examined as a strong indicator in the context of sustainability (Han et al., 2019; Hwang & Kim, 2019; Meng et al., 2020; Rezvani et al., 2017). In addition, customer intention to use has been extensively studied as an important predictor of actual behavior in the field of technology adoption (Han et al., 2019; Hwang et al., 2019; Hwang et al., 2019; Okumus & Bilgihan, 2014; Trang, Lee, & Han, 2019). In particular, many studies have adopted the technology acceptance model, and customer intention to use has been widely utilized to measure consumer adoption and willingness to try new technology once it becomes available. For example, Okumus and Bilgihan (2014) tested intention to use to investigate the utilization of smartphone applications as innovative tools for promoting healthy eating behavior among customers when ordering food in restaurants.

## Hypothesis development

### Effect of INELOC on anticipated emotions

Extant studies on eco-friendly behavior/intention are abundant, and the significant association between consumer belief in environmental protection and emotions has been demonstrated in a number of studies. For example, Moons and De Pelsmacker (2012) centered on sustainable value of electric car usage and recognized the different customer segment. They categorized 1,202 Belgians into environmental concern, environmental behavior, opinion leadership, and personal values and the significant role of emotions was illustrated in more environment-oriented groups. Koenig-Lewis et al. (2014) tested consumer emotional evaluations of ecologically responsible packaging, and their results confirmed that individual concern with environmental issues significantly influences positive emotions. Powell and Bullock (2014) explored the relationships among individuals' predispositions toward nature, emotional experience, and conservation mindedness based on 408 pieces of data collected in New York. They measured individuals' propensity toward nature by questioning their level of interests and commitment to conservation efforts or organization and reported the significant association between the dispositions of the people toward nature and emotional responses. Jang et al. (2015) conducted an online survey among 312 coffee customers to examine the moderating role of green consciousness. Their results found that consumers with a high degree of green consciousness displayed stronger emotional attachment and responded more positively to green practices than those who are less environmentally conscious. Xie et al. (2015) examined the role of the individual characteristics of people, such as demonstrating a high value for social justice and having high morals in the relationship between green activities conducted by enterprises and consumer emotional reactions. The results reliably supported the notion that individuals who are empathetic tend to exude positive emotions that induce favorable word of mouth. Gravante and Poma (2016) carried out a series of in-depth interviews with four self-organized environmental groups in Mexico. Their findings described how and where environmental activism begins and its association with emotions which influence the organizational choices. Han et al. (2018) studied biospheric value, environmental concern, awareness of consequences, and ascription of responsibility as cognitive drivers of green behavioral intention. Their analysis results based on 302 samples from US cruise passengers indicated the statistically supported association between cognitive triggers and affective drivers (i.e., positive and negative anticipated emotions) in an environmentally responsible manner. In particular, these researchers confirmed that environmental concerns influence the formation of positive and negative anticipated emotions. Chiang et al. (2019) reported the significant relationship between locus of control and emotional stability in inducing customer pro-environmental behavior through an empirical analysis of 473 responses collected in Taiwan. The aforementioned studies imply that INELOC is related to anticipated emotions, and the following hypotheses are proposed accordingly.

Hypothesis 1a: Green consumers will relate significantly to positive anticipated emotions toward eco-friendly drone food delivery services.

Hypothesis 1b: Green consumers will relate significantly to negative anticipated emotions toward eco-friendly drone food delivery services.

Hypothesis 2a: Activists will relate significantly to positive anticipated emotions toward eco-friendly drone food delivery services.

Hypothesis 2b: Activists will relate significantly to negative anticipated emotions toward eco-friendly drone food delivery services.

Hypothesis 3a: Advocates will relate significantly to positive anticipated emotions toward eco-friendly drone food delivery services.

Hypothesis 3b: Advocates will relate significantly to negative anticipated emotions toward eco-friendly drone food delivery services.

Hypothesis 4a: Recyclers will relate significantly to positive anticipated emotions toward eco-friendly drone food delivery services.

Hypothesis 4b: Recyclers will relate significantly to negative anticipated emotion toward eco-friendly drone food delivery services.

### Effect of anticipated emotions on intention to use

Numerous attempts have been made to involve positive and negative anticipated emotions in explaining customer pro-environmental behavior/intention; and significant associations have been supported by empirical evidence (Koenig-Lewis et al., 2014; Perugini & Bagozzi, 2001; Piçarra & Giger, 2018). Mellers and McGraw (2001) examined how anticipated emotions are

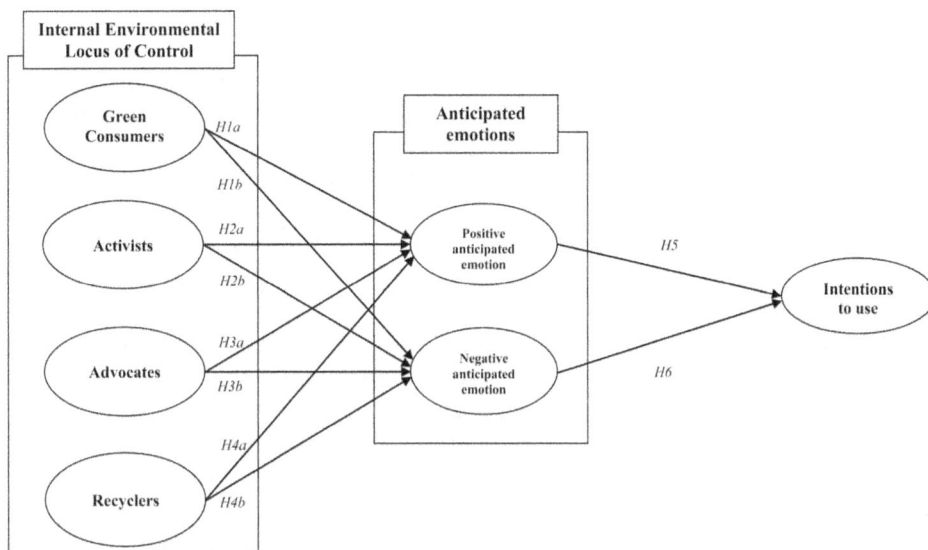

**Figure 1.** Proposed conceptual model.

related to the choices people make through laboratory and real-world studies. They found a significant relationship between anticipated emotions and decision-making, and they explained that people tend to choose the option with greater average pleasure. Kals and Maes (2002) tested the association between emotions and behavioral intention; they demonstrated the prominent role of emotions in predicting an individual's pro-environmental willingness. In particular, their data analysis based on 281 German samples indicated that anger and affinity are related to behavioral intention. Baumgartner et al. (2008) explored the impact of future-oriented emotions on motivating goal-directed behavior. Their results provided evidence that positive and negative anticipated emotions influence behavioral intention. Ha (2018) built a framework based on regret theory and analyzed 805 samples to predict consumer behavior in the field of new technology. The analysis results indicated that anticipated regret affected customer adoption level of innovative technology. Ahn and Kwon (2020) focused on green practices in the hotel industry and confirmed the salient impact of positive and negative anticipated emotions on environment-friendly behavioral intention toward green hotels. Accordingly, we present the following hypotheses.

Hypothesis 5: Positive anticipated emotions will relate significantly to intention to use eco-friendly drone food delivery services.

Hypothesis 6: Negative anticipated emotions will relate significantly to intention to use eco-friendly drone food delivery services.

### Research model

By integrating the aforementioned theoretical background and empirical evidence, the current study developed a conceptual model for exploring the associations among INELOC, anticipated emotions, and intention to use eco-friendly drone food delivery services (Figure 1).

### Methodology

#### Measurement items

The present study used multiple-item scales adopted from prior studies. First, INELOC included four sub-dimensions: green consumers, environmental activists, environmental advocates, and recyclers. These subdimensions were measured using 16 items from Cleveland et al. (2005) and Cleveland et al. (2012). Second, anticipated emotions consisted of two subdimensions: positive and negative anticipated emotions. These subdimensions were measured using six items from Perugini and Bagozzi (2001, 2004) and Hwang et al. (2019). Lastly, intention to use was measured using three items from Hwang and Lyu (2018) and Zeithaml et al. (1996). All measurement items were measured with a seven-point Likert-type scale ranging from 1 (strongly disagree) to 7 (strongly agree).

The questionnaire was thoroughly reviewed by two groups of experts: (1) three faculty members with a major research focus on the restaurant industry and (2) individuals holding a remote pilot certificate. After confirming that the questionnaire items exhibited no problems, a pretest was conducted among 50 restaurant customers through an online survey in South Korea. To enable the respondents to clearly understand the importance of drone food delivery services in protecting the environment, a 2 min newspaper article on the environmental benefits of drone-based delivery services was provided to the respondents before the survey began. The data analysis results indicated that the Cronbach's alpha values for all the constructs were greater than 0.70, suggesting a high level of reliability (Nunnally, 1978).

## Data collection

Data collection was achieved using an online survey company, which is one of the biggest companies in South Korea. This study is related to food delivery services, so the survey was conducted among restaurant customers who had used food delivery services within the last 6 months. The respondents filled the survey form after reading the newspaper article that was used on the pretest before starting the survey. The newspaper article used in the pretest explained the eco-friendly role of the drone food delivery service compared to the current delivery services. The survey company sent an invitation email to 4,525 panelists. Among them, 442 participated in the survey. It showed a participation rate of about 10%. In addition, 37 samples were deleted due to multicollinearity problems and visual inspections. In particular, because these outliers distort the statistical results (Agresti & Finlay, 2009), we exclude them from statistical analysis. Finally, 405 samples were used for statistical analysis.

## Data analysis

### Profile of survey respondents

Table 1 provides the profile of the survey respondents. Among the respondents, 200 were males (49.4%) and 205 were females (50.6%). In terms of monthly household income, 27.7% of the respondents indicated that their household income ranged from US$1,001 to US$2,000. In addition, 50.4% of the respondents were single and 48.6% were married. With regard to education level, most of the respondents held a bachelor's degree (63.7%, $n = 258$). Lastly, mean age was 37.81 years old.

### Confirmatory factor analysis (CFA)

Table 2 presents the CFA results. CFA was used to evaluate composite reliability, convergent validity, and discriminant validity. The CFA model provided a suitable fit for the data, i.e., $\chi^2 = 383.849$, df $= 168$, $\chi^2$/df $= 2.285$, $p < 0.001$, normed fit index (NFI) $= 0.960$, incremental fit index (IFI) $= 0.977$, comparative fit index (CFI) $= 0.977$, Tucker–Lewis index (TLI) $= 0.971$, and root mean square error of approximation (RMSEA) $= 0.056$. All factor loadings were $\geq 0.813$ ($p < 0.001$).

**Table 1.** Profile of survey respondents ($n = 405$).

| Variable | $n$ | Percentage |
|---|---|---|
| **Gender** | | |
| Male | 200 | 49.4 |
| Female | 205 | 50.6 |
| **Monthly household income** | | |
| US$6,001 and over | 23 | 5.7 |
| US$5,001–US$6,000 | 14 | 3.5 |
| US$4,001–US$5,000 | 36 | 8.9 |
| US$3,001–US$4,000 | 51 | 12.6 |
| US$2,001–US$3,000 | 109 | 26.9 |
| US$1,001–US$2,000 | 112 | 27.7 |
| Under US$1,000 | 60 | 14.8 |
| **Marital status** | | |
| Single | 204 | 50.4 |
| Married | 197 | 48.6 |
| Widowed/Divorced | 4 | 1.0 |
| **Education level** | | |
| Less than High school diploma | 48 | 11.9 |
| Associate's degree | 62 | 15.3 |
| Bachelor's degree | 258 | 63.7 |
| Graduate degree | 37 | 9.1 |
| **Mean age = 37.92 years old** | | |

As presented in Table 3, the composite reliability values ranged from 0.930 to 0.964, which were higher than the recommended level of 0.600 (Bagozzi & Yi, 1988). Thus, all the constructs have acceptable internal consistency. In addition, the average variance extracted (AVE) values fell within the range of 0.792–0.899; thus, they were greater than the suggested cutoff of 0.500 (Hair et al., 2006), signifying high convergent validity levels. Lastly, the AVE values for all the constructs were greater than all of the squared correlations ($R^2$) between all possible pairs of constructs, indicating adequate discriminant validity.

### Structural equation modeling (SEM)

SEM was conducted to validate the proposed hypotheses. The overall evaluation of model fit exhibited an acceptable fit of the model to the data ($\chi^2 = 410.410$, df $= 173$, $\chi^2$/df $= 2.372$,

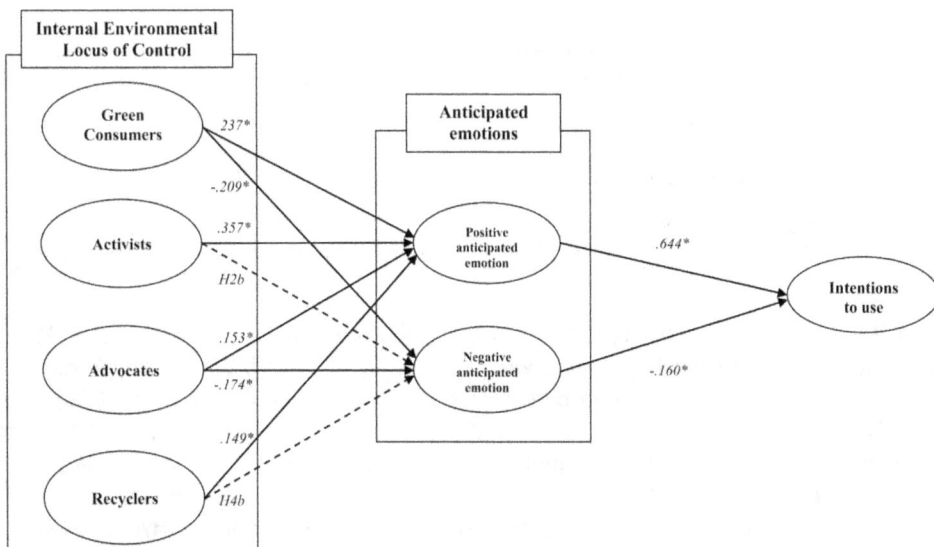

**Figure 2.** Standardized theoretical path coefficients.

**Table 2.** Confirmatory factor analysis: Items and loadings.

| Construct and scale items | Standardized Loading[a] |
| --- | --- |
| **Green consumers** | |
| The sooner consumers start buying greener products, the sooner companies will transform to respond to their demands. | .929 |
| The more I buy 'green' products, the more I help persuade companies to become 'friendlier' to the environment. | .941 |
| By buying greener products, I can make a difference in helping the environment. | 887 |
| **Activists** | |
| Any donation to environmental groups helps it attain its goals. | .878 |
| The efforts deployed by environmental groups have an impact on the end result of many ecological challenges. | .930 |
| By making donations to pro-environmental groups, I can help make a positive difference on the state of the environment. | .920 |
| **Advocates** | |
| I am able to convince a friend to change his/her conservation habits. | .926 |
| I am able to convince some of my friends to take some kind of action with regards to environmental challenges. | .913 |
| If willing, people can generally influence their friends' transportation habits. | .827 |
| **Recyclers** | |
| By recycling, I am helping to reduce pollution. | .939 |
| By recycling, I am doing my part to help the state of the environment. | .953 |
| By recycling, I am saving valuable natural resources. | .952 |
| **Anticipated emotions** | |
| **Positive anticipated emotion** If I use an environmentally friendly way, such as drone food delivery services, I will feel ... | |
| Excited | .870 |
| Delighted | .930 |
| Happy | .931 |
| **Negative anticipated emotion** | |
| If I use an environmentally friendly way, such as drone food delivery services, I will feel ... | |
| Disappointed | .813 |
| Depressed | .944 |
| Uncomfortable | .947 |
| **Intentions to use** | |
| I will use drone food delivery services when ordering food. | .947 |
| I am willing to use drone food delivery services when ordering food. | .940 |
| I am likely to use drone food delivery services when ordering food. | .952 |

Goodness-of-fit statistics: $\chi^2 = 383.849$, df $= 168$, $\chi^2/df = 2.285$, $p < .001$, NFI $= .960$, IFI $= .977$, CFI $= .977$, TLI $= .971$, RMSEA $= .056$.
Notes 1: [a]All factors loadings are significant at $p < .001$.
Notes 2: NFI = Normed Fit Index, CFI = Comparative Fit Index, TLI = Tucker-Lewis Index, RMSEA = Root Mean Square Error of Approximation.

$p < 0.001$, NFI $= 0.957$, IFI $= 0.975$, CFI $= 0.975$, TLI $= 0.969$, RMSEA $= 0.058$). The SEM results showed that eight of the ten hypotheses were statistically accepted, but two hypotheses were rejected (see Figure 2 and Table 4). In particular, green consumers ($\beta = 0.237$, $p < 0.05$), activists ($\beta = 0.357$, $p < 0.05$), advocates ($\beta = 0.153$, $p < 0.05$), and recyclers ($\beta = 0.149$, $p < 0.05$) exerted a positive effect on positive anticipated emotions. Thus, Hypotheses 1a, 2a, 3a, and 4a were supported. In addition, green consumers ($\beta = -0.209$, $p < 0.05$) and advocates ($\beta = -0.174$, $p < 0.05$) exerted a negative effect on negative anticipated emotions; hence, Hypotheses 1 b and 3 b were supported. However, activists and recyclers had no effect on negative anticipated

**Table 3.** Descriptive statistics and associated measures.

| | No. of Items | Mean (SD) | AVE | (1) | (2) | (3) | (4) | (5) | (6) | (7) |
|---|---|---|---|---|---|---|---|---|---|---|
| (1) Green consumers | 3 | 5.39 (1.00) | .845 | **.942**[a] | .643[b] | .498 | .663 | .635 | −.363 | .561 |
| (2) Activists | 3 | 5.08 (1.04) | .827 | .413[c] | **.935** | .424 | .548 | .651 | −.273 | .528 |
| (3) Advocates | 3 | 4.67 (1.03) | .792 | .248 | .180 | **.919** | .536 | .492 | −.330 | .507 |
| (4) Recyclers | 3 | 5.34 (1.05) | .899 | .440 | .300 | .287 | **.964** | .581 | −.327 | .467 |
| (5) Positive anticipated emotion | 3 | 4.48 (1.12) | .830 | .403 | .424 | .242 | .338 | **.936** | −.253 | .677 |
| (6) Negative anticipated emotion | 3 | 2.80 (1.27) | .816 | .132 | .075 | .109 | .107 | .064 | **.930** | −.323 |
| (7) Intentions to use | 3 | 4.51 (1.30) | .896 | .315 | .279 | .257 | .218 | .458 | .104 | **.963** |

Notes 1: SD = Standard Deviation, AVE = Average Variance Extracted.
Notes 2: a. Composite reliabilities are along the diagonal, b. Correlations are above the diagonal, c. Squared correlations are below the diagonal.

**Table 4.** Standardized parameter estimates for structural model.

| | | | Standardized Estimate | t-value | Hypothesis |
|---|---|---|---|---|---|
| H1a Green consumers | ⟶ | Positive anticipated emotion | .237 | 3.970* | Supported |
| H1b Green consumers | ⟶ | Negative anticipated emotion | −.209 | −2.713* | Supported |
| H2a Activists | ⟶ | Positive anticipated emotion | .357 | 6.676* | Supported |
| H2b Activists | ⟶ | Negative anticipated emotion | −.017 | −.254 | Not supported |
| H3a Advocates | ⟶ | Positive anticipated emotion | .153 | 3.241* | Supported |
| H3b Advocates | ⟶ | Negative anticipated emotion | −.174 | −2.840* | Supported |
| H4a Recyclers | ⟶ | Positive anticipated emotion | .149 | 2.737* | Supported |
| H4b Recyclers | ⟶ | Negative anticipated emotion | −.086 | −1.215 | Not supported |
| H5 Positive anticipated emotion | ⟶ | Intentions to use | .644 | 14.485* | Supported |
| H6 Negative anticipated emotion | ⟶ | Intentions to use | −.160 | −3.988* | Supported |

Goodness-of-fit statistics: $\chi^2 = 410.410$, df = 173, $\chi^2/df = 2.372$, $p < .001$, NFI = .957, IFI = .975, CFI = .975, TLI = .969, RMSEA = .058.
Notes 1: *$p < .05$.
Notes 2: NFI = Normed Fit Index, CFI = Comparative Fit Index, TLI = Tucker-Lewis Index, RMSEA = Root Mean Square Error of Approximation.

emotions. Therefore, Hypotheses 2b and 4b were not supported. Lastly, the data analysis results also indicated that positive anticipated emotions ($\beta = 0.644$, $p < 0.05$) and negative anticipated emotions ($\beta = -0.160$, $p < 0.05$) help enhance intention to use, supporting Hypotheses 5 and 6.

## Discussions and implications

Building a sustainable environment in the food service industry has apparently become a social issue, and it is addressed as a primary concern among today's increasingly knowledgeable and mature consumers (Koenig-Lewis et al., 2014; Siddiqui et al., 2018). Home cooking can disappear by 2030, and most meals will instead be ordered by consumers by making arrangements with central kitchens or restaurants to have their food delivered (UBS, 2018). Advancements in innovative technologies have enabled us to consume food conveniently and wisely in many aspects. In particular, drones have emerged as a disruptive innovation in the field of food delivery services, and many experiments have proven the premium value of drones in various areas, particularly in the environment (Goodchild & Toy, 2018; Koiwanit, 2018; Park et al., 2018). Accordingly, drone-based food delivery services have been recognized as a possible solution to achieving the highest HSEQ standards that incorporate the principles of sustainability in modern times.

Customer propensity is prominent, and INELOC has been examined as a salient indicator for explaining customer pro-environmental behavioral intention (Aguilar et al., 2008; Cleveland et al., 2012). In this regard, understanding how INELOC is linked to customer responses in an eco-friendly manner in drone food delivery services is a worthy undertaking. The present study is the first attempt to explore the associations among INELOC, anticipated emotions, and intention to undertake pro-environmental behavior when ordering food delivery services. Moreover, this work focused on drone-based delivery, which is an emerging issue as a disruptive innovation in food delivery services. A relatively smaller number of studies have connected INELOC and anticipated emotions; however, these two variables have not been used to articulate customer ecological behavior/intention in the context of drone food delivery services. Given that minimal information is available, the present study provides rich theoretical originality and lays the foundation for future research in the field of eco-friendly drone food delivery services. To be specific, a total of ten hypotheses were examined in the current study and eight proposed links were statistically supported. Green consumers were found to be significantly related to positive and negative anticipated emotions which, in turn influence intentions to use. It is interpreted that individuals who have a strong confidence in their pro-environmental actions to improve sustainability consider drone food delivery services in building the optimal emotional experience through increased positive anticipated emotions and decreased negative anticipated emotions, and consequently inducing intention to use. Similarly, advocates showed the significant associations with both positive and negative anticipated emotions which subsequently influence intention to use drone-based food delivery services. That is, drone food delivery services are easily regarded for advocates to motivate the people around them and that lead to the perfect emotional status in increasing intention to use. On the other hand, activists exhibited the significant associations with positive anticipated emotions but not with negative anticipated emotions in the formation of intention to use. These results are construed that, because activities are committed to develop the public morals and create systems to improve sustainability rather than the personal ecological practices, they are not necessarily affecting the negative anticipated emotions in using drone food delivery services. Recyclers displayed the similar results regarding the impact on anticipated emotions, which the influence on positive anticipated emotions was statically supported and the effect on negative anticipated emotions was rejected. We interpret it is due to the fact that drone-based food delivery services are not directly linked to recycling behaviors.

The present study provides empirical evidence for the relationships between multifaceted INELOC and anticipated emotions in the context of drone food delivery services. Bohlen et al. (1993) emphasized the importance of identifying consumers with green consciousness for organizations to communicate their environmental effort to the appropriate group of people. In addition, consumer segmentation depending on environment-specific variables is more adequate and stable than that based on demographic criteria (Cleveland et al., 2012; Straughan & Roberts, 1999), suggesting the superiority of INELOC in consumer segmentation in sustainability. By contrast, Hwang and Hyun (2017) asserted that understanding how different personality traits evoke varying affective responses is helpful for practitioners in developing marketing strategies. In this regard, our work is distinguished from previous studies by determining how customers' multifaceted perceptions of control over the environment are connected with customers' anticipated emotions as a result of their future behavior. These findings are similar to those of prior studies (Chiang et al., 2019; Jang et al., 2015; Koenig-Lewis et al., 2014), suggesting that the close relationship between INELOC and anticipated emotions includes positive and negative anticipated emotions. Hence, our study validated the high levels of rationality for using INELOC to explain an individual's anticipated emotions in the domain of eco-friendly drone food delivery services. Furthermore, our results will be helpful for practitioners in establishing marketing strategies depending on each distinct customer propensity in implementing drone food delivery services. Similarly, food service companies should understand the different propensity of consumers and establish differentiated marketing strategies depending on various segments in contributing to the environmental footprint with greater value return.

Green consumers relate significantly to positive and negative anticipated emotions in using drone food delivery services (Hypotheses 1a and 1b). A green consumer is described as a believer who has a decisive effect on environmental improvement through his/her eco-friendly behavior. Such individuals abide by environmental ethics and tend to refuse harmful processes or products to the environment (Cleveland et al., 2012; Moisander & Pesonen, 2002). Our results indicated that green consumers increased their positive anticipated emotions, but decreased their negative anticipated emotions through drone-based food delivery services, as exemplified by an environment-friendly mode of food delivery. Thus, using drones for food delivery is appropriate for green consumers. At present, a substantial number of food delivery orders are made via mobile applications. Thus, we suggest that a default option of drone-based delivery be used rather than a menu selection of delivery tools for green consumers. An option to use other delivery modes, such as motorcycles, can be provided, but these choices should be available with an additional charge that will be considered a penalty or cost to reduce environmental damage caused by the selected alternative delivery tool. That is, using an eco-friendly delivery method, namely, drone food delivery services, should be the norm among green consumers, simultaneously enhancing their anticipated positive emotions and reducing their anticipated negative emotions.

The results indicated that activists and positive anticipated emotions ($\beta = 0.357$, $p < 0.05$) are closely related (Hypothesis 2a). By contrast, activists did not exhibit a correlation with negative anticipated emotions (Hypothesis 2b). That is, activists relate to anticipated positive emotions toward drone food delivery services as an ecologically responsible behavior, but they do not relate to anticipated negative emotions. We inferred that these results are attributed to the characteristics of activists. Activists are typically depicted as individuals who influence people at large to improve infrastructure or systems (Dono et al., 2010; Fielding et al., 2008), and they may be insensitive to the direct negative consequences of relatively small actions. However, activists are still related to positive anticipated emotions through drone food delivery services; thus, food service companies should encourage activists to contribute to inspiring public support and confidence to use drone food delivery services. For example, food delivery service companies can organize informative campaigns that are supported by activists to raise awareness of the environmental roles of drones in food delivery services. Furthermore, collecting donations from activists to roll out complete drone-based food delivery services for small-scale or startup food service companies is recommended as a viable alternative. Thus, activists will be able to reduce their environmental footprint and enhance their anticipated positive emotions by using eco-friendly drone food delivery services.

Advocates and anticipated emotions are deeply involved with each other. In particular, advocates are related significantly to positive and negative anticipated emotions (Hypotheses 3a and 3b). Hence, practitioners should direct advocates to inspire people around them to preserve the environment by using drone-based food delivery services and avoid the use of environmentally harmful delivery tools. When advocates practice "working green" in food delivery services, they should be continuously encouraged to remind people of green deeds and convince those around them to use eco-friendly drone food delivery services. For example, practitioners in food delivery services may run a reward program for advocates who post influential stories or videos on their social networking sites. Meanwhile, the analysis results confirmed that a close association ($\beta = 0.149$, $p < 0.05$) exists between recyclers and positive anticipated emotions (Hypothesis 4a). By contrast, the relationship between recyclers and negative anticipated emotions was not statistically supported (Hypothesis 4b). Drone-based food delivery services are not directly linked to recycling behavior, possibly explaining why recyclers are not related to negative anticipated emotions. However, consumers with recycling habits exhibit a higher probability of positive anticipated emotions toward ecological food delivery services, and food service companies should pay attention to enhancing this relationship. As previously explained, food delivery services are growing substantially, and an increasing number of customers are using food delivery services for their daily food consumption. Therefore, we suggest that consumers should be led to believe that recyclers' routine participation in drone food delivery services is part of the effort to protect the environment. That is, having recyclers become fully aware of the environmental roles of drone food delivery services is recommended to build their practices.

Lastly, anticipated emotions exerted a salient influence on intention to use (Hypotheses 5 and 6). These results supported those of previous studies (Hwang et al., 2019; Piçarra & Giger, 2018), suggesting the positive impact of positive anticipated emotions and the negative impact of negative anticipated emotions on intention to use. As we postulated, when individuals anticipate positive emotions, such as excitement and delight, they demonstrate favorable behavioral intention toward drone food delivery services. By contrast, when people anticipate negative emotions, such as disappointment and depression, their intention to use drone food delivery services is reduced. Our study enhanced the understanding of such relationship by providing an empirical evidence that increasing positive anticipated emotions and reducing negative anticipated emotions will build overall high intention to use drone food delivery services. Therefore, food service companies should understand that people anticipate the emotions that they might experience in the future as a result of the choices they make and recognize the influence of anticipated emotions on customer decisions. Moreover, an effort to increase positive anticipated emotions is necessary. Environmental certifications are regarded as useful tools because they are related to customer behavior in the hospitality context (Martínez García de Leaniz et al., 2018). Similarly, acquiring certifications to formalize the environmental roles of drones in food delivery services may be considered to induce positive anticipated emotions.

## Limitations and future study

Although the objectives of this study were successfully accomplished through extensive literature review and empirical analyses, the findings of this research exhibit the following limitations. First, applying the findings of this study to other regions is difficult because this study collected data only from South Korea. Second, given that drone food delivery services have not been fully commercialized, we failed to measure actual behavior. Pro-environmental behavior is frequently considered not directly connected with intention, and future studies are suggested to measure actual behavior once services become fully available in the market. Another limitation of the current research is that no control variables were included in the data analysis. A number of studies addressed that individuals' attitudes in adopting novel technology were depending on their demographic characteristics (e.g. Hwang & Kim, 2019; Hwang et al., 2019), and therefore it is recommended to include the demographic profiles such as gender and age as control variables in examining the relationships among proposed constructs to increase the predicting power. Last, technical barriers or potential risks are other areas of concern when a new innovative technology is adopted (Okumus & Bilgihan, 2014). Thus, effort to understand consumer unfavorable perception and challenges in accepting drones as food delivery tools will be meaningful.

## Disclosure statement

No potential conflict of interest was reported by the author(s).

# References

Agresti, A., & Finlay, B. (2009). *Statistical methods for the social sciences* (4th ed.). Pearson Prentice Hall.

Aguilar, O. M., Waliczek, T. M., & Zajicek, J. M. (2008). Growing environmental stewards: The overall effect of a school gardening program on environmental attitudes and environmental locus of control of different demographic groups of elementary school children. *HortTechnology, 18*(2), 243–249. https://doi.org/10.21273/HORTTECH.18.2.243

Ahn, J., & Kwon, J. (2020). Green hotel brands in Malaysia: perceived value, cost, anticipated emotion, and revisit intention. *Current Issues in Tourism, 23*(12), 1559–1516. https://doi.org/10.1080/13683500.2019.1646715

Bagozzi, R. P., Belanche, D., Casaló, L. V., & Flavián, C. (2016). The role of anticipated emotions in purchase intentions. *Psychology & Marketing, 33*(8), 629–645.

Bagozzi, R. P., & Yi, Y. (1988). On the evaluation of structural equation models. *Journal of the Academy of Marketing Science, 16*(1), 74–94. https://doi.org/10.1007/BF02723327

Bamburry, D. (2015). Drones: Designed for product delivery. *Design Management Review, 26*(1), 40–48.

Baumgartner, H., Pieters, R., & Bagozzi, R. P. (2008). Future-oriented emotions: Conceptualization and behavioral effects. *European Journal of Social Psychology, 38*(4), 685–696. https://doi.org/10.1002/ejsp.467

Bohlen, G., Schlegelmilch, B. B., & Diamantopoulos, A. (1993). Measuring ecological concern: A multi-construct perspective. *Journal of Marketing Management, 9*(4), 415–430. https://doi.org/10.1080/0267257X.1993.9964250

Bradley, G. L., & Sparks, B. A. (2002). Service locus of control: Its conceptualization and measurement. *Journal of Service Research, 4*(4), 312–324. https://doi.org/10.1177/1094670502004004008

CBS News. (2018, February 15). Winter Olympics in Pyeongchang are the highest-tech games yet. Retrieved from https://www.cbsnews.com/news/winter-olympics-in-pyeongchang-are-the-highest-tech-games-yet/

Chiang, Y. T., Fang, W. T., Kaplan, U., & Ng, E. (2019). Locus of control: The mediation effect between emotional stability and pro-environmental behavior. *Sustainability, 11*(3), 820. https://doi.org/10.3390/su11030820

Chua, B. L., Kim, H. C., Lee, S., & Han, H. (2019). The role of brand personality, self-congruity, and sensory experience in elucidating sky lounge users' behavior. *Journal of Travel & Tourism Marketing, 36*(1), 29–42. https://doi.org/10.1080/10548408.2018.1488650

Cleveland, M., Kalamas, M., & Laroche, M. (2005). Shades of green: Linking environmental locus of control and pro-environmental behaviors. *Journal of Consumer Marketing, 22*(4), 198–212. https://doi.org/10.1108/07363760510605317

Cleveland, M., Kalamas, M., & Laroche, M. (2012). It's not easy being green": Exploring green creeds, green deeds, and internal environmental locus of control. *Psychology & Marketing, 29*(5), 293–305. https://doi.org/10.1002/mar.20522

Clore, G. L., Ortony, A., & Foss, M. A. (1987). The psychological foundations of the affective lexicon. *Journal of Personality and Social Psychology, 53*(4), 751–766. https://doi.org/10.1037/0022-3514.53.4.751

CNET. (2019). Manna's 5G drone delivery deal will help you track your airborne pizza. Retrieved from https://www.cnet.com/news/manna-5g-drone-delivery-deal-will-help-you-track-your-airborne-pizza/

Colebrook-Claude, C. (2019). *Development and validation of the adolescent internal environmental locus of control scale* [Doctoral dissertation]. Fielding Graduate University.

Dono, J., Webb, J., & Richardson, B. (2010). The relationship between environmental activism, pro-environmental behaviour and social identity. *Journal of Environmental Psychology, 30*(2), 178–186. https://doi.org/10.1016/j.jenvp.2009.11.006

Doole, M., Ellerbroek, J., & Hoekstra, J. (2018). Drone delivery: Urban airspace traffic density estimation. In 8th SESAR Innovation Days, Salzburg, Austria.

Fielding, K. S., McDonald, R., & Louis, W. R. (2008). Theory of planned behaviour, identity and intentions to engage in environmental activism. *Journal of Environmental Psychology, 28*(4), 318–326. https://doi.org/10.1016/j.jenvp.2008.03.003

Figliozzi, M. A. (2017). Lifecycle modeling and assessment of unmanned aerial vehicles (drones) CO2 e emissions. *Transportation Research Part D: Transport and Environment, 57*, 251–261. https://doi.org/10.1016/j.trd.2017.09.011

Forbes. (2018, June 26). Millennials are ordering more food delivery, but are they killing the kitchen, too? Retrieved from https://www.forbes.com/sites/andriacheng/2018/06/26/millennials-are-ordering-food-for-delivery-more-but-are-they-killing-the-kitchen-too/#70cdcac2393e

Forbes. (2019, October 28). First look: Uber unveils new design for Uber Eats delivery drone. Retrieved from https://www.forbes.com/sites/bizcarson/2019/10/28/first-look-uber-unveils-new-design-for-uber-eats-delivery-drone/#37cfdce778f2

Goodchild, A., & Toy, J. (2018). Delivery by drone: An evaluation of unmanned aerial vehicle technology in reducing CO2 emissions in the delivery service industry. *Transportation Research Part D: Transport and Environment, 61*, 58–67. https://doi.org/10.1016/j.trd.2017.02.017

Gravante, T., & Poma, A. (2016). Environmental self-organized activism: Emotion, organization and collective identity in Mexico. *International Journal of Sociology and Social Policy, 36*(9/10), 647–661. https://doi.org/10.1108/IJSSP-11-2015-0128

Guagnano, G. A. (1995). Locus of control, altruism and agentic disposition. *Population and Environment, 17*(1), 63–77. https://doi.org/10.1007/BF02208278

Ha, Y. (2018). Expectations gap, anticipated regret, and behavior intention in the context of rapid technology evolvement. *Industrial Management & Data Systems, 118*(3), 606–617. https://doi.org/10.1108/IMDS-02-2017-0045

Hair, J. F., Jr., Black, W. C., Babin, B. J., Anderson, R. E., & Tatham, R. L. (2006). *Multivariate data analysis* (6th ed.). Prentice-Hall.

Han, H., Chua, B. L., & Hyun, S. S. (2019). Consumers' intention to adopt eco-friendly electric airplanes: The moderating role of perceived uncertainty of outcomes and attachment to eco-friendly products. *International Journal of Sustainable Transportation*, 1–15. https://doi.org/10.1080/15568318.2019.1607957

Han, H., & Hwang, J. (2017). What motivates delegates' conservation behaviors while attending a convention? *Journal of Travel & Tourism Marketing, 34*(1), 82–98. https://doi.org/10.1080/10548408.2015.1130111

Han, H., Hwang, J., & Lee, M. J. (2017). The value–belief–emotion–norm model: Investigating customers' eco-friendly behavior. *Journal of Travel & Tourism Marketing, 34*(5), 590–607. https://doi.org/10.1080/10548408.2016.1208790

Han, H., Olya, H. G., Kim, J., & Kim, W. (2018). Model of sustainable behavior: Assessing cognitive, emotional and normative influence in the cruise context. *Business Strategy and the Environment, 27*(7), 789–800. https://doi.org/10.1002/bse.2031

Han, H., Yu, J., & Kim, W. (2018). Youth travelers and waste reduction behaviors while traveling to tourist destinations. *Journal of Travel & Tourism Marketing, 35*(9), 1119–1131. https://doi.org/10.1080/10548408.2018.1435335

Han, H., Yu, J., & Kim, W. (2019). Environmental corporate social responsibility and the strategy to boost the airline's image and customer loyalty intentions. *Journal of Travel & Tourism Marketing, 36*(3), 371–383. https://doi.org/10.1080/10548408.2018.1557580

Hines, J. M., Hungerford, H. R., & Tomera, A. N. (1987). Analysis and synthesis of research on responsible environmental behavior: A meta-analysis. *The* Journal of Environmental Education, *18*(2), 1–8. https://doi.org/10.1080/00958964.1987.9943482

Hwang, J., Cho, S. B., & Kim, W. (2019). Consequences of psychological benefits of using eco-friendly services in the context of drone food delivery services. *Journal of Travel & Tourism Marketing, 36*(7), 835–846. https://doi.org/10.1080/10548408.2019.1586619

Hwang, J., & Hyun, S. S. (2017). First-class airline travelers' tendency to seek uniqueness: how does it influence their purchase of expensive tickets? *Journal of Travel & Tourism Marketing, 34*(7), 935–947. https://doi.org/10.1080/10548408.2016.1251376

Hwang, J., & Kim, H. (2019). Consequences of a green image of drone food delivery services: The moderating role of gender and age. *Business Strategy and the Environment, 28*(5), 872–884. https://doi.org/10.1002/bse.2289

Hwang, J., Kim, H., & Kim, W. (2019). Investigating motivated consumer innovativeness in the context of drone food delivery services. *Journal of Hospitality and Tourism Management, 38*, 102–110. https://doi.org/10.1016/j.jhtm.2019.01.004

Hwang, J., Lee, J. S., & Kim, H. (2019). Perceived innovativeness of drone food delivery services and its impacts on attitude and behavioral intentions: The moderating role of gender and age. *International Journal of Hospitality Management, 81*, 94–103. https://doi.org/10.1016/j.ijhm.2019.03.002

Hwang, J., & Lyu, S. O. (2018). Understanding first-class passengers' luxury value perceptions in the US airline industry. *Tourism Management Perspectives, 28*, 29–40. https://doi.org/10.1016/j.tmp.2018.07.001

Iyer, E. S., & Kashyap, R. K. (2007). Consumer recycling: Role of incentives, information, and social class. *Journal of Consumer Behaviour, 6*(1), 32–47. https://doi.org/10.1002/cb.206

Jang, Y. J., Kim, W. G., & Lee, H. Y. (2015). Coffee shop consumers' emotional attachment and loyalty to green stores: The moderating role of green consciousness. *International Journal of Hospitality Management, 44*, 146–156. https://doi.org/10.1016/j.ijhm.2014.10.001

Kals, E., & Maes, J. (2002). Sustainable development and emotions. In P. Schmuck, & W. P. Schultz (Eds.), *Psychology of sustainable development* (pp. 97–122). Kluwer Academic Publications.

Kim, J. J., & Hwang, J. (2020). Merging the norm activation model and the theory of planned behavior in the context of drone food delivery services: Does the level of product knowledge really matter? *Journal of Hospitality and Tourism Management, 42*, 1–11. https://doi.org/10.1016/j.jhtm.2019.11.002

Koenig-Lewis, N., Palmer, A., Dermody, J., & Urbye, A. (2014). Consumers' evaluations of ecological packaging – Rational and emotional approaches. *Journal of Environmental Psychology, 37*, 94–105. https://doi.org/10.1016/j.jenvp.2013.11.009

Koiwanit, J. (2018). Analysis of environmental impacts of drone delivery on an online shopping system. *Advances* in Climate Change Research, *9*(3), 201–207. https://doi.org/10.1016/j.accre.2018.09.001

Kornilaki, M., Thomas, R., & Font, X. (2019). The sustainability behaviour of small firms in tourism: The role of self-efficacy and contextual constraints. *Journal of Sustainable Tourism, 27*(1), 97–117. https://doi.org/10.1080/09669582.2018.1561706

Larson, L. R., Stedman, R. C., Cooper, C. B., & Decker, D. J. (2015). Understanding the multi-dimensional structure of pro-environmental behavior. *Journal of Environmental Psychology*, *43*, 112–124. https://doi.org/10.1016/j.jenvp.2015.06.004

Lefcourt, H. M. (1991). Locus of control. In J. P. Robinson, P. R. Shaver, & L. S. Wrightsman (Eds.), *Measures of personality and social psychological attitudes* (pp. 413–499). Academic Press.

Martínez García de Leaniz, P., Herrero Crespo, Á., & Gómez López, R. (2018). Customer responses to environmentally certified hotels: The moderating effect of environmental consciousness on the formation of behavioral intentions. *Journal of Sustainable Tourism*, *26*(7), 1160–1177. https://doi.org/10.1080/09669582.2017.1349775

McCarty, J. A., & Shrum, L. J. (2001). The influence of individualism, collectivism, and locus of control on environmental beliefs and behavior. *Journal of Public Policy & Marketing*, *20*(1), 93–104.

Mellers, B. A., & McGraw, A. P. (2001). Anticipated emotions as guides to choice. *Current Directions in Psychological Science*, *10*(6), 210–214. https://doi.org/10.1111/1467-8721.00151

Meng, B., Ryu, H. B., Chua, B. L., & Han, H. (2020). Predictors of intention for continuing volunteer tourism activities among young tourists. *Asia Pacific Journal of Tourism Research*, *25*(3), 261–273. https://doi.org/10.1080/10941665.2019.1692046

Moisander, J., & Pesonen, S. (2002). Narratives of sustainable ways of living: Constructing the self and the other as a green consumer. *Management Decision*, *40*(4), 329–342. https://doi.org/10.1108/00251740210426321

Moon, H., & Han, H. (2019). Tourist experience quality and loyalty to an island destination: The moderating impact of destination image. *Journal of Travel & Tourism Marketing*, *36*(1), 43–59.

Moons, I., & De Pelsmacker, P. (2012). Emotions as determinants of electric car usage intention. *Journal of Marketing Management*, *28*(3-4), 195–237. https://doi.org/10.1080/0267257X.2012.659007

Nunnally, J. C. (1978). *Psychometric theory*. McGraw-Hill.

Okumus, B., & Bilgihan, A. (2014). Proposing a model to test smartphone users' intention to use smart applications when ordering food in restaurants. *Journal of Hospitality and Tourism Technology*, *5*(1), 31–49. https://doi.org/10.1108/JHTT-01-2013-0003

Oliver, R. L. (1997). *Satisfaction: A behavioral perspective on the consumer*. Irwin-McGraw-Hill.

Onwezen, M. C., Antonides, G., & Bartels, J. (2013). The Norm Activation Model: An exploration of the functions of anticipated pride and guilt in pro-environmental behaviour. *Journal of Economic Psychology*, *39*, 141–153. https://doi.org/10.1016/j.joep.2013.07.005

Park, J., Kim, S., & Suh, K. (2018). A comparative analysis of the environmental benefits of drone-based delivery services in urban and rural areas. *Sustainability*, *10*(3), 888. https://doi.org/10.3390/su10030888

Perugini, M., & Bagozzi, R. P. (2001). The role of desires and anticipated emotions in goal-directed behaviors: Broadening and deepening the theory of planned behavior. *The British Journal of Social Psychology*, *40*(1), 79–98. https://doi.org/10.1348/014466601164704

Perugini, M., & Bagozzi, R. P. (2004). The distinction between desires and intentions. *European Journal of Social Psychology*, *34*(1), 69–84. https://doi.org/10.1002/ejsp.186

Piçarra, N., & Giger, J. C. (2018). Predicting intention to work with social robots at anticipation stage: Assessing the role of behavioral desire and anticipated emotions. *Computers in Human Behavior*, *86*, 129–146. https://doi.org/10.1016/j.chb.2018.04.026

Powell, D. M., & Bullock, E. V. (2014). Evaluation of factors affecting emotional responses in zoo visitors and the impact of emotion on conservation mindedness. *Anthrozoös*, *27*(3), 389–405. https://doi.org/10.2752/175303714X13903827488042

Rahimi, M. (1995). Merging strategic safety, health and environment into total quality management. *International Journal of Industrial Ergonomics*, *16*(2), 83–94. https://doi.org/10.1016/0169-8141(94)00074-D

Ray, A., Dhir, A., Bala, P. K., & Kaur, P. (2019). Why do people use food delivery apps (FDA)? A uses and gratification theory perspective. *Journal of Retailing and Consumer Services*, *51*, 221–230. https://doi.org/10.1016/j.jretconser.2019.05.025

Reuters. (2019, April 9). Alphabet's drone delivery service takes off in Australia. Retrieved from https://www.reuters.com/article/us-alphabet-wing/alphabets-drone-delivery-service-takes-off-in-australia-idUSKCN1RL16U

Rezvani, Z., Jansson, J., & Bengtsson, M. (2017). Cause I'll feel good! An investigation into the effects of anticipated emotions and personal moral norms on consumer pro-environmental behavior. *Journal of Promotion Management*, *23*(1), 163–183. https://doi.org/10.1080/10496491.2016.1267681

Rotter, J. B. (1966). Generalized expectancies for internal versus external control of reinforcement. *Psychological Monographs: General and Applied*, *80*(1), 1–28. https://doi.org/10.1037/h0092976

Shavarani, S. M., Nejad, M. G., Rismanchian, F., & Izbirak, G. (2018). Application of hierarchical facility location problem for optimization of a drone delivery system: A case study of Amazon prime air in the city of San Francisco. *The International Journal of Advanced Manufacturing Technology*, *95*(9-12), 3141–3153. https://doi.org/10.1007/s00170-017-1363-1

Siddiqui, M. W., Rahman, M. S., & Wani, A. A. (Eds.). (2018). *Innovative packaging of fruits and vegetables: strategies for safety and quality maintenance*. Apple Academic Press, NY.

Stolaroff, J. K., Samaras, C., O'Neill, E. R., Lubers, A., Mitchell, A. S., & Ceperley, D. (2018). Energy use and life cycle greenhouse gas emissions of drones for commercial package delivery. *Nature Communications*, *9*(1), 409. https://doi.org/10.1038/s41467-017-02411-5

Straughan, R. D., & Roberts, J. A. (1999). Environmental segmentation alternatives: A look at green consumer behavior in the new millennium. *Journal of Consumer Marketing*, *16*(6), 558–575. https://doi.org/10.1108/07363769910297506

Trang, H. L. T., Lee, J. S., & Han, H. (2019). How do green attributes elicit pro-environmental behaviors in guests? The case of green hotels in Vietnam. *Journal of Travel & Tourism Marketing*, *36*(1), 14–28. https://doi.org/10.1080/10548408.2018.1486782

UBS. (2018, June 18). Is the kitchen dead? Retrieved from https://www.ubs.com/global/en/investment-bank/in-focus-/2018/dead-kitchen.html

Van Riper, C. J., & Kyle, G. T. (2014). Understanding the internal processes of behavioral engagement in a national park: A latent variable path analysis of the value-belief-norm theory. *Journal of Environmental Psychology*, *38*, 288–297. https://doi.org/10.1016/j.jenvp.2014.03.002

Xie, C., Bagozzi, R. P., & Grønhaug, K. (2015). The role of moral emotions and individual differences in consumer responses to corporate green and non-green actions. *Journal of the Academy of Marketing Science*, *43*(3), 333–356. https://doi.org/10.1007/s11747-014-0394-5

Yang, X., & Weber, A. (2019). Who can improve the environment - Me or the powerful others? An integrative approach to locus of control and pro-environmental behavior in China. *Resources, Conservation and Recycling*, *146*, 55–67. https://doi.org/10.1016/j.resconrec.2019.03.005

Zeithaml, V. A., Berry, L. L., & Parasuraman, A. (1996). The behavioral consequences of service quality. *The Journal of Marketing*, *60*(2), 31–46. https://doi.org/10.2307/1251929

# The impact of the Middle East Respiratory Syndrome coronavirus on inbound tourism in South Korea toward sustainable tourism

Yunseon Choe, Junhui Wang and HakJun Song

**ABSTRACT**

Despite the declaration of the end of the Middle East Respiratory Syndrome coronavirus (MERS) outbreak in December 2015 in South Korea that the epidemic lasted for 2 months, the depressed domestic economy and tourism sector did not immediately restore. Thus, it is important to explore how much of an impact the disease had and how long the damage lasted. Using quantitative time-series models, the present study explored the influence of MERS on inbound tourism in South Korea and estimates the concrete impact of MERS on the market in 2015. Monthly international tourist arrival data were provided by the Korea Tourism Organization from January 2009 to December 2015. The results showed that the contagious disease was statistically and negatively significant for inbound tourists visiting South Korea. During the time of MERS, from June 2015 to September 2015, the total effect was estimated to be −1,968,765 tourists with a loss of 3.1 billion USD in receipts. This study can not only better estimate the impact on tourism number of inbound tourist arrivals, but also supports policy-makers in their attempts to establish proper policies to assure tourists of their safety in such crises.

## Introduction

Tourism is highly vulnerable, but resilient to external events; economic development trends; environmental, political, and socio-cultural contexts, and international situations as such events can positively or negatively alter the flow of tourism (Biggs et al., 2012; Carney et al., 1999; Chambers & Conway, 1992; Pratt et al., 2011). The growth of tourism and the significance of the tourism industry have impacted tourism destinations causing sustainability practices to move to the forefront of continuing international tourism growth (Dodds, 2003; Hunter, 2002; Mastny, 2002; Murphy & Price, 1998; Pratt et al., 2011; Pryce, 2001; Stipanuk, 1996). Sustainability can be threatened by epidemics, which also threaten tourists, tourism destinations, and locals living in tourism areas. In addition, epidemics can cause tour cancellations, temporary facility closures, changes in air routes and cruise routes, and loss of access to tourism destination (Capo et al., 2007; Han & Hyun, 2018; Manning, 2004). In particular, sustainable consumption can be impeded by epidemics that result in decreases in tourist numbers, employment, and hotel occupancies; increases in tourists impacted by the negative side effects caused by the epidemics; and increases in tourists who fear to travel to destinations, change plans, and/or avoid public places to lessen their risks. These disruptions have devastating effects on sustainable tourism behaviors, supply chain tourism, and the economy (Jung et al., 2016).

Despite these epidemic impacts, the results for the tourism sector may still be substantial in many countries and subsectors (Glaesser, 2011). The impact of epidemics could last for months or years as tourists' perceptions of the risks related to destinations could last for that time (Manning, 2004). In order to determine how the tourism industry can move toward sustainability, one must first determine the barriers to sustainable tourism as well as develop strategies to mitigate the types of crises discussed above (Bramwell & Lane, 2011; De Sausmarez, 2007; Graci, 2007). Toward this goal, the aim of this research is to identify the impact of Middle East Respiratory Syndrome coronavirus (hereafter MERS) on inbound tourism in South Korea (hereafter Korea). In this study, MERS was selected as the case to be studied because it heavily devastated the Korean economy in 2015, especially in regard to the inbound tourism sector due to the risk factors related to the spread of the infection through contact with an infected person's respiratory secretions. Despite the declaration of the end of MERS outbreak in December 2015, the depressed domestic economy and tourism sector did not immediately recover. Thus, it is important to explore how much of an impact the disease had and how long the impact lasted so as to provide policy-makers and tourism businesses with evidence related to critical implications of epidemics on sustainable tourism development and guidance for sustainable consumption of tourism in the economy (Brown, 1998; Jung et al., 2016; Sharpley, 2003).

Recently, several studies have revealed the effects of epidemics, such as SARS (Cooper, 2005; McKercher & Chon, 2004; Min, 2005), the Chikungunya virus (Mavalankar et al., 2009; Parola et al., 2006), the Dengue virus (Choe et al., 2017; Mavalankar et al., 2009), and the Ebola virus (Baker, 2015), on the tourism industry. However, little estimation has been conducted related to the impact of epidemics on inbound tourism demand (Song & Lee, 2010), and only a few studies have shown the impact of MERS on inbound tourism using a rigorous methodology (Shi & Li, 2017). Instead, studies have examined the impact of MERS with a particular focus such as on the Chinese tourists to Korea (Shi & Li, 2017); on the number of tourists to Jeju Island (Song, 2016); and on casino visitors to Korea (Suh & Kim, 2018). Although these studies emphasized the negative impact of MERS on the tourism demand, rather less attention has been paid to the overall impact of MERS on Korea's inbound tourism demands using forecasting methods throughout the MERS period. Therefore, the purpose of this study is to concretely reveal the impact of MERS on tourism demand in Korea as well as estimate tourism data for the period affected by MERS using various forecasting methods (e.g., the autoregressive integrated moving average (ARIMA) model, winters exponential smoothing model, and stepwise autoregressive model). By assessing the impact of MERS on inbound tourism demands, the findings of this study may be helpful to the tourism industry as it attempts to prepare for and predict tourist demand in the face of other external tourism crises, especially in regard to creating crises coping strategies and the resulting necessary enhanced level sustainable consumption of tourism resources. Furthermore, this study could provide South Korea's practical strategic actions to prepare for the tourism industry to keep its business afloat from a previous MERS coronavirus outbreak into recovery.

## Literature review

### *MERS outbreak and its impact on Korea*

An external event can be defined as any unpredicted incident that influences travelers' confidences in tourism destinations and impedes continued normal operations (Luhrman, 2003). Among the various external events that can affect tourists' flow the influence of infectious diseases seems to be powerful (Choe et al., 2017; Richter, 2003). For instance, SARS occurred in 32 countries, produced 8,422 cases, and caused 916 deaths worldwide, with a loss of 60 billion USD in receipts (Glaesser, 2011; World Health Organization [WHO], 2003). In Taiwan, it only produced 674 cases and 84 deaths, but it caused Taiwan's inbound tourism numbers to decline by 50% for the second quarter of 2013 (Min, 2005). Epidemics, such as SARS, the Avian Flu, malaria, cholera, and other diseases that can be transmitted by international tourism, have caused worldwide concern for the WHO (Kuo et al., 2008). According to the WHO (2018), MERS can be transmitted from animal to human through close contact (e.g., contact with dromedary camels was the main cause of MERS in humans) as well as through interpersonal transmission. The typical case of MERS is characterized by fever, cough, diarrhea, and shortness of breath (WHO, 2018).

Since the first case was identified in Saudi Arabia in June 2012, MERS has been identified in over 27 countries, with 2,260 cases and 803 deaths, including in the Middle East, Europe, Asia and the United States (WHO, 2018). In May 2015, the first case of MERS in Korea was identified in a 68-year-old male. From the first case to the official end of the MERS crisis on December 23, 2015, MERS spread quickly through household and hospital contacts between healthcare workers and patients. During this period, 186 Korean people were infected, 38 of whom died (Korea Centers for Disease Control and Prevention [KCDC], 2018). MERS significantly and negatively affected Korea's inbound tourism including domestic tourism although the Korean government, city of Seoul, and the Korea Association of Travel Agents attempted to make recovery efforts by offering MERS insurance for international visitors as well as undertaking campaigns to deal with MERS crisis on the Korean economy and inbound tourism.

According to the Korea Culture and Tourism Institute (2015), it is briefly estimated that the MERS outbreak caused a decrease in about half of the country's 332 travel agencies, with 1.53 million fewer international visitors compared to June through September of the previous year. In addition, few academic studies have investigated the impacts of the MERS on the tourism industry in Korea (Shi & Li, 2017; Song, 2016; Suh & Kim, 2018). For example, Shi and Li (2017) examined the impact of MERS on the arrival of international tourists from China to Korea. This study aimed at discovering the effects of MERS on four different types of tourism demand (i.e., tour, business, official, other types). The study utilized an autoregressive distributed lag model, using quarterly data collected by the Korea Tourism Organization. The results showed that MERS had a negative influence on the total number of inbound tourist arrivals and Chinese tourist arrivals to Korea, but did not find significant impacts on business, official, and other types of tourism demands.

Song (2016) analyzed the number of tourists to Jeju Island based on the purpose of their travel (i.e., rest and sightseeing, leisure and sport, conference and business). He fit the SARIMA model to the monthly international tourist arrival data from 2005 to 2016. The results showed that, while the leisure and sport and conference and business groups were significantly affected by MERS in June, resulting in a decreased number of tourists by 30%-40%, the rest and sightseeing groups were barely affected. Recently, Suh and Kim (2018) estimated the macroeconomic or financial variables related to the number of international visitors to Korean casinos arriving from four of Korea's major tourism markets (i.e., Japan, China, Hong Kong, Taiwan). The author used monthly international tourist arrival data from 2006 to 2016, applying a SVAR model. This study found that the impact of MERS had a statistically significant impact on all four countries, causing a 40%-85% decrease in casino visitors.

However, previous studies on the impact of MERS on the tourism industry were not conclusive because most studies employed a SARIMA model (Song, 2016), ARIMA model (Sung, 2016), or SVAR (Suh & Kim, 2018). The methods can be limited to calculate the specific effect sizes for MERS (Jung & Sung, 2017). As such, to resolve these issues, this research provides more integrated methodologies focused on the impact of MERS on Korea's inbound tourism demands employing a variety of rigorous forecasting methods (i.e., ARIMA, Winters Exponential Smoothing, Stepwise Autoregressive Model). This study is timely as Korea recently reported a second MERS case in September 2018 that has caused the Korean government to isolate 21 people and identify 399 others who may have been exposed to MERS (KCDC, 2018). This situation implies that infectious diseases, such as MERS, will continue to impact the tourism industry for the foreseeable future. The processes used within and the results of this study will help countries better predict the impact of such diseases on tourism numbers so as to establish appropriate tourism policies when similar infectious diseases occur (Mai & Smith, 2015).

### *Tourism and epidemics*

Recent empirical studies have measured the effects of external crises on international tourism demand (Enders & Sandler, 1991; Huang & Min, 2002). To date, according to Maditinos and Vassiliadis (2008), the majority of the research on how external events impact tourism falls into four categories: (1) natural disasters (e.g., earthquakes, avalanches, floods, hurricanes); (2) terrorism; (3) political instability and war (e.g., wars, oil shocks, energy crises, economic crises, armed conflicts); and (4) epidemics (e.g., SARS, foot-and-mouth disease, swine flu, the Ebola virus, MERS, COVID-19). Among the external crises, the external events of epidemics seem to be the more important and serious issues because epidemics can affect not only the areas where they occur, but also the surrounding areas, while the influence of other external events can be limited to single regions and the durations of other external events are often shorter than that of epidemics (Rodway-Dyer & Shaw, 2005). Moreover, the effect of epidemics may be greater than expected in modern societies where exchanges between countries are high (Jamal & Budke, 2020). Fatal epidemics have scared people because of unknown preventions and cures, which has, in turn, influenced tourism. Specifically, the spread of epidemics is closely related to living in a globalized world, not only limits individual's activities, but also results in a country's economic loss (Jung & Sung, 2017). Prior research has dedicated sensitive economic sectors such as inbound tourism, hotel, and airlines. For example, the research on the epidemic outbreaks of SARS in 2003 has focused on the impact of SARS on China's tourism industry (Wen et al., 2005; Zhang & Wei, 2003) and the hotel industry (Chien & Law, 2003). Liu et al. (2011) demonstrated that the outbreak of SARS in Taiwan had a significant influence on the loss of inbound tourists (−64.31%), compared to other crisis such as the impact of 9/11 attacks in 2001 was −24.59%, the earthquake in 2001 was −17.07%, and Asian financial crisis in 1997 was −2.54%. Also, Chi and Baek (2013) proposed that SARS epidemic had a long-term influence on air transportation due to a lack of demand.

In Korea, Choe et al. (2017) examined the overall incidence of four infectious diseases between 2003 and 2012. This study analyzed the relationship among international travel, certain infectious diseases (i.e., salmonellosis, shigellosis, malaria, dengue), and seasonal variability. The authors obtained data from the National Notifiable Disease Surveillance System for 75 cases of infectious diseases and from the Korea Tourism Organization for the number of inbound and outbound travelers to Korea. The results of the study showed that international travel was associated with incidences of imported infectious diseases in Korea between 2003 and 2012. Several researchers have conducted econometric studies to determine the effects of epidemics on the tourism industry. For instance, the effect of SARS on Taiwanese foreign visitor arrivals was examined by applying a SARIMA model (Min, 2005). This research compared the predicted and actual values of Taiwanese inbound tourists. The results showed that the number of international tourists to the country decreased until November 2003. While the number of inbound tourists was not considerably lower in March and April 2003, the biggest value difference of 224,178 visitors occurred in May.

Yim and Sohn (2007) measured the impact of SARS on Korean inbound tourists. While the figure of inbound tourists was not substantially lower in March and April 2003, the prevalent difference of 208,341 tourists was found in May 2003. The occurrence of the Ebola virus in West

Africa in February 2014 and its spread to Guinea, Liberia, Nigeria, and Sierra Leone, affected the African region's economies, including the tourism industry (Baker, 2015). Due to a decrease in trade, borders were closed, foreign investment was reduced, and flights were cancelled. According to the World Bank Group, the Ebola virus epidemic resulted in an economic loss of least 1.6 billion USD in Guinea, Liberia and Sierra Leone (Baker, 2015). Various forecast methods have been used to determine the effect of these events on the tourism industry. Goh and Law (2002) investigated Hong Kong tourists' patterns related to number of events, using forecasting techniques. They found that the SARIMA and ARIMA intervention models are better to used when forecasting than other time-series models. Since the majority of the forecasting research has focused on estimating coefficients as an effect of external events on tourism demand, limited research exists on the estimation of the impact of external events on inbound tourism demand (Song & Lee, 2010). Moreover, not many researchers have studied how long it will take for the impact of epidemics to dissipate. Within this strand of research, we aim to examine the estimation of the impact of epidemics on inbound tourism demand, employing three forecasting techniques.

## Method

### Measurements

We obtained the monthly international tourist arrival data from the Korean Tourism Statistics from the Korea Tourism Organization (2017). Monthly international tourist arrivals are the number of visitors by air and/or ships, which includes tourists, cruise passengers, crew members, and arrivals of nationals residing abroad (Korea Tourism Organization, 2017; Manosuthi et al., 2020). Monthly international tourist arrival from a country to Korea used in this study was provided by the Korea Tourism Organization and runs from January 2009 to December 2015. As monthly international tourist arrivals can be considered the most intuitive variable to show the impact of MERS, it was selected to be the main variable in this research. In this study, a univariate time-series model was employed in order to evaluate the effect of MERS on tourism. Previous studies on tourism and crises have shown that the impact of epidemics (e.g., SARS, Ebola virus, MERS) on international tourism has been serious. The spread of epidemics leads to a decrease in the arrival of tourists, sometimes severe (Baker, 2015; Min, 2005; Yim & Sohn, 2007). Between July and August 2015, the number of inbound tourists who booked accommodations and other tickets fell by 82.1% in South Korea, with this damage valued at 95.8 million USD according to the Korea Association of Travel Agents (2015).

### ARIMA and the ARIMA intervention model

In terms of understanding the ARIMA, it is necessary to understand the basic model of ARIMA (i.e., the Autoregressive Moving Average Model (ARMA)). As the ARMA model can only be used with stationary data series, it must be expanded into more general model (i.e., ARIMA model) by considering the process of differencing which refers to subtracting each observation from its previous observation in the time-series to make the time-series data stationary (Makridakis et al., 1998).

Furthermore, the ARIMA model can become more complete by considering exogenous events called interventions (e.g., incidents of terrorism or wars, outbreaks of contagious diseases, economic fluctuations, energy crises) (Box & Tiao, 1975). In order words, the univariate ARIMA model can be altered to become the ARIMA intervention model by additionally permitting the dependent variable to be affected by the time path of the binary independent variable called the intervention (Enders, 1995; Makridakis et al., 1998). Bonham et al. (2006) stated that interventions in the ARIMA model are able to decrease uncertainty and increase model accuracy. The formula for the ARIMA intervention model can be expressed as follows (Box & Tiao, 1975):

$$Y_t = f(I_t) + N_t \quad t = 1,\ 2,\ \ldots,\ T \tag{1}$$

(1)where $Y_t$ = dependent variable, $f(I_t)$ = intervention component of the model, and $N_t$ = stochastic disturbance assumed to be the ARIMA model.

Thus, $f(I_t)$ indicates the crucial component between an intervention(s) and the time-series. Given the assumption of multiplicative seasonality, the time structure of $N_t$ can be expressed with a general ARIMA $(p,d,q)(P,D,Q)_s$ process. Under these assumptions, the disturbance term can be formulated as follows (Enders, 1995):

$$N_t = \frac{\Theta_Q(B^S)\theta_q(B)}{\Phi_P(B^S)\varphi_p(B)(1-B)^d(1-B)^D} e_t \tag{2}$$

where $B$ = back shift operator such that $Y_t = Y_{t-1}$, $d$ = d-order non-seasonal difference operator, $D$ = D-order seasonal difference operator, $\varphi_p(B)$ = p-order non seasonal AR (auto regressive) model, $\Phi_P(B^S)$ = P-order seasonal AR (auto regressive) model, $\theta_q(B)$ = q-order non seasonal MA (moving average) model, $\Theta_Q(B^S)$ = Q-order seasonal MA (moving average) model, and $e_t$ = error term $\sim$ IID $(0,\ \sigma^2)$.

Briefly, $Y_t$, the response or target variable, is a synthesized function of the crucial intervention indicator and pre-intervention ARIMA disturbance (Nelson, 2000). A dummy variable is usually employed to express the intervention indicator in the model. In the ARIMA intervention model, two types of dummy variables exist: a pulse or temporary intervention lasting for only one period, which can be represented by a one-time dummy variable, and a step or permanent intervention, representing a level shift at the time period of the intervention. The analyzing process of the ARIMA intervention model is not very different from that of the ARIMA model. The analyzing process usually consists of four steps (i.e., stationary process, model identification, estimation, diagnostic checking). The ARIMA intervention model can be distinguished from the ARIMA model by the fact that it is required to take into account the component of the intervention(s) within the process of the pre-intervention, which inspects the portion of the data set that happened before the intervention by using the ARIMA model (Enders, 1995).

A stationary process indicates a process to alter non-stationary time-series data to stationary time-series data, which will reveal the data changes around a constant mean, independent of time, and the variance of the data exists principally constant over time. The most prevalent way of making the time-series data stationary is to employ the approach of differencing by subtracting one observation from the previous observation in the time-series data (Makridakis et al., 1998). A seasonal stationary process should be also considered for seasonal time-series data and seasonal differencing needs to be performed for the time-series data seasonally non-stationary. Through the process of model identification, the ACF (autocorrelation function) and PACF (partial autocorrelation function) are used to tentatively identify the specific ARIMA model type.

While an ACF, a computation of the correlation of the observed data with consecutive lags of the data, is employed in order to estimate the order of the MA model, the PACF, a computation of the partial correlation of the observed data with consecutive lags of the data, is used to estimate the order of the AR model. In the process of estimating and undergoing diagnostic checking, the tentative model is evaluated in regard to whether the model is properly developed, which will confirm that all of the coefficients are significant and that the error term is not different from white noise (i.e., a purely random process), which will indicate that the error term of the tentative model can be considered to have a zero mean, have a constant variance, and be serially uncorrelated (Enders, 1995). The pre-intervention process, one of the remarkable differences between the ARIMA and ARIMA intervention models, indicates the process of model identification with some part of a given data set occurring before the intervention(s) in the ARIMA intervention model. Once the specific order of the ARIMA as a pre-intervention model has been successfully developed, the intervention(s) can be considered a binary dummy variable in the process of the post-intervention in order to explore the significance or impact of the intervention(s).

## Winters model

As one of the typical exponential smoothing methods, the Winters model is appropriate when the time-series data has a seasonal pattern. The forecast equation and three smoothing equations (i.e., level, trend, seasonality equations) are used to extrapolate seasonal time-series data in the Winters model (Makridakis et al., 1998). Specifically, two types of Winters model methods exist (i.e., additive model and multiplicative method) and the type of Winters model is decided by the nature of the seasonal variation. While the additive method is available when the seasonal variations are roughly constant, the multiplicative method is preferred when the seasonal variations are proportional to the level of the series. As the data related to tourism demand has a feature that causes the extent of the seasonal pattern to increase as the value of the data increases, the Winters multiplicative model was used in this study. The formula of the Winters multiplicative model can be suggested as follows (Makridakis et al., 1998):

Level:

$$Lt = \alpha \frac{Yt}{St-s} + (1-\alpha)(Lt-1 + bt-1),$$ (3)

Trend:

$$bt = \beta(Lt-Lt-1) + (1-\beta)bt-1,$$ (4)

Seasonal:

$$St = \gamma \frac{Yt}{Lt} + (1-\gamma)St-s,$$ (5)

and
Forecast:

$$Ft + m = (Lt + btm)St-s+m$$ (6)

(6)where $Y_t$ = actual (observed) values, $L_t$ = the single smoothed values (the level of the series), $b_t$ = the trend, $S_t$ = the seasonal component, $s$ = the length of seasonality, $F_{t+m}$ = the forecast for $m$ period ahead, and $\alpha$, $\beta$, $\gamma$ are coefficients to be estimated that range from 0 to 1 and used to minimize mean squared errors.

### 3.3. Stepwise autoregressive model

Two sub-models (i.e., the trend and autoregressive models) form the stepwise autoregressive model. It implies that, in the first stage, the predicted value is forecasted based on the trend model for the long-term trend and then the error terms are predicted from the autoregressive model in order to cover short-term fluctuations in the second stage. Using the method of ordinary least squares, the proper type of trend model is decided. After this process, the coefficients of the lagged variables are considered in order to explain the part of the error terms from the trend model by increasing the order of the autoregressive model. The stepwise autoregressive model is usually expressed as follows (SAS Publishing, 2004):

$$Y_t = TR_t + e_t,$$ (7)

$$e_t = b_1e_{t-1} + b_2e_{t-2} + \ldots\ldots + b_te_{t-p} + \varepsilon_t,$$ (8)

**Table 1.** Unit root test for data.

| Variable | Model specification | Non-seasonal unit root[a] | Seasonal unit root[b] | Lag length[c] |
|---|---|---|---|---|
| Tourist | None | 0.658 | 0.541 | 4 |
| | constant | −1.531 | −0.151 | 4 |
| | constant, trend | −5.456*** | −0.394 | 4 |
| ΔTourists | none | −4.549*** | 0.181 | 4 |
| | constant | −4.674*** | 0.182 | 4 |
| | constant, trend | −4.656*** | 0.184 | 4 |
| $\Delta_{12}$Tourists | none | −1.136 | −5.917*** | 4 |
| | constant | −2.69 | −9.013*** | 4 |
| | constant, trend | −2.918* | −8.973*** | 4 |
| $\Delta\Delta_{12}$Tourist | none | −4.130*** | −7.743*** | 4 |
| | constant | −4.072** | −7.683*** | 4 |
| | constant, trend | −7.732*** | −7.732*** | 4 |

Note: Δ indicates the first non-seasonal differencing of the data.
$\Delta_{12}$ indicates the first seasonal differencing of the data.
[a]H0: data have a non-seasonal unit root.
[b]H0: data have a seasonal unit root.
[c]lag length is decided by Ng and Perron (1995)'s general-to-specific method.

**Table 2.** Estimation of ARIMA intervention model for detecting MERS.

| ARIMA(2,1,0)(0,1,0) Parameters | | Estimate | Standard error | t-value | Sig. |
|---|---|---|---|---|---|
| AR (1) | | 0.452 | 0.105 | 4.30 | <0.0001 |
| AR (2) | | −0.496 | 0.109 | −4.57 | <0.0001 |
| MERS | Impact ($\omega_1$) | −142,383.8 | 55,717.9 | −2.56 | 0.011 |
| | Slope($\delta$) | −0.71474 | 0.254 | −2.82 | 0.005 |
| Box–Ljung Q | | Chi-Square | | Sig. | |
| Lag 6 | | 5.13 | | 0.275 | |
| Lag 12 | | 7.70 | | 0.658 | |
| Lag 18 | | 11.69 | | 0.765 | |

where $Y_t=$ tourism demand, $TR_t=$ time period, $e_t=$ error term of trend model, $e_{t-p}=$ lag values of error term, and $\varepsilon_t=$error term of trend model's errors.

## Results

### The statistical significance of the MERS

In this research, the ARIMA intervention model was employed to confirm the statistical significance of the MERS because the model has been beneficial to proving the existence of outside events, such as terrorism and diseases (Nelson, 2000).

*Stationary process*: Non-seasonal and seasonal augmented Dickey–Fuller (ADF) tests were performed in order to logically confirm the appearance of the non-seasonal and seasonal stationarity of the tourism data (Dickey & Fuller, 1979). As shown in Table 1, the monthly tourism data were found to be non-seasonally or seasonally non-stationary based on the individual ADF tests since the null hypotheses for the non-seasonal and seasonal unit roots could not be rejected ($p > 0.05$). After both of the non-seasonal and seasonal differencing, it was checked again to decide whether it was seasonally and non-seasonally stationary at $p < 0.05$.

*Consideration of the MERS in the model*: The ARIMA intervention model was utilized in order to confirm whether the MERS significantly decreased inbound tourism to Korea. First, based on prior knowledge and a literature review, the specific time frame in which MERS occurred was determined.

*Model identification*: The ACF and PACF from the seasonal and non-seasonal differenced series proposed two tentative ARIMA intervention models: $(2,1,0)(0,1,0)_{12}$ and $(2,1,0)(1,1,0)_{12}$. The Akaike information criterion (AIC) (Akaike, 1974) and Schwarz Bayesian criterion (SBC) (Schwarz, 1978) were employed in order to select the best ARIMA intervention model. The ARIMA intervention

**Table 3.** Estimation of three forecasting models for inbound tourists to Korea.

| Models Variables | ARIMA model $(0,1,0)(1,1,0)_{12}$ Coefficients (t-value) | SE | Winters exponential smoothing model Coefficients (t-value) | SE | Stepwise autoregressive model Coefficients (t-value) | SE |
|---|---|---|---|---|---|---|
| SAR(1) | −0.413** (−3.085) | 0.134 | – | – | – | – |
| Linear trend | – | – | – | – | 8749*** (7.19) | 1217 |
| Alpha (level) | – | – | 0.383*** (7.927) | 0.048 | – | – |
| Gamma (trend) | – | – | 0.001 (0.058) | 0.017 0.149 | | |
| Delta (season) | | | 0.952*** (6.385) | | | |
| AR1 of residual | – | – | – | – | −0.633*** (−7.04) | 0.090 |
| Constant | – | – | – | – | 568695*** (10.31) | 55186 |
| $R^2$ | 0.932 | | 0.928 | | 0.837 | |
| MAPE | 4.352 | | 5.311 | | 8.310 | |

$**p < 0.01$, $*p < 0.05$.

model $(2,1,0)(0,1,0)_{12}$ with the lowest AIC (1929.487) and SBC (1838.538) values was determined to be the best model for the current study (Table 2).

*Estimation*: Table 2 shows the results of the estimation for the ARIMA intervention model $(2,1,0)(0,1,0)_{12}$. The coefficients of the AR models were statistically significant at $p < 0.0001$ or less. The coefficients of the intervention variable (MERS) were also statistically significant, as expected.

*Diagnosis*: Using the residual PACF and ACF plot and the Box-Ljung Q-statistic (Table 2), it was confirmed that the probability of the residual autocorrelations not to be white noise was less than 5% (Dharmaratne, 1995). As they fulfill the diagnostic criteria, the ARIMA intervention model $(2,1,0)(0,1,0)_{12}$ was theoretically satisfactory and the effect of the MERS for international tourists coming to Korea was statistically and negatively significant (McCleary & Hay, 1980).

### *Impact of MERS on international tourism to korea*

The estimated effect was calculated after the significance of the MERS was statistically confirmed. The impact of MERS on inbound tourists to Korea was evaluated using three forecasting models (i.e., Stepwise Autoregressive, Winters Exponential Smoothing, ARIMA models) by comparing each model's forecasted number of tourists (under the assumption that MERS did not exist in the country) with the actual number of tourists. As shown in Table 3, in the ARIMA model $(0,1,0)(1,1,0)_{12}$, the seasonal AR term was statistically significant at $p < 0.01$ with the $R^2$ at 0.932. The ARIMA model appeared accurate within the sample forecasts with a MAPE of 4.4% (Lewis, 1982). In the Winters Exponential Smoothing model, the sum of the squared errors was minimized when the parameters were $\alpha = 0.383$, $\gamma = 0.001$, and $\delta = 0.952$.

The Winters model appeared accurate within the sample forecasts with a MAPE of 5.3%. The Stepwise Autoregressive model was statistically significant in that all of the parameters were significant at $p < 0.05$ or less with a high $R^2$ of 0.837 and the model was accurate with a MAPE of 8.3%. The three forecasting models were accurate according to Lewis (1982) since the values of their MAPEs were less than 10%. As such, any of the forecasting models could be used to predict tourism demand under the assumption that no MERS was present. Different forecasting models may generate different forecasts for the same periods since they have their own advantages and limitations. Therefore, taking average of the combined approaches (i.e., all three models) may provide some useful information not conveyed by any single forecasting model (Var & Lee, 1993). As shown in Figure 1, the three forecasting models predicted the tourism demand

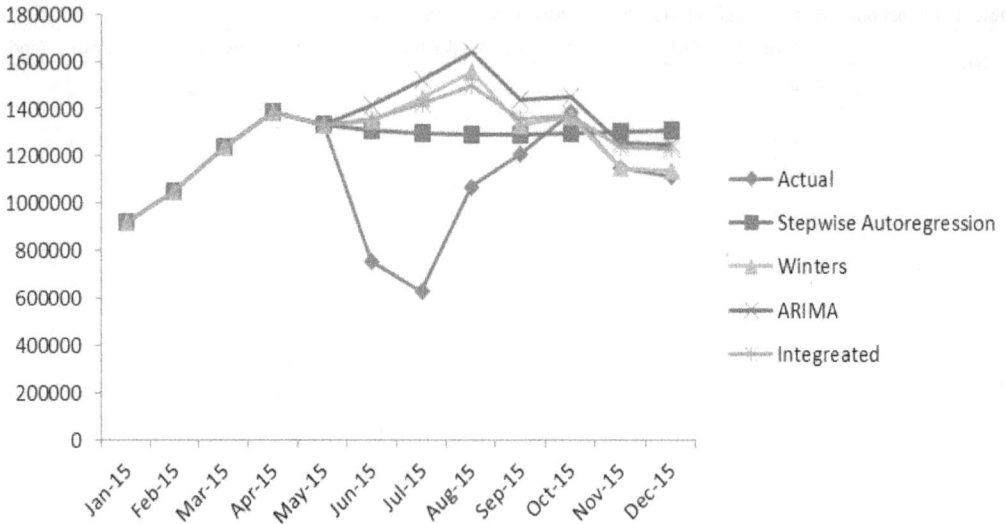

**Figure 1.** Impact of MERS on inbound tourists to Korea

**Table 4.** Impact of MERS on inbound tourists to Korea.

| Time period | Actual (A) | ARIMA model (0,1,0) $(1,1,0)_{12}$ | Winters Smoothing model | Stepwise Auto-regressive model | Integrated average (B) | Impact of tourists C=(A-B) | Proportion (C/B)*100 |
|---|---|---|---|---|---|---|---|
| Jun-15 | 750,925 | 1,411,922 | 1,347,723 | 1,309,273 | 1,356,306 | −605,381 | −44.6 |
| Jul-15 | 629,737 | 1,519,890 | 1,442,194 | 1,296,690 | 1,419,591 | −789,854 | −55.6 |
| Aug-15 | 1,069,314 | 1,641,713 | 1,555,355 | 1,291,930 | 1,496,333 | −427,019 | −28.5 |
| Sep-15 | 1,206,764 | 1,436,018 | 1,331,682 | 1,292,125 | 1,353,275 | −146,511 | −10.8 |
| Total | 3,656,740 | 6,009,543 | 5,676,954 | 5,190,019 | 5,625,505 | −1,968,765 | −35.0 |
| MAPE | | 4.352 | 5.311 | 8.310 | 4.854 | | |

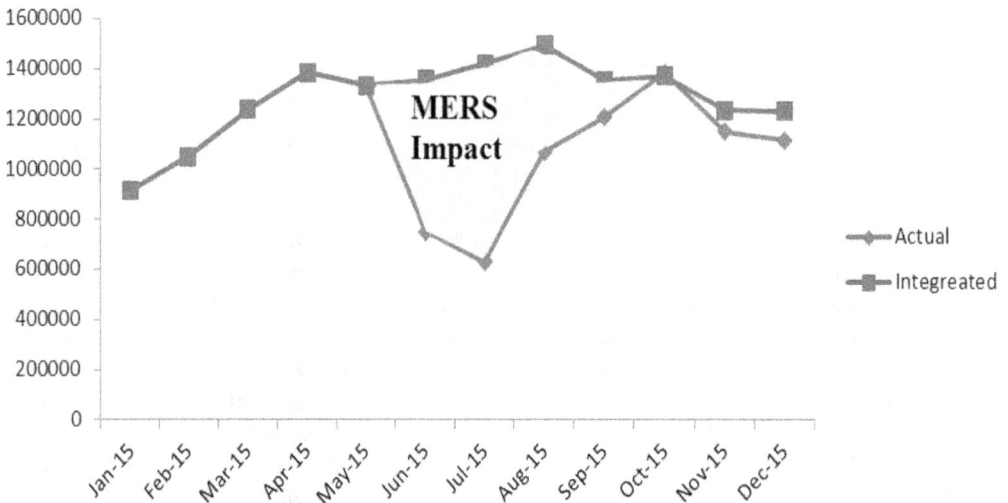

**Figure 2.** MERS impact derived from the results of current study

similarly depicting some overlapping, where the average values from these models produced moderate forecasts. These findings suggest that various forecasting models needed to be used and compared as multiple models can offer more advantages than single models.

**Table 5.** Economic impact of MERS on inbound tourists to Korea.

| Time period | Impact of MERS (tourists) | Per capita expenditure of Korean inbound tourist (USD) | Impact of MERS (tourism receipts million USD) |
|---|---|---|---|
| Jun-15 | −605,381 | 1,605.5[a] | −971.9 |
| Jul-15 | −789,854 | | −1,268.1 |
| Aug-15 | −427,019 | | −685.6 |
| Sep-15 | −146,511 | | −235.2 |
| Total | −1,968,765 | | −3,160.9 |

[a]Korea Tourism Organization (2015).

Utilizing multiple forecasting models delivers researchers and policy-makers a range of impact of MERS which cannot be conveyed by any single model. While the actual tourist arrivals data contained the impact of MERS, the average values of the tourism demand by the three forecasting models were predicted under the assumption that no MERS was present. Thus, the differences between the actual tourist arrivals and the average values indicated the impact of MERS. As shown in Table 4 and Figure 2, the MERS impact in June 2015 was relatively small at −605,381 tourists. The MERS impact increased in July 2015 to −789,854 tourists and gradually decreased in August 2015 by −427,019 tourists. The impact of MERS showed the smallest value at −146,511 tourists in September 2015.

As shown in Table 5, the total effect of MERS between June 2015 to September 2015 was estimated to be −1,968,765 tourists with −3,160.9 million USD receipts. Specifically, economic impact of MERS (tourism receipts of −3,160.9 million USD) was calculated by multiplying the number of international tourist arrivals to Korea (-1,968,765) by per capita expenditure of Korean inbound tourist (1,605.5 USD).

## Discussion and conclusions

It is not easy to properly forecast time-series data, as it contains external events, which can cause significant structure changes (Glass, 1972). External events (e.g., natural disasters, terrorism, political instability, wars, epidemics) negatively affect internal tourism demand. In the tourism industry, during the occurrence of external events, it is important to have proper decision-making based on precise effect estimations as such estimations can help in the creation and preparation of appropriate strategies to avoid shortages or excesses related to tourism demand (Burger et al., 2001; Huang & Min, 2002). Estimating the impact of an external event is valuable when attempting to choose and perform proper marketing strategies, especially as similar external events will likely occur again in the future (Lee, Oh, & O'Leary, 2004).

In this regard, the current study aimed at estimating the impact of MERS, an external event, which had a large impact on Korean inbound tourism in 2015. Specifically, the present study considered two research goals (i.e., identifying the influence of MERS, and estimating the specific impact of MERS on inbound tourism in Korea). With regard to the first issue, identifying the negative influence of MERS, the current study revealed that the ARIMA intervention model was proper to confirm the existence of external events that could describe the first issue (Nelson, 2000). The results of the current study revealed that MERS was statistically significant at $p < 0.05$ with a negative sign, indicating that MERS had a strong negative impact on Korea's inbound tourism industry in 2015. The results were similar to that of previous studies, which showed that fatal epidemics had significant effects on international tourism demand (Min, 2005; Shi & Li, 2017; Song, 2016; Suh & Kim, 2018; Yim & Sohn, 2007).

In terms of the second issue, estimating the specific impact of MERS on inbound tourism in Korea, three popular time-series models were employed (i.e., the ARIMA, Winters Exponential Smoothing, the Stepwise Autoregressive Model). The value of MAPE was 4.352 for the ARIMA,

5.311 for the Winters Exponential Smoothing, and 8.31 for the Stepwise Autoregressive model, showing a high level of forecast accuracy for all three models (Lewis, 1982). The differences between the actual tourist arrivals and forecasted tourist arrivals on average from these forecasting models were considered to be the impact of MERS.

With this approach, the total impact of MERS was calculated as −1,968,765 tourists. Specifically, the impact pattern of MERS was the smallest in June 2015 (-605,381 tourists), representing a 44.6% decrease in demand for that month. The peak of the impact was in July 2015 (-789,854 tourists), but the impact continually decreased after that point. By the end of September 2015, the impact was down to −146,511 tourists, representing a 10.8% decrease in demand for the month. Viewed in this light, the results from this study demonstrate that, within two months, the latent influence of MERS had peaked and, in an existing tourism market sensitive to fatal epidemics, the emergence of MERS caused a financial loss. Based on these figures, the loss of tourism receipts was estimated at 3,161 million USD. The tourism industry implemented sustainability initiatives and developed policies to overcome the MERS disaster in tourism, such as offering free insurance and promoting a special loan program for tourism companies in order to minimize the impact of external interventions. Specifically, during the period of MERS, the Korean government offered free insurance to compensate international visitors in case they were diagnosed with MERS during their visits to Korea. The Korean Ministry of Tourism also waived visa fees for group travelers from China and Southeast Asia, and the ministry supported a special loan program worth 64.4 million USD to help tourism-related businesses, including hotels and travel agencies. Moreover, the city of Seoul spent about 25 million USD to attract international tourists; this money was spent on developing free promotional tours and holding large concerts.

Practically, this study identified the impacts of MERS on inbound tourism and found evidence of the total effect was estimated to be −1,968,765 tourists with a loss of 3.1 billion USD in receipts from June 2015 to September 2015. To make up for the loss and increase tourist confidence when travelling, the Korean government implemented revitalization policies to promote sustainable tourism and recommendation for recovery efforts (e.g., visa fee waive for travelers, special loan program to the industry, and a monetary support (or subsidy package) by a regional government). In order to assist accelerate recovery in tourism sector, the government and/or public health authorities such as KCDC could focus on communication and sharing information on the prevention of epidemics in major tourist attractions for domestic tourists to make up the loss caused by a decrease of inbound tourism but also inbound tourists' perceptions of the risks and health security toward a destination as counties close borders, quarantine new arrivals, and cease international travel. Specifically, domestic tourism has been increased for three months after the end of MERS, and it is expected that the quarantine activities of tourist attractions will serve as an important criterion for selecting tourism destinations in relation to seasonal infectious disease. Also, using mobile positioning data, managers and practitioners could provide crowding information and develop the traffic congestion guidance system for the purpose of dispersing tourists by region during peak seasons. Destination management organizations (DMOs) can discover unknown tourist spots, direct tourists to less crowded areas, and raise awareness of open spaces. DMOs should address recovery in line with target markets to help restore and integrate with product diversification to reduce tourists' fear to travel to destinations. It is necessary to develop marketing plans for healing with recreational and outdoor activities such as nature (river, sea, mountain, lake), parks (national park, urban park, botanical garden, arboretum, recreation forest,), and leisure activities (fishing, camping).

In addition to these policies, if an infectious disease such as MERS epidemic or pandemic occurred again in the future, new effective policies need to be implemented in the tourism sector during the period impacted by the disease. While epidemic is an outbreak of a sudden disease on a community, geographical area or several countries, affecting many individuals, pandemic that has a worldwide spread of an epidemic on many countries or continents, affecting a large percent of the population (US Centers for Disease Control & Prevention, 2020). As both epidemic and pandemic affect a wider geographical area, infect a larger number of individuals, create loss at the national and global level, pre-crisis planning should be discussed in order to prepare for similar external epidemic or pandemic in the future as such planning will lead to more effective management during a crisis. In this case, MERS caused larger damage to Korea than expected because the perception of MERS was very low and the preparation for it was not enough (Lee & Ki, 2015). Thus, the government should prepare for the next potential emergency such as pandemic in order to reduce economic crises across all sectors, including tourism (Ritchie, 2004). In the situation of an external event, stakeholder collaboration is important (Choe & Schuett, 2020). Countries need to prepare internationally unified regulation standards and emergency management plans for tourists' safety and security in order to minimize damage by taking prompt response action plan via international collaboration (Eslami et al., 2019). The disease crisis will bring uncertainty into the lives of the tourists and the country will need to rebuild a safe image (Min, 2005; Moon et al., 2017) as the disease outbreak can have long-term impacts on the image of the tourism destination and perceptions of risk and safety toward a destination will influence people's willingness to travel there (Sonmez & Graefe, 1998).

Due to globalization and increasing travel abroad, national and global stakeholders must work together to develop public health guidelines for travelers visiting high risk destinations so as to mitigate global health and economic impacts (Arjona-Fuentes et al., 2019). National and international stakeholders can collaborate to bring back foreign travelers through arranging MICE, concerts and performances, and familiarization tours for target countries (Song, Kim, & Choe, 2019). It is a good idea for celebrities to promote tourism safety by meeting major travel agencies and persuading tourists. Special legislation or government policies mandating can be also useful when attempting to deliver effective communication channels and government administrative control tower operate properly to support on-site capability (Kim, 2015). It is important to note that, this research is limited because, by using the univariate time-series model, the research does not consider various macroeconomic variables as control variables or intervener variables, such as the GDP, exchange rate, and consumer price index. Employing multivariate time-series models would be useful in regard to explaining the complicated relationships between international tourism demand and socio-economic variables in future research. In addition, even when ideal estimation models are found, due to complex tourist behaviors, other factors may be associated with the impact of the rebound, such as beginning to fade in people's memories (Huang & Min, 2005), motivations, or external variables (Petrevska, 2017). Researchers can further test whether the variable acts as a confounder (Huang & Min, 2005). Practitioners can be prepared to revise the previously identified models and adapt them to upcoming similar crises (Petrevska, 2017). This study is based on the number of international tourist arrivals to Korea. It is also highly likely that some differences may occur in terms of the purpose of travel or estimating the impact of promotion or recovery status in large markets. It would be interesting to compare the purpose of travel to the inbound tourism demand during a crisis: tour, business, official, and others. Hence, policy-makers and marketing managers can easily identify target markets where travel purpose or nationality would be the most sensitive to external events in relation to inbound tourism demand. Finally, future research should be conducted in various settings at the national, regional, and international levels to consider external events affected locally or globally.

## Disclosure statement

No potential conflict of interest was reported by the authors.

## Funding

This work was supported by the Paichai University Research grant in 2020.

## ORCID

*Yunseon Choe* http://orcid.org/0000-0001-5527-2263
*Junhui Wang* http://orcid.org/0000-0001-6849-5896

## References

Akaike, H. (1974). A new look at the statistical model identification. *IEEE Transactions on Automatic Control, 19*(6), 716–723. https://doi.org/10.1109/TAC.1974.1100705

Arjona-Fuentes, J. M., Ariza-Montes, A., Han, H., & Law, R. (2019). Silent threat of presenteeism in the hospitality industry: Examining individual, organisational and physical/mental health factors. *International Journal of Hospitality Management, 82*, 191–198. https://doi.org/10.1016/j.ijhm.2019.05.005

Baker, D. M. A. (2015). Tourism and the health effects of infectious diseases: Are there potential risks for tourists?. *International Journal of Safety and Security in Tourism and Hospitality, 1*(12), 1–17.

Biggs, D., Hall, C. M., & Stoeckl, N. (2012). The resilience of formal and informal tourism enterprises to disasters: Reef tourism in Phuket. *Journal of Sustainable Tourism, 20*(5), 645–665. https://doi.org/10.1080/09669582.2011.630080

Bonham, C., Edmonds, C., & Mak, J. (2006). The impact of 9/11 and other terrible global events on tourism in the United States and Hawaii. *Journal of Travel Research, 45*(1), 99–110. https://doi.org/10.1177/0047287506288812

Box, G. E., & Tiao, G. C. (1975). Intervention analysis with applications to economic and environmental problems. *Journal of the American Statistical Association, 70*(349), 70–79. https://doi.org/10.1080/01621459.1975.10480264

Bramwell, B., & Lane, B. (2011). Crises, sustainable tourism and achieving critical understanding. *Journal of Sustainable Tourism, 19*(1), 1–3. https://doi.org/10.1080/09669582.2010.535743

Brown, D. O. (1998). Debt-funded environmental swaps in Africa: Vehicles for tourism development? *Journal of Sustainable Tourism, 6*(1), 69–79. https://doi.org/10.1080/09669589808667302

Burger, C., Dohnal, M., Kathrada, M., & Law, R. (2001). A practitioner's guide to time-series methods for tourism demand forecasting: A case study of Durban. *Tourism Management, 22*(4), 403–409. https://doi.org/10.1016/S0261-5177(00)00068-6

Capo, J., Font, A. R., & Nadal, J. R. (2007). Dutch disease in tourism economies: Evidence from the Balearics and the Canary Islands. *Journal of Sustainable Tourism, 15*(6), 615–627. https://doi.org/10.2167/jost698.0

Carney, D., Drinkwater, M., Rusinow, T., Neefjes, K., Wanmali, S., Singh, N. (1999). *Livelihoods approaches compared.* London: DFID. Retrieved March 10, 2019, from http://www.start.org/Program/advanced_institute3_web/p3_documents_folder/Carney_etal.pdf.

Chambers, R., Conway, G. (1992). *Sustainable rural livelihoods: Practical concept for the 21st century.* IDS Discussion Paper No. 296. Brighton: IDS. Retrieved March 10, 2019, from https://opendocs.ids.ac.uk/opendocs/bitstream/handle/123456789/775/Dp296.pdf?sequence=1.

Chi, J., & Baek, J. (2013). Dynamic relationship between air transport demand and economic growth in the United States: A new look. *Transport Policy, 29*, 257–260. https://doi.org/10.1016/j.tranpol.2013.03.005

Chien, G. C., & Law, R. (2003). The impact of the severe acute respiratory syndrome on hotels: a case study of Hong Kong. *International Journal of Hospitality Management, 22*(3), 327–332. https://doi.org/10.1016/S0278-4319(03)00041-0

Choe, Y. J., Choe, S. A., & Cho, S. I. (2017). Importation of travel-related infectious diseases is increasing in South Korea: An analysis of salmonellosis, shigellosis, malaria, and dengue surveillance data. *Travel Medicine and Infectious Disease, 19*, 22–27. https://doi.org/10.1016/j.tmaid.2017.09.003

Choe, Y., & Schuett, M. A. (2020). Stakeholders' perceptions of social and environmental changes affecting Everglades National Park in South Florida. Environmental Development, 100524.

Cooper, M. (2005). Japanese tourism and the SARS epidemic of 2003. *Journal of Travel & Tourism Marketing, 19*(2-3), 117–131. https://doi.org/10.1300/J073v19n02_10

De Sausmarez, N. (2007). Crisis management, tourism and sustainability: The role of indicators. *Journal of Sustainable Tourism, 15*(6), 700–714. https://doi.org/10.2167/jost653.0

Dharmaratne, G. S. (1995). Forecasting tourist arrivals in Barbados. *Annals of Tourism Research, 22*(4), 804–818. https://doi.org/10.1016/0160-7383(95)00022-3

Dickey, D. A., & Fuller, W. A. (1979). Distribution of the estimators for autoregressive time series with a unit root. *Journal of the American Statistical Association, 74*(366), 427–431. https://doi.org/10.2307/2286348

Dodds, R. (2003). *Developing new markets for traditional destinations: Is sustainable tourism policy a successful option for creating new markets?* TTRA Canada Conference, October. St.John, NB, October 2-5, 2003.

Enders, W. (1995). *Applied Econometric Time Series* (1st ed.). John Wiley & Sons.

Enders, W., & Sandler, T. (1991). Causality between transnational terrorism and tourism: The case of Spain. *Terrorism, 14*(1), 49–58. https://doi.org/10.1080/10576109108435856

Eslami, S., Khalifah, Z., Mardani, A., Streimikiene, D., & Han, H. (2019). Community attachment, tourism impacts, quality of life and residents' support for sustainable tourism development. *Journal of Travel & Tourism Marketing, 36*(9), 1061–1079. https://doi.org/10.1080/10548408.2019.1689224

Glaesser, D. (2011). Toward a safer world: The travel, tourism and aviation sector. Retrieved July 20, 2017, from http://webunwto.s3.amazonaws.com/imported_images/41552/unwtotowardasaferworld.pdf.

Glass, G. V. (1972). Estimating the effects of intervention into a nonstationary time series. *American Educational Research Journal, 9*(3), 463–477. https://doi.org/10.3102/00028312009003463

Goh, C., & Law, R. (2002). Modeling and forecasting tourism demand for arrivals with stochastic nonstationary seasonality and intervention. *Tourism Management, 23*(5), 499–510. https://doi.org/10.1016/S0261-5177(02)00009-2

Graci, S. (2007). Accommodating green: Examining barriers to sustainable tourism development. Paper presented at the TTRA Canada conference, Montebello, Quebec.

Han, H., & Hyun, S. (2018). Eliciting customer green decisions related to water saving at hotels: Impact of customer characteristics. *Journal of Sustainable Tourism, 26*(8), 1437–1452. https://doi.org/10.1080/09669582.2018.1458857

Huang, J. H., & Min, J. C. H. (2002). Earthquake devastation and recovery in tourism: The Taiwan case. *Tourism Management, 23*(2), 145–154. https://doi.org/10.1016/S0261-5177(01)00051-6

Hunter, C. (2002). Sustainable tourism and the touristic ecological footprint. *Environment, Development and Sustainability, 4*(1), 7–20. https://doi.org/10.1023/A:1016336125627

Jamal, T., & Budke, C. (2020). Tourism in a world with pandemics: Local–global responsibility and action. *Journal of Tourism Futures.* https://doi.org/10.1108/JTF-02-2020-0014

Jung, E., & Sung, H. (2017). The Influence of the Middle East Respiratory syndrome outbreak on online and offline markets for retail sales. *Sustainability, 9*(3), 411–412. Retrieved May 2020, from https:// https://doi.org/10.3390/su9030411

Jung, H., Park, M., Hong, K., & Hyun, E. (2016). The impact of an epidemic outbreak on consumer expenditures: An empirical assessment for. *Sustainability, 8*(5), 454–412. Retrieved September 2019, from https://doi.org/10.3390/su8050454

Kim, D. H. (2015). Structural factors of the Middle East respiratory syndrome coronavirus outbreak as a public health crisis in Korea and future response strategies. *Journal of Preventive Medicine and Public Health = Yebang Uihakhoe Chi, 48*(6), 265–270. https://doi.org/10.3961/jpmph.15.066

Korea Association of Travel Agents (2015). Travel statistics. Retrieved September 12, 2016, from https://www.kata.or.kr.

Korea Centers for Disease Control and Prevention (2018). MERS Statistics. Retrieved July 20, 2020, from http://www.cdc.go.kr/contents.es?mid=a30329000000

Korea Culture and Tourism Institute (2015). Impacts and countermeasures of MERS-CoV in tourism industry. Retrieved July 20, 2018, from http://www.kcti.re.kr/web/board/boardContentsView.do.

Korea Tourism Organization (2015). Visitor Arrivals, Korean Departures, International Tourism Receipts & Expenditures. https://kto.visitkorea.or.kr/eng/tourismStatics/keyFacts/KoreaMonthlyStatistics.kto

Korea Tourism Organization. (2017). *Korean tourism statistics.* https://kto.visitkorea.or.kr/eng/tourismStatics/keyFacts/KoreaMonthlyStatistics.kto

Kuo, H. I., Chen, C. C., Tseng, W. C., Ju, L. F., & Huang, B. W. (2008). Assessing impacts of SARS and Avian Flu on international tourism demand to Asia. *Tourism Management, 29*(5), 917–928. https://doi.org/10.1016/j.tourman.2007.10.006

Lee, C., & Ki, M. (2015). Strengthening epidemiologic investigation of infectious diseases in Korea: Lessons from the Middle East Respiratory Syndrome outbreak. *Epidemiology and Health, 37*, e2015040. https://doi.org/10.4178/epih/e2015040

Lee, S., Oh, C., & O'Leary, J. T. (2004). Estimating the impact of the September 11 terrorist attacks on the US air transport passenger demand and using intervention analysis. *Tourism Analysis, 9*(4), 355–361. https://doi.org/10.3727/108354205789807238

Lewis, C. D. (1982). *Industrial and business forecasting methods*. Butterworth.

Liu, Y. V., Massare, M. J., Barnard, D. L., Kort, T., Nathan, M., Wang, L., & Smith, G. (2011). Chimeric severe acute respiratory syndrome coronavirus (SARS-CoV) S glycoprotein and influenza matrix 1 efficiently form virus-like particles (VLPs) that protect mice against challenge with SARS-CoV. *Vaccine, 29*(38), 6606–6613. https://doi.org/10.1016/j.vaccine.2011.06.111

Luhrman, D. (2003). *Crisis guidelines for the tourism industry. Asia-Pacific Ministerial Summit on Crisis Management*.

Maditinos, Z., Vassiliadis, C. (2008). Crises and disasters in tourism industry: Happen locally – affect globally. MIBES ebook 2008 (pp. 67–76). Retrieved December 20, 2016, from http://mibes.teilar.gr/e-books/2008/maditinos_vasiliadi s%2067-76.pdf

Mai, T., & Smith, C. (2015). Addressing the threats to tourism sustainability using systems thinking: A case study of Cat Ba Island, Vietnam. *Journal of Sustainable Tourism, 23*(10), 1504–1528. https://doi.org/10.1080/09669582.2015.1045514

Makridakis, S., Wheelwright, S. C., & Hyndman, R. J. (1998). *Forecasting: Methods and applications* (3rd ed.). Wiley.

Manning, T. (2004). Indicators of sustainable development for tourism destinations. *Indicators of sustainable development for tourism destinations*. A Guidebook.

Manosuthi, N., Lee, J., & Han, H. (2020). Impact of distance on the arrivals, behaviors and attitudes of international tourists in Hong Kong: A longitudinal approach. *Tourism Management , 78*, 103963. https://doi.org/10.1016/j.tourman.2019.103963

Mastny, L. (2002). Redirecting international tourism. In L. Starke (Ed.) *State of the. World 2002* (pp. 101–126). Worldwatch Institute, Worldwatch.

Mavalankar, D. V., Puwar, T. I., Murtola, T. M., & Vasan, S. S. (2009). Quantifying the impact of Chikungunya and Dengue on tourism revenues of Gujarat (India), Malaysia and Thailand, Working Paper, Indian Institute of Management, Ahmedabad 380 015, India, February (2009). Retrieved December 20, 2016, from http://vslir.iima.ac.in:8080/jspui/bitstream/11718/17086/1/2009-02-03Mavalankar.pdf.

McCleary, M., & Hay, R. A. (1980). *Applied time series analysis for the social sciences*. Sage Publication.

McKercher, B., & Chon, K. (2004). The over-reaction to SARS and the collapse of Asian Tourism. *Annals of Tourism Research, 31*(3), 716–719. https://doi.org/10.1016/j.annals.2003.11.002

Min, J. C. H. (2005). SARS devastation on tourism: The Taiwan case. *Journal of American Academy of Business, 6*(1), 278–284.

Moon, H., Yoon, H., & Han, H. (2017). The effect of airport atmospherics on satisfaction and behavioral intentions: Testing the moderating role of perceived safety. *Journal of Travel & Tourism Marketing, 34*(6), 749–763. https://doi.org/10.1080/10548408.2016.1223779

Murphy, P. E., & Price, G. G. (1998). Tourism and sustainable development. In W.F. Theobald (Eds.), *Global tourism* (pp. 167–193). Butterworth-Heinemann.

Nelson, J. P. (2000). Consumer bankruptcies and the Bankruptcy Reform Act: A time-series intervention analysis, 1960–1997. *Journal of Financial Services Research, 17*(2), 181–200. https://doi.org/10.1023/A:1008166614928

Ng, S., & Perron, P. (1995). Unit root tests in ARMA models with data-dependent methods for the selection of the truncation lag. *Journal of the American Statistical Association, 90*(429), 268–281. https://doi.org/10.1080/01621459.1995.10476510

Parola, P., de Lamballerie, X., Jourdan, J., Rovery, C., Vaillant, V., Minodier, P., Brouqui, P., Flahault, A., Raoult, D., & Charrel, R. N. (2006). Novel Chikungunya virus variant in travelers returning from Indian Ocean islands. *Emerging Infectious Diseases, 12*(10), 1493–1499. https://doi.org/10.3201/eid1210.060610

Petrevska, B. (2017). Predicting tourism demand by ARIMA models. *Economic Research-Ekonomska Istraživanja, 30*(1), 939–950. https://doi org/pdf/10 1080/1331677X 2017 1314822 https //doi org/10 1080/1331677X 2017 1314822

Pratt, L., Rivera, L., Bien, A. (2011). Tourism: Investing in energy and resource efficiency. United Nations Environment Programme. Retrieved March 10, 2019, from www.unep.org/greeneconomy/Portals/88/ … /GER_11_Tourism.pdf

Pryce, A. (2001). Sustainability in the hotel industry. *Travel and Tourism Analyst, 6*, 3–23.

Richter, L. K. (2003). International tourism and its global public health consequences. *Journal of Travel Research, 41*(4), 340–347. https://doi.org/10.1177/0047287503041004002

Ritchie, B. W. (2004). Chaos, crises and disasters: A strategic approach to crisis management in the tourism industry. *Tourism Management, 25*(6), 669–683. https://doi.org/10.1016/j.tourman.2003.09.004

Rodway-Dyer, S., & Shaw, G. (2005). The effects of the foot-and-mouth outbreak on visitor behaviour: The case of Dartmoor National Park, South-West England. *Journal of Sustainable Tourism, 13*(1), 63–81. https://doi.org/10.1080/17501220508668473

SAS Publishing (2004). *SAS/ETS 9.1 user's guide*. SAS Institute.

Schwarz, G. (1978). Estimating the dimension of a model. *The Annals of Statistics, 6*(2), 461–464. https://doi.org/10.1214/aos/1176344136

Sharpley, R. (2003). Tourism, modernisation and development on the island of Cyprus: Challenges and policy responses. *Journal of Sustainable Tourism, 11*(2/3), 246–265. https://doi.org/10.1080/09669580308667205

Shi, W., & Li, K. X. (2017). Impact of unexpected events on inbound tourism demand modeling: evidence of Middle East Respiratory Syndrome outbreak in South Korea. *Asia Pacific Journal of Tourism Research, 22*(3), 344–356. https://doi.org/10.1080/10941665.2016.1250795

Song, H. J., & Lee, C. K. (2010). Impact assessment of external shocks on international tourism demand. *Korean Journal of Hospitality Administration, 19*(5), 1–16.

Song, J. (2016). A study on demand forecasting for Jeju-bound tourists by travel purpose using seasonal ARIMA-intervention model. *Journal of the Korean Data and Information Science Society, 27*(3), 725–732. https://doi.org/10.7465/jkdi.2016.27.3.725

Song H. J., Kim, M. C., & Choe, Y. (2019). Structural relationships among mega-event experiences, emotional responses, and satisfaction: Focused on the 2014 Incheon Asian Games. *Current Issues in Tourism, 22*(5), 575–581. https://doi.org/10.1080/13683500.2018.1462310 https://doi.org/10.1080/13683500.2018.1462310

Sönmez, S. F., & Graefe, A. R. (1998). Determining future travel behavior from past travel experience and perceptions of risk and safety. *Journal of travel research, 37*(2), 171–177. https://doi.org/10.1177/004728759803700209

Stipanuk, D. M. (1996). The US lodging industry and the environment: An historical view. *Cornell Hotel and Restaurant Administration Quarterly, 37*(5), 39–45. https://doi.org/10.1177/001088049603700522

Suh, H., & Kim, S. B. (2018). The macroeconomic determinants of international casino travel: Evidence from South Korea's top four inbound markets. *Sustainability, 10*(2), 554. https://doi.org/10.3390/su10020554

Sung, H. G. (2016). Impacts of the outbreak and proliferation of the Middle East Respiratory Syndrome on rail transit ridership in the Seoul metropolitan city. *Journal of Korea Planning Association, 51*(3), 163–179. https://doi.org/10.17208/jkpa.2016.06.51.3.163

US Centers for Disease Control and Prevention (2020). Principles of epidemiology in public health practice. Retrieved May 12, 2020, from https://www.cdc.gov/csels/dsepd/ss1978/lesson1/section11.html

Var, T., & Lee, C. K. (1993). Tourism forecasting: State-of-the-art techniques. In M. A. Khan, M. D. Olsen, & T. Var (Eds.), *VNR's Encyclopedia of hospitality and tourism* (pp. 679–696). Van Nostrand Reinhold.

Wen, Z., Huimin, G., & Kavanaugh, R. R. (2005). The impacts of SARS on the consumer behaviour of Chinese domestic tourists. *Current Issues in Tourism, 8*(1), 22–38. https://doi.org/10.1080/13683500508668203

World Health Organization (2003). Retrieved August 12, 2017, from http://www.who.int/csr/sars/en/

World Health Organization (2018). Middle East respiratory syndrome coronavirus (MERS-CoV). Author.

Yim, E. S., & Sohn, T. H. (2007). Effect of the SARS illness on demand for tourism: An application of ARIMA model. *Journal of Tourism Sciences, 31*(1), 365–381

Zhang, G., & Wei, X. (2003). *The impact of SARS on China's tourism industry and the recovery strategy*. Social Science Documentation Publishing House.

# Exploring preferences and sustainable attitudes of Airbnb green users in the review comments and ratings: a text mining approach

Laura Serrano, Antonio Ariza-Montes⊕, Martín Nader, Antonio Sianes and Rob Law

**ABSTRACT**

The sharing economy platforms for accommodation have emerged as a disruptive model that has revolutionized the tourist lodging sector and modified tourist consumption patterns. Airbnb is the paradigm of this business model, with approximately 300 million users and more than 7 million listings worldwide. However, studies that have attempted to explore the consumer experiences of green Airbnb users are still scarce, especially those using big data approaches. The present study explores the preferences and attitudes of green Airbnb users by analyzing the online review published in Inside Airbnb using text mining and sentiment analysis. Findings reveal six latent aspects among which the "sustainability" predominates in the online opinions of green Airbnb users, situating sustainability in relation to the most positive emotion of the Plutchik model. Moreover, the results of the analysis suggest a positivity bias in the online reviews of green Airbnb users. Managerial implications for sustainability, consumer behavior management, and marketing fields are discussed.

## Introduction

The irruption of the sharing economy in the tourism industry has introduced a new model that added value by making all idle resources available for direct transactions between tourists and local communities (Toni et al., 2018). At the beginning of this phenomenon, a certain aura of romanticism was observed to the point that Cheng and Jin (2019) attributed part of the success of this new model to guests' search for greater sustainability in their consumption actions.

The extension of the sharing economy has favored the emergence of an informed and demanding consumer, who evaluates the economic value of products and services consumed and requires such products and services to be respectful of the great challenges facing the world in terms of sustainability. In this way, the number of tourists who prefer environmentally-friendly accommodations has grown along with the number of consumers who declare their preference for green or sustainable products (Han et al., 2017; Han & Yoon, 2015). This paradigm shift means that ensuring the sustainability of the tourism industry is not only a strategy for the future but also a challenge for the survival of the sector.

The present work focuses on one of the activities at the heart of the business: the tourist lodging sector. For years, academia has attempted to identify the critical attributes of traditional hotel establishments that contribute to customer satisfaction (Sthapit & Jiménez-Barreto, 2018). For obvious reasons, research on these issues within the sharing economy framework is much recent and therefore limited. However, the obtained evidence highlights attributes that are similar to those identified in traditional hotel establishments. For example, Cheng and Jin (2019) indicate that the three most significant attributes that determine Airbnb users' experiences are the location of the accommodation, the service facilities, and the interaction with hosts. Although the most materialistic attributes continue to dominate the most relevant study results, some studies have marginally revealed other attributes linked to concern for social and environmental sustainability or the importance of human interactions (Guttentag et al., 2018; Tussyadiah & Zach, 2015).

This change in research approach reflects the main objective of the current work, which is to focus on tourists who are more committed to social and environmental sustainability and have greater ecological awareness. This interest in green consumers finds justification in a general trend toward sustainability throughout the tourism and hospitality sector. This trend causes the appearance of green consumers, who focus on incorporating all dimensions of sustainability into their tourism experiences (Midgett et al., 2018). According to these authors, this market segment is entirely new and requires more academic research.

Airbnb was chosen as the object because it is the leader in the peer-to-peer accommodation sector and approximately 72% of users appear to choose this alternative because of ecological convictions (Airbnb, 2017). From a broad perspective, Böcker and Meelen (2017) point out that those who engage in the sharing economy frequently act so because they wish to take care of the ecosystem and the environment and one can expect these individuals to be environmentally conscientious in their actions. This characteristic of the sharing economy is especially visible in the case of Airbnb. Midgett et al. (2017) highlight that Airbnb is evidence of the sharing economy's level of environmental sustainability; the overall theme of sustainability is embedded in the minds of many hosts and users of the service.

This study has a dual objective. On the one hand, it seeks to identify the preferences revealed by "green tourists" on Airbnb, the sharing economy website for accommodation, in relation to the attributes they value the most. On the other hand, it seeks to understand how the emotions of these green tourists toward such attributes determine their final assessment of an establishment. These objectives are intended to address the research gap identified by Yadav et al. (2019), who states that deepening the understanding of tourists' preferences and their relationship with satisfaction can contribute to a deeper understanding of green consumption in the tourist accommodation sector.

## Literature review

### The sharing economy in the lodging sector

Although the concept of the collaborative economy is a current topic of academic debate (Gil, 2019), its origins should be understood on the idea of extending the useful life of underused resources (Botsman & Rogers, 2010). From this point of view, the use of services offered through digital platforms can be perceived in terms of a sustainability factor because they reduce the amount of resources required, thus limiting the environmental impact (Wu & Zhi, 2016). This idea has led different authors, such as Cheng and Jin (2019), to affirm that the germ of the collaborative economy is the desire for consumer sustainability.

Short-term rental is one of the economic activities most affected by the development of the sharing economy (Quaglieri-Domínguez & Sánchez- Bergara, 2019). Airbnb represents the paradigm in this research area (Sthapit & Jiménez-Barreto, 2018). In slightly more than a decade, Airbnb has been viewed as the leading tourist accommodation company worldwide (Sigala, 2018). The dizzying rate of Airbnb's growth has not been free of criticism because of the tension that it has caused in some areas (Higgins-Desbiolles, 2018), such as its disruptive effect on the hotel sector.

Airbnb offers guests added value compared with traditional accommodation services by providing authentic experiences (Adamiak, 2019; Quaglieri-Domínguez & Sánchez-Bergara, 2019). These experiences are composed of a series of dimensions or attributes that determine consumer behavior when choosing this type of accommodation (Cheng & Jin, 2019). Precisely, one of the most important factors in the purchasing behavior of Airbnb users is their concern for sustainability and the company's commitment to reducing its negative impact on the environment and supporting the local economy (Gunter, 2018).

The academic interest in Airbnb is evident in the growing number of scientific disciplines that study its model of success. From different academic areas, researchers have attempted to understand the effects of Airbnb on the tourism industry and the behavior of tourists and the local effects that Airbnb generates in urban contexts (Cocola-Gant & Gago, 2019; Gil & Sequera, 2018). Some comparative studies have also been carried out among cities and countries (Adamiak, 2019; Liu et al., 2019; Yang & Mao, 2019). Following Sthapit and Jiménez-Barreto (2018), other topics of interest related to Airbnb studies include regulations (Dredge et al., 2016), changes in production (Gil, 2019), and the explanatory factors of consumer behavior, which are in line with the objectives of this work.

This work adds to the literature on the factors that determine consumer behavior and its relationship with consumers' level of satisfaction but focuses on users characterized as green consumers who present different characteristics as described in the following section.

### Green consumerism in the lodging sector

Since the early 2000s, two revolutionary phenomena have marked the evolution of consumer behavior, namely, the exponential growth of Internet use (Buhalis & Law, 2008) and concern for sustainability as a new paradigm (Han et al., 2017; Meng et al., 2020; Serrano et al., 2019). On the one hand, social networks, digital travel platforms, and websites have reached a point of development that has made them high-priority communication channels between tourists and professionals in the sector (Hu et al., 2017). These information and communication technologies are sources of information with a high degree of influence on the decision-making process (Zhang et al., 2016). On the other hand, the growing concern over the effects generated by economic activity on the environment is especially significant in the tourism industry, particularly in the hotel sector. In this way, the effort of companies in the tourism sector to increase sustainability and reduce the environmental impact of their processes have been important (Chi & Han, 2020; Han & Hyun, 2018) and may be required especially by consumers who express a preference for companies with sustainable practices (Han et al., 2018; Presenza et al., 2019).

Since the emergence of sustainability as a paradigm, the figure of the pro-environmental, sustainable, or green consumer has emerged as an object of exploration and study within the framework of consumer behavior. Thus, the investigation of sustainable consumer behavior constitutes an amalgamation of different theories and methodologies, each of which contributes to the complex objective of outlining the attributes that define the behavior of the sustainable consumer (Antonides, 2017). In this sense, the underlying factors in pro-environmental behavior have been explored from different theoretical perspectives, and the conclusion has been reached that proactivity toward sustainability depends, above all, on an individual's previous engagement in sustainable practices (Dharmesti et al., 2020; Eom & Han, 2019; Han et al., 2009).

The sharing economy fits perfectly into this new paradigm of sustainability. Nadler (2014) argues that the sharing economy has the potential to increase access to goods and services while reducing the need for investment in resources and infrastructure; therefore, it contributes to improving consumer welfare and reducing social costs. Belk (2014) also ensures that the sharing economy offers a sustainable alternative to the traditional consumption model. For example, many users of digital platforms on which this economy is developed believe that collaborative consumption translates into high income for local people, which in turn enables to contribute to the improvement of the well-being of the community (Guttentag et al., 2018).

Airbnb is an optimal research object to analyze green consumer behavior in the lodging sector. Different researchers have attempted to gain a better understanding of the attributes that determine the behavior of users of this type of lodging and their level of satisfaction with the experience (Cheng & Jin, 2019). Most studies have given limited attention to the sustainability attributes of accommodation (Cheng, 2016). In-depth analysis of the results obtained by Cheng and Jin (2019) suggests that the sustainability factor does not constitute a critical element in the purchase decision of the user. Thus, developing new research and gaining a deeper apprehension of the intentions of green Airbnb users when choosing accommodation are needed (Yadav et al., 2019), which justifies the purpose of the present study.

The scarcity of empirical research on this issue in the field of the sharing economy is possibly because of the dispersion and heterogeneity of information available on consumer behavior, which has not been feasible until the recent appearance of innovative study methodologies. The next section will verify that big data analysis represents a useful tool that can overcome limitations associated with the most classic analysis methods.

## Big data as a tool to evaluate tourist satisfaction in the sharing economy

In contrast to the classical methodologies that are mainly based on deductive reasoning, big data analysis is based on inductive reasoning and a large amount of statistical data processing without the need for prior theory. This approach can be seen as an a posteriori hypothesis generator (Olteanu et al., 2019) in contrast to the hypothesis tests that are characteristic of classical science. The big data approach expands the sources of data extraction, offering new perspectives on the analysis and how researchers address research questions (Li et al., 2018); this approach also allows the researcher to analyze large databases with information generated practically in real-time and collected without spatial limitations, making going from anecdote- to evidence-based data possible (Song et al., 2015). Briefly, big data analysis complements the conventional methodologies used in the scientific research to overcome their limitations to the point that many researchers have identified it as a new paradigm in the field of investigation (Xiang et al., 2017).

Big data analysis has multiple advantages in the context of tourism research. First, the industry generates data intensively (Hlee et al., 2018). Consequently, the emergence of big data has allowed the processing of data into relevant information, helping academics and professionals to understand better the different variables and constructs that have a decisive effect on the success of the tourism sector (Xiang et al., 2015a). The field of tourism constitutes an exemplary field of study for the implementation of big data analysis because it provides guidance for new research approaches and allows for the analysis of unconventional data sources and the processing of a huge volume of information (Cheng & Jin, 2019; Li et al., 2018). These factors provide new opportunities for the research on tourism (Xu et al., 2020), introducing a range of possibilities, such as the visualization of behavioral patterns in tourists' consumption and mobility, the identification of tourist saturation zones, and the identification of the circuits frequented by visitors in specific territories.

The use of big data in conjunction with content analysis (as is the case with sentiment analysis) is being increasingly utilized by researchers exploring the field of tourism (Fuchs et al., 2014; Li et al., 2013, 2015, 2018; Liu et al., 2013). Online comments posted by tourists on social networks generate millions of data points that, when examined through the lens of sentiment analysis, allow us to understand better their perceptions and satisfaction and that of other tourism industry players (Alaei et al., 2019). The content of these comments reflects tourists' free and spontaneous opinions on their lived experience and constitutes a highly valuable source of information. In an increasingly competitive industry in which new disruptive business models, such as Airbnb, have been implemented, the use of knowledge is essential for companies' survival and competitiveness.

Although big data analysis has gained popularity in tourism research, few studies have adopted this methodology as the focus (Cheng & Jin, 2019), and fewer studies have applied this methodology in the context of the sharing economy for accommodation with players, such as Airbnb. The present investigation attempts to shed light on this matter through the development of empirical investigation in the following sections.

## Research methodology

After data collection and before the data analysis, the following text processing steps similar to those adopted in previous studies (Guo et al., 2017; Luo & Tang, 2019) were carried out: data preprocessing to clean the data collected of all irrelevant elements and filter the resulting database, aspect identification to identify dimensions from reviews online and extract seed words, sentiment detection to determine the sentiment behind online reviews, and latent sentiment rating of each aspect implementing the Latent Rating Aspect Analysis (LARA).

## Data collection and preprocessing

The database used for this study was retrieved from Inside Airbnb, a website that gathers all comments on Airbnb listings from 83 cities around the world. A total of 176,852,704 comments were retrieved. Inside Airbnb collects information yearly and in all languages. Thus, data cleansing and removing of duplicates and comments not written in English are required. The final database comprised 13,181,297 unique comments. Then, a data preprocessing step was carried out to clean the collected data. First, all the letters that make up the words were converted to lowercase; punctuation marks, numbers, special characters (#, @, "",/, and\), and words that can distort the analysis (stop words) because of their frequency of appearance were eliminated.

Subsequently, comments from Airbnb users who could be classified as "green users" had to be filtered through the words they used in their online comments. Following Hurley (2011), green consumers are those who use words that are associated with sustainable lifestyles. Therefore, words that made specific reference to issues related to sustainability were selected. This research strategy is consistent with the proposals of other authors (Cheng, 2016; Han & Yoon, 2015; Kim & Han, 2010, Han et al., 2009; Pizam, 2009), who consider that the term "green" can be understood as an alternative for "sustainable" and that this term is synonymous and interchangeable with concepts such as "environmentally responsible" or "environmentally friendly." To compile the list of search terms, an iterative process was conducted, in which the researchers tested different terms to evaluate how they function and are used publicly (Kim et al., 2013). This process for the search and identification of terms has been the norm in content analysis in which a large amount of unstructured data is used, such as the data collected on social networks or user-generated comments. Thus, all the words previously identified by several studies related to green consumer behavior were tried. However, only "sustainable," "sustainability," and "organic" produced satisfactory results in terms of obtaining comments and evaluations highlighting the dimensions related to green consumption. The comments that employed these terms were selected. As a result, 10,488 review comments were retained for data analysis, which means only 0.079% of the customer comments posted on Airbnb express a preference for factors related to sustainability.

The information provided in the review comments by green Airbnb users resulted in a matrix with almost 1.3 million records of Airbnb listings. This information was processed using a script written in MySQL language. An interface of R statistical programming language was used to process the text data. Different packages were applied for the sentiment analysis phase: "Tm" and "Tidytext" packages (Feinerer et al., 2008) were used to pre-process the text and obtain clean summary information of the unstructured data to facilitate the handling of data; "Topicmodels" package (Grün and Hornik, 2011) was used to implement simply Latent Dirichlet allocation (LDA) to identify how many topics exist in the review online database and the words that each topic contains, and it reports the proportion of each topic in each text too; "Sentimentr" used Bayesian classifiers for positive, negative, and emotion classification at the sentence level.

## Data analysis

Sentiment analysis is the methodology used to achieve the first objective of the present study, which is to explore the preferences revealed by green Airbnb users during their stay. This method employs data mining to extract and process the subjective information provided by users in review comments and understand their "emotions" toward a specific topic (Chang & Wang, 2018; Xiang et al., 2015a, 2015b; Ye et al., 2009; Zheng & Ye, 2009).

The present research was based on the approaches to sentiments analysis and emotions analysis through the lens of computational linguistics established by Kim and Klinger (2018). Under this approach, the computational analysis of emotions may deepen understanding of and help overcome the "positive" and "negative" dichotomy of traditional sentiment analysis and facilitate the exploration of more detailed emotional dimensions with the National Research Council (NRC) Emotion Lexicon (Mohammad & Turney, 2013) and Plutchik's Wheel of Emotions (Plutchik, 1994). NRC Emotion Lexicon is an emotion estimation lexicon widely used in the sentiment analysis with more than 14,000 tagged words in English, corresponding associations with Plutchik's eight basic emotions (anger, fear, anticipation, confidence, surprise, sadness, joy and disgust) and two feelings (negative and positive). Each word is reviewed with several different types of emotion, which allows the contextualization of terms and gives valuable flexibility. All the terms were made manually by individual voluntary annotation.

Sentiment analysis was carried out in several stages. The first stage consisted of an initial approach to preferences of green Airbnb users through the identification of prominent themes when they post reviews online spontaneously and freely. For this purpose, on the one hand, the terms most frequently mentioned by guests were identified and represented in a bar chart. On the other hand, Leximancer software was used to obtain a conceptual map with latent semantic dimensions identifying the terms' hidden relationships with the labeled themes. The second stage included an analysis of emotions, which was performed by identifying the predominant emotions in each review comment through text mining. The Plutchik model and the NRC Emotion Lexicon were proposed as a reference framework, as was commented previously. The final stage implemented preference analysis using LDA, which is a widely utilized method that aids in the identification of the latent aspects in a large dataset of review comments (Li et al., 2018). The LDA method, originally formulated by Blei et al. (2003), is a mechanism used for topic extraction that is developed in three steps. The algorithm estimates the latent themes, then proceeds to classify each word to a theme, and finally assigns the theme for each word. Once the

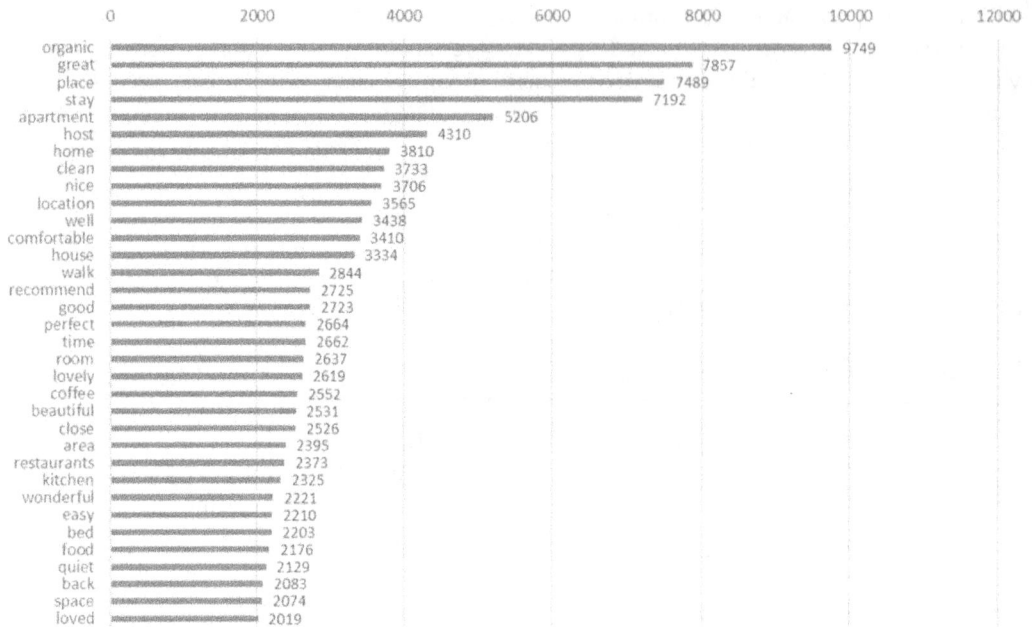

**Figure 1.** Most frequently used terms.

latent aspects were obtained, their theta coefficients were calculated, specifically, the probability that a text is associated with a specific aspect. After these coefficients were established, their average was calculated, thereby shedding light on the preferences of green Airbnb users expressed in their review comments (Luo & Tang, 2019).

The second objective of the present study is to determine the extent to which the final assessment made by green Airbnb users of their accommodations is conditioned by the polarity understood as the intensity of the feelings implicit in the comments. To reach that goal, LARA was used in a manner similar to those employed in other studies (Luo & Tang, 2019; Wang et al., 2010, 2011). LARA is a methodological approach composed of two parts. The first part explores the latent dimensions in portions of the text (in this case, the comments of green Airbnb users). The second part involves the estimation of a regression model (latent rating regression) that seeks to establish the extent to which the latent emotional intensity in textual pieces generates effects on global numerical ratings. The operation of the model is based on three basic assumptions: (1) the attitudes expressed in users' evaluations of a product or service are positive or negative (Solomon, 2009), as defined by the emotional dimension, (2) each review comment expresses different emotions with variable intensities, and (3) the emotion contained in the review comments have some effect on the rating scores given by the users.

## Results

### Sentiment analysis

In the first phase of the sentiment analysis, a descriptive analysis was performed to explore the preferences of green Airbnb users. The terms of the dataset were unified using a thesaurus built ad hoc for this research. Then, every term with a frequency equal to or greater than 2,000 appearances was selected, with a total result of 34 keywords.

Figure 1 presents a bar chart with the 34 keywords. The chart provides a hierarchical visualization approach (Hunt et al., 2014) and provides high detail with the exact number of frequency by term. "Organic" is the preponderant term, demonstrating the relevance of this attribute for

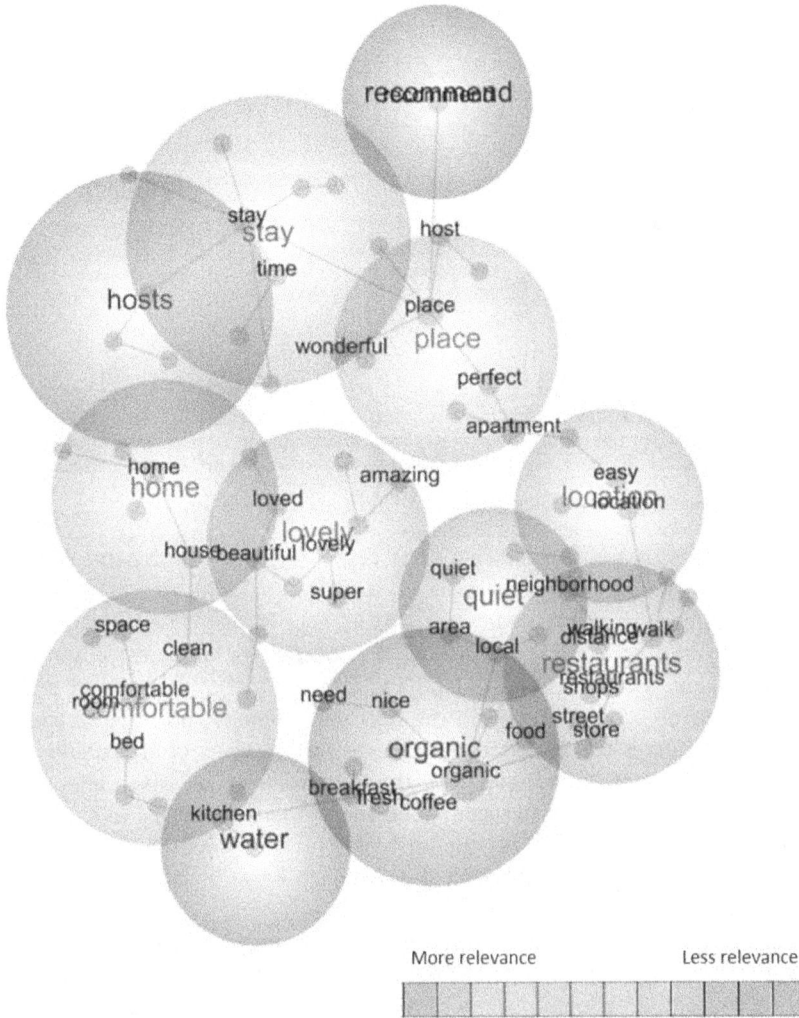

More relevance                          Less relevance

**Figure 2.** Conceptual map of the themes and concepts.

green Airbnb users. Subsequently, after a set of terms focusing on lodging attributes ("place" and "apartment"), other words appear to be related to activities carried out during the stay ("walk," "restaurants," and "store"), the features of the property ("kitchen" and "clean"), and the characteristics of its location ("quiet" and "close"). A profusion of positive emotions emerges ("great," "nice," "lovely," or "wonderful") in contrast to the absence of negative emotion terms.

Next, a relational semantic analysis was performed using Leximancer software, allowing us to obtain a conceptual map of the latent relevant themes. This technique is especially useful for big data analysis (Brochado et al., 2017; Cheng & Jin, 2019), and researchers give it coherence by interpreting the results. Figure 2 presents a heat map in which the color indicates the highest or lowest preponderance of themes and the size shows the number of concepts that belong to the theme represented by each circle. Figure 2 shows that the three most prominent dimensions are "organic," "restaurants," and "comfortable."

"Organic" reflects the importance of the sustainability factor in the comments of green Airbnb users. On the one hand, the concepts clustered under the theme described concerns such as healthy eating ("food," "fresh," "breakfast," and "coffee"), connecting to the theme "restaurants."

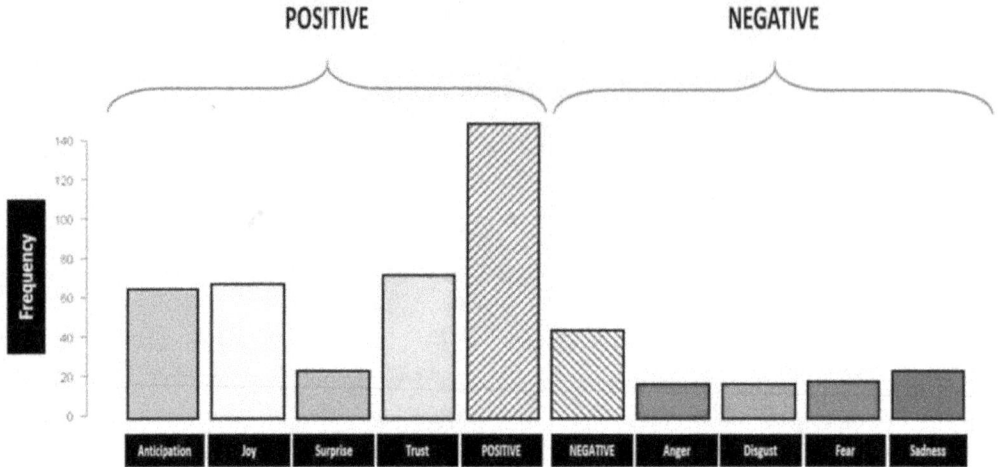

**Figure 3.** Distribution of emotional labels according to Plutchik's theory of basic emotions.

This theme is also near others that indicate concerns, such as the sustainable use of resources ("water") or forms of tourism with less impact on local communities ("quiet") that are reflected widely in the literature (Fuentes-Moraleda et al., 2019).

"Restaurants" includes concepts that refer to the location of these establishments, thereby connecting them with "location" and "quiet," and others that connect them with "organic." "Location" appears to be a bridge between the dominant themes and other latent dimensions. This connectivity anticipates the importance for green Airbnb users of an experience related to their demands for sustainability and links with the local community, which affects their assessment and final decision on whether to recommend the property to other users.

The prevalence of the "comfortable" theme highlights the importance attached by green Airbnb users to property amenities although it appears to be an important issue connected to the dimension of feelings in the experience.

Once the relational semantic analysis was explored, sentiment analysis was used to deepen the knowledge to the emotions contained in the online review comments of green Airbnb users. To this purpose, the NRC Emotional Lexicon previously described was used. The results obtained show that positive sentiments are predominant and that positive emotions ("trust," "anticipation," "joy," and "surprise") are more common than negative emotions ("fear," "anger," "sadness," and "disgust"). Figure 3 shows that "trust" was the most prevalent emotion. Based on Lahnos' (2001) definition of trust as a rational belief in the behavior of another person, the prevalence of this emotion indicates that green Airbnb users are aware of what they consume because of the strong rational component in that decision.

Figure 4 shows the terms associated with the different emotions of Plutchik's model. This figure shows an emotional map of Airbnb's green users, which could be an alternative tool for characterizing them. Based on the fact that they express their reviews online through emotions, identifying different types of Airbnb users is possible. Some terms with a strong sentimental charge were repeated in almost all emotions, such as "lovely," "wonderful," "perfect," and "friendly" among the positive ones or "feeling" and "treat" among the negative ones. Notably, some recurring terms in all types of stay, including those of green Airbnb users, are linked closely to a single emotion, such as "safe" to "trust" or "change" to "fear." The terms linked to sustainability elements, such as "garden," "farm," or "food," are fundamentally associated with positive emotions, as well as "organic", which falls directly and exclusively into the positive category.

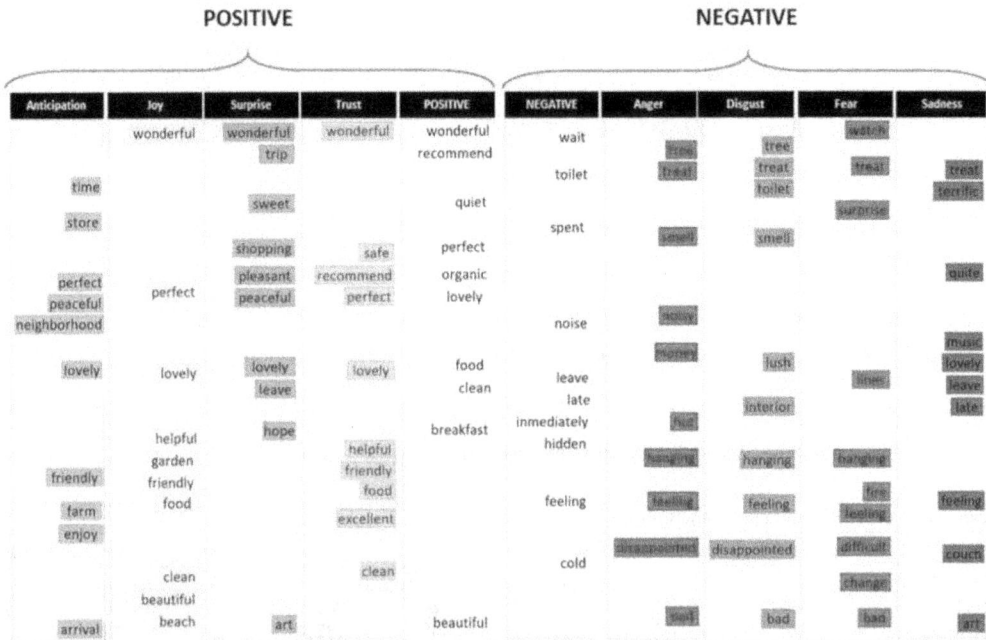

**Figure 4.** Terms associated with each emotion in the Plutchik model.

## Aspect-based sentiment analysis

To deepen the sentiment analysis with greater methodological rigor, the LDA algorithm (Xiang et al., 2017) was used to extract the latent aspects of online comments of green Airbnb users. A document and term matrix was created, and the parameters for the identification of latent aspects were established using the Gibbs sampling strategy. Generally, the researcher decides how many latent aspects will be extracted (Wang et al., 2010). In this case, the review corpus suggested six major aspects because six is the number of characteristics Airbnb users' rate in the digital platform after their stay. The beta coefficients of each term were calculated to assign to each of the six latent aspects extracted, as shown in Figure 5. A holistic interpretation of the seed terms included in each aspect allowed them to be labeled because the preferences expressed by green Airbnb users in their comments were known.

The first latent aspect, "amenities," describes the characteristics and resources of the property (room, kitchen, and bed). The second aspect, "sustainability," includes the seed terms related to the "organic" attribute. The third aspect, "experience," is linked in the literature with why a property is recommended. The fourth latent aspect is "facilities," which is defined as the general house environment and other amenities, such as a garden. The fifth latent aspect is the "host" and everything related to users' treatment (stay, time, and arrival). The last aspect is "location," which highlights the proximity of the accommodation to valuable items, including restaurants and markets, and the quality of the neighborhood.

The corresponding theta coefficients were calculated to rank the preferences revealed by green Airbnb users, where a higher average value indicates the latent aspect to which green Airbnb users devote more attention in their comments. The coefficients obtained were as follows: "amenities" (0.17), "sustainability" (0.22), "experience" (0.18), "accommodation" (0.13), "host" (0.21), and "location" (0.09). Significantly, the most relevant latent aspect for green Airbnb users was "sustainability"; therefore, their experience is highly mediated by this aspect.

The last step in the analysis is to apply aspect-based sentiment analysis (ABSA) to link the six latent aspects identified with the polarity of the sentiments awakened in green Airbnb users

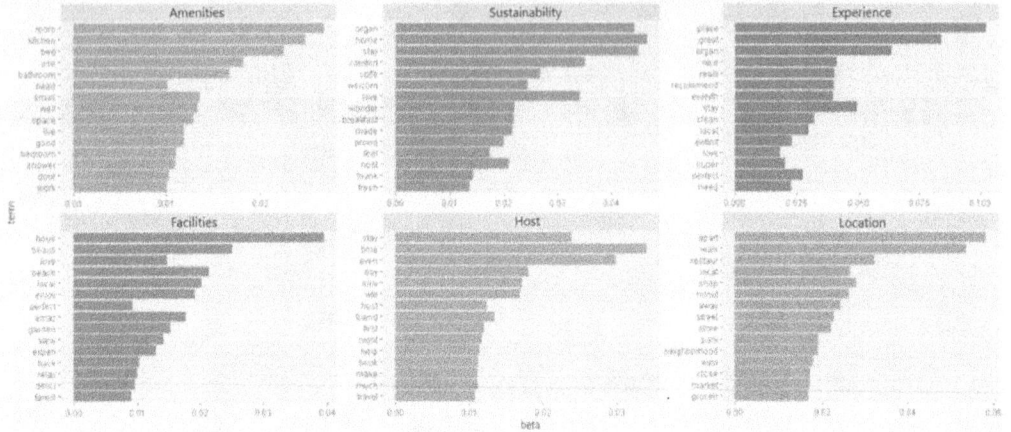

**Figure 5.** Latent aspects in review comments from green Airbnb users.

**Table 1.** Aspect-based sentiment analysis results of each aspect.

| Aspect | Anticipation | Joy | Surprise | Trust | Anger | Disgust | Fear | Sadness |
|---|---|---|---|---|---|---|---|---|
| Amenities | 3,9366 | 4,7634 | 1,8173 | **4,9716** | 0,4622 | 0,3780 | 0,5970 | 1,1583 |
| Sustainability | 3,4840 | **5,0442** | 1,6839 | 4,2324 | 0,4083 | 0,3635 | 0,4856 | 0,9253 |
| Experience | 3,8707 | 4,5532 | 1,7344 | **5,2305** | 0,7285 | 0,5818 | 0,8175 | 1,4857 |
| Facilities | 2,0971 | 2,9962 | 0,9283 | **3,2440** | 0,1575 | 0,1133 | 0,1861 | 0,5958 |
| Host | 2,6215 | **4,1269** | 1,4502 | 3,8455 | 0,2245 | 0,2101 | 0,2616 | 0,8167 |
| Location | 2,8553 | 3,5625 | 1,3032 | **3,8175** | 0,2683 | 0,1898 | 0,3183 | 0,8891 |

(Zhao et al., 2019). Logically, each of these aspects generates mixed sentiments, distributing their preeminence in the eight basic emotions described by Plutchik. However, as the previous general sentiment analysis anticipated, the feelings aroused by each aspect in green Airbnb users are concentrated on positive emotions, particularly "joy," "trust," and "anticipation" (see Table 1).

Some significant results must be highlighted. The latent aspect of "sustainability" is fundamentally experienced with the emotion "joy," as is the case with the aspect of great human interaction, which is the relationship with the "host." The rest of the aspects, which are more focused on the physical aspects of the home ("amenities," "facilities," "location," and "experience"), are fundamentally experienced through the "trust" emotion. The joint analysis of these results shows a green Airbnb user who experiences human enjoyment and the aspects of sustainability that endow the stay with value and a user who confidently knows what to seek and acknowledges when expectations are satisfied.

The results obtained have made it possible to explore the preferences revealed by green Airbnb users. All modes of analysis conducted by the study show a centrality of the elements related to sustainability, which is not only the most employed term but also the term that awakens in users the most positive emotion, such as "joy." This finding does not detract from the findings of the users' preferences and attitudes regarding other aspects of their stay, which are also important and produce satisfactory emotions. From these results, the first objective of the present investigation is fulfilled. Next, the second objective is addressed, that is, understanding the extent to which the intensity of the feelings of green Airbnb users in their review comments influences the evaluation and numerical assessment they make of Airbnb accommodations.

### Attitudes of sustainable airbnb consumers

Following Luo and Tang (2019) model, to explore the attitudes of green Airbnb users toward the accommodations they selected, a structural equation model was designed and estimated using a

**Table 2.** Adjustment indicators and a simple linear regression model estimated using the Bayesian approach.

| Aspects rating | Average | Beta | 95% Lower bound | 95% Upper bound | PPP | DIC |
|---|---|---|---|---|---|---|
| Accuracy | 4.993.996 | −0.004 | −0.004 | −0.015 | 0.50 | 9.99 |
| Cleanliness | 497.141 | 0.000 | −0.045 | 0.001 | 0.50 | 9.98 |
| Check-in | 4.996.596 | 0.000 | −0.005 | 0.010 | 0.50 | 9.98 |
| Communication | 4.995.997 | 0.000 | −0.002 | 0.016 | 0.50 | 9.96 |
| Location | 4.994.758 | 0.000 | −0.014 | 0.008 | 0.51 | 9.96 |
| Value | 4.984.656 | 0.000 | −0.015 | 0.020 | 0.50 | 10.00 |
| Overall | 4.818.417 | 0.135 | 0.135 | 0.113 | 5.00 | 9.99 |

Bayesian approach by applying the LARA model. Unlike frequentist statistics, Bayesian statistics are designed to identify the best scenario to respond to uncertainty. In the case at hand, Bayesian statistics are used to determine how the emotional intensity of the comments, the polarity according to Jockers (2017), affects different overall ratings made by guests.

For this purpose, the emotional polarity of each of the review comments was established using Sentimentr package. The calculation of polarity was carried out using an algorithm, in which all the unique terms in the "bag of words" were extracted from each comment. Those words were compared with a dictionary of polarized terms, valued numerically, and modified if appropriate by the accompanying amplifier, de-amplifier, denier, and adversary terms. The polarity of the paragraph was calculated as the last step. AMOS 26.0 software was employed to estimate the test model fit and test the causal hypothesized relationships of the two variables of the study. Two variables were created, namely, an independent variable associated with the average of the intensity of feelings called "polarity" (Liu, 2015) and a dependent variable that included the numerical assessment of the properties made by green Airbnb users. The independent variable was called "av_sent," and the dependent variable was called "eval." The latter includes the overall numerical ratings given by green Airbnb users for each of the six dimensions evaluated (accuracy in the description, check-in, communication with the host, cleanliness of the place, location, and added value) and the overall rating of their experience.

The model estimation method was the Markov Chains Monte Carlo method (Asparouhov & Muthén, 2014), which is an increasingly popular technique that responds to Bayesian statistics. This method includes a class of algorithms to estimate by simulating the posterior distribution of the parameter selected for the study. It is used to fit a model and extract samples from the joint posterior distribution of the model parameters. In Bayesian statistics, the probability is represented by a distribution because it is based on the idea that the probability of something happening is influenced by the previous assumption of probability and the probability of something happening as indicated by the data. The initial values were the ratings according to the polarity analysis that the users are expected to give according to their experience (latent rating). The a posteriori figures are the actual assessments made by the users on each dimension and their overall experience (effective rating). The results obtained by applying the model estimation method showed that the trajectories for each dimension and global assessment are relatively stable, which is desirable in assuming robustness in this type of analysis.

The prediction of the model consists of estimating the extent to which the emotionality associated with each review comment was transferred to a numerical rating and how much the evaluation should have changed based on the intensity of the emotions shown by green Airbnb users. LARA, which evaluates the opinions, emotions, and feelings expressed in each review comment at the aspect level, establishes the latent rating for each comment and the relative importance given by users to these aspects when forming their general opinion on the accommodation. The results of this analysis are presented in Table 2, which contains seven rows: one row for each of the six aspect ratings that Airbnb asks users to give in their comments and the last row for the overall rating.

Table 2 shows that all the models present a good fit because the discrepancy (PPP) and good-ness of fit (DIC) indicators have values of approximately 0.5 and 10, respectively, as suggested in the literature (Cain & Zhang, 2019).

The Bayesian inference shows that sentiments affect the numerical rating of each of the dimensions because all the credibility intervals indicate that none of the values in the a posteri-ori distribution is zero, which leads to the rejection of the null hypothesis. The a posteriori distri-butions, which express the values that each of the numerical evaluations should assume if the emotional intensity associated with the texts is considered, revealed that the ratings made by green Airbnb users are consistent with the emotional expressions present in each of their com-ments. This result means that, if the emotional polarity present in the comment is positive, the numerical rating will also be positive.

However, considering the beta coefficient in Table 2, a small difference is observed between the latent and effective ratings. Among all the elements evaluated, the one with the higher effect is the overall rating of the property. Specifically, if the global assessments of properties made by sustainable Airbnb users are considered and adjusted for the presence of sentiments in their comments, the expected value of their rating should be slightly more than one-tenth lower than the assessment that was made.

In summary, the effective ratings made by green Airbnb users after their experience are only relatively affected by the feeling that this experience has awakened in them. Moreover, this affectation occurs more in the overall rating than in the rating for each aspect. This finding is consistent with the previous literature on Airbnb, which will be discussed in the follow-ing section.

## Discussion and implications

Using big data analytics through sentiment analysis has provided significant evidence on the preferences of green Airbnb users, including similarities and differences with the findings previ-ously established by the literature. Our study has proven an evident positivity bias in comments of green Airbnb users. This bias could be explained by several factors. On the one hand, the overall inflation of the ratings as a result of the peer evaluation system of Airbnb has already been stated (Lee & Kim, 2019). On the other hand, customers tend to evaluate their experience more positively when they are in an emotional state in which positive emotions are prevalent (Isen, 1987). This finding points in the same direction as the trend determined by the most recent academic literature for comments from most Airbnb users (Cheng & Jin, 2019; Luo & Tang, 2019).

Another relevant finding derived from the descriptive analysis and the ABSA was that green Airbnb users exhibit a clear preference for sustainability. Consistent with previous empirical stud-ies, the location factor is the aspect with the greatest weight in the preferences revealed by green Airbnb users because it is for overall Airbnb users (Cheng & Jin, 2019; Young et al., 2017). However, our results also show how green Airbnb users are deeply connected with the origins of the sharing economy. Their choices are based on the search for greater sustainability, that is, they consider a more environmentally and socially sustainable alternative to the consumption model proposed by the traditional economy (Guttentag et al., 2018).

The study also identified that the host plays a critical role in the experience perceived by green Airbnb users, which is a conclusion that differs from previous studies (Cheng & Jin, 2019; Tussyadiah & Park, 2018). These studies have acknowledged that Airbnb hosts play a more important role than conventional hotel workers but have not confirmed whether the users of this platform seek any special relationship with the host. Therefore, the common Airbnb user perceives the host as only a facilitator of the experience, whereas green Airbnb users perceive that the host plays a central role as a builder of social relationships. This circumstance could explain the evolution of Airbnb toward a traditional economic model, where the host is limited to be a figure who manages the accommodation in a professional manner (Gil & Sequera, 2018) and moves away from the role originally assigned to hosts. Our findings regarding green Airbnb users support the philosophy with which Airbnb was started, which reinforces the idea that green Airbnb users respond to preferences closer to the non-commercial model of the sharing economy.

The findings of ABSA also show that "sustainability" and "host," the main mediators in the experience of green Airbnb users, are the aspects related to the emotion "joy." This fact could indicate that these two aspects provide the satisfaction with the experience to green Airbnb users. The results also reveal that the prominent emotion in four of the six aspects is "trust." This result reveals how aware of their preferences green Airbnb users are. Such awareness includes all the products and services offered to them (background) and is reflected in their a posteriori review comments when their expectations are adequately satisfied.

The second stage of the study analyzed the extent to which the numerical ratings granted by green Airbnb users to the accommodations are conditioned by the emotions and feelings they experienced. The results obtained indicate that, in green consumer attitudes toward Airbnb properties, the emotional intensity associated with their comments is reflected in the numerical values assigned in general and in each of the six specific aspects (cleanliness, location, communication, added value, the accuracy of the description, and registration). On the one hand, this finding is consistent with that of Luo and Tang (2019), who noted that the intensity of the emotion is reflected in the numerical evaluation. On the other hand, it is also consistent with "trust" being revealed as the predominant emotion in most latent aspects. Both pieces of evidence could indicate that these users know what they want and that their stay confirms their appreciation, explaining a relationship among the different emotional expressions in each of the latent aspects that were revealed, the qualifications of each aspect, and the relative weights in the formation of a numerical overall rating.

In summary, in the different techniques applied to the dataset, all elements related to sustainability play a leading role in the expression of opinions and evaluations by green Airbnb users. All of these factors were also revealed to be linked to the most positive emotion of all the options in Plutchik's model, "joy." The preferences revealed by green Airbnb users differ from those of other users of the platform who are not defined by a sustainability profile. These preferences are connected with the original characteristics of the Airbnb model, which are closer to the real sharing economy than to the conventional economy. Finally, this study shows that the a posterori ratings made by green Airbnb users are not conditioned by their a priori feelings contained in the comments nor by the information they had before staying at the establishment.

This study has several useful managerial implications for professionals and researchers in hospitality. The results of this study delve into the factors that determine the behavior pattern of users with a sustainable profile, which informs the sector on how to improve the service it offers and thus increase overall satisfaction. In this sense, responding to the preferences identified in the study, digital accommodation platforms could design and implement automated establishment recommendation systems for potential green Airbnb users based on the preferences identified in this study. Combining search categories established a priori (e.g., price, location, experiences, and amenities) with the information contained in the online review comments and the ratings granted by these users, faster and more precise and reliable decisions could be made by future users. Moreover, the study reaffirms the centrality of sustainability and the prominent role of the host for this type of Airbnb user. The most compelling aspects for green Airbnb users to stay in an Airbnb accommodation are the host and sustainability facilities. Therefore, Airbnb hosts should focus on providing a description of themselves and the home according to the desire for sustainability that moves these guests. The private sector plays a fundamental role in promoting sustainable consumption by establishing a corporate culture linked to sustainable development that translates into a change in production models that incorporates sustainability criteria. Thus, although the Airbnb platform includes sustainable properties that are available, implementing a classification system for its services with sustainability criteria to facilitate users' selection process is important.

## Limitations and directions for future research

Despite the significant contributions of this study, concluding with a series of necessary observations about the study's limitations, which are understood as a starting point for further future lines of work, is mandatory. First, the sample contains only online review comments from cities included on the Inside Airbnb website. Thus, expanding the sample to other cities where Airbnb operates and using other reliable sources of data, such as AirDNA, would be interesting for future research.

Second, texts that were not written in English were excluded from this study, thereby denying the inclusion of other cultural and demographic perspectives that could provide depth and richness to the findings obtained. This approach represents a powerful line of research for future studies.

Third, the analysis of online review comments from green Airbnb users could incorporate information extracted from the descriptions made by the hosts of the properties. Such information, when combined with other reference variables, such as the price per night and geolocation, would contribute to improving the results of the study. Similarly, exploring the relationship between user behavior and sustainable consumption from the perspective of the host is highly recommended for future studies. According to the study findings, the question is not to classify whether a consumer is green but rather to the proposal of the lifestyle in the property that the host makes to the user. Future studies could explore how to eliminate dissonances between Airbnb hosts and guests. This topic would point to another topic that is closely related to the objective of studying the differences between one and the other and the barrier that technology creates between them and how technology should be sought for the solution to it.

## Disclosure statement

No potential conflict of interest was reported by the authors.

## ORCID

*Antonio Ariza-Montes* (iD) http://orcid.org/0000-0002-5921-0753

## References

Adamiak, C. (2019). Current state and development of Airbnb accommodation offer in 167 countries. *Current Issues in Tourism*, 0(0), 1–19. https://doi.org/10.1080/13683500.2019.1696758

Airbnb. (2017). *Informe sobre la comunidad de Airbnb en España*.

Alaei, A. R., Becken, S., & Stantic, B. (2019). Sentiment analysis in tourism: Capitalizing on Big Data. *Journal of Travel Research*, 58(2), 175–191. https://doi.org/10.1177/0047287517747753

Antonides, G. (2017). Sustainable consumer behaviour: A collection of empirical studies. *Sustainability (Switzerland)*, 9(10), 1686. https://doi.org/10.3390/su9101686

Asparouhov, T., & Muthén, B. (2014). Multiple-group factor analysis alignment. *Structural Equation Modeling*, 21(4), 495–508. https://doi.org/10.1080/10705511.2014.919210

Belk, R. (2014). Sharing versus pseudo-sharing in web 2.0. *The Anthropologist*, 18(1), 7–23. https://doi.org/10.1080/09720073.2014.11891518

Blei, D. M., Ng, A. Y., & Jordan, M. I. (2003). Latent dirichlet allocation. *Journal of Machine Learning Research*, 3(4–5), 993–1022. https://doi.org/10.1016/b978-0-12-411519-4.00006-9

Böcker, L., & Meelen, T. (2017). Sharing for people, planet or profit? Analysing motivations for intended sharing economy participation. *Environmental Innovation and Societal Transitions*, 23, 28–39. https://doi.org/10.1016/j.eist.2016.09.004

Botsman, R., & Rogers, R. (2010). *What's mine is yours: The rise of collaborative consumption*. Harper Collins.

Brochado, A., Troilo, M., & Shah, A. (2017). Airbnb customer experience: Evidence of convergence across three countries. *Annals of Tourism Research*, 63, 210–212. https://doi.org/10.1016/j.annals.2017.01.001

Buhalis, D., & Law, R. (2008). Progress in information technology and tourism management: 20 years on and 10 years after the internet: The state of eTourism research. *Tourism Management*, 29(4), 609–623. https://doi.org/10.1016/j.tourman.2008.01.005

Cain, M. K., & Zhang, Z. (2019). Fit for a Bayesian: An evaluation of PPP and DIC for structural equation modeling. *Structural Equation Modeling: A Multidisciplinary Journal*, 26(1), 39–50. https://doi.org/10.1080/10705511.2018.1490648

Chang, W. L., & Wang, J. Y. (2018). Mine is yours? Using sentiment analysis to explore the degree of risk in the sharing economy. *Electronic Commerce Research and Applications*, 28, 141–158. https://doi.org/10.1016/j.elerap.2018.01.014

Cheng, M. (2016). Sharing economy: A review and agenda for future research. *International Journal of Hospitality Management*, 57, 60–70. https://doi.org/10.1016/j.ijhm.2016.06.003

Cheng, M., & Jin, X. (2019). What do Airbnb users care about? An analysis of online review comments. *International Journal of Hospitality Management*, 76, 58–70. https://doi.org/10.1016/j.ijhm.2018.04.004

Chi, X., & Han, H. (2020). Exploring slow city attributes in Mainland China: Tourist perceptions and behavioral intentions toward Chinese Cittaslow. *Journal of Travel & Tourism Marketing*, 37(3), 361–379. https://doi.org/10.1080/10548408.2020.1758286

Cocola-Gant, A., & Gago, A. (2019). Airbnb, buy-to-let investment and tourism-driven displacement: A case study in Lisbon. *EPA: Economy and Space*, 0(0),1–18. https://doi.org/doi: https://doi.org/10.1177/0308518X19869012

Dharmesti, M., Merrilees, B., & Winata, L. (2020). "I'm mindfully green": Examining the determinants of guest proenvironmental behaviors (PEB) in hotels. *Journal of Hospitality Marketing and Management*, 29, 1–18. https://doi.org/10.1080/19368623.2020.1710317

Dredge, D., Gyimóthy, S., Birkbak, A., Jensen, T. E., & Madsen, A. K. (2016). *The impact of regulatory approaches targeting collaborative economy in the tourism accommodation sector: Barcelona, Berlin, Amsterdam and Paris*. Impulse Paper No. 9. European Commission.

Eom, T., & Han, H. (2019). Community-based tourism (TourDure) experience program: A theoretical approach. *Journal of Travel & Tourism Marketing*, 36(8), 956–968. https://doi.org/10.1080/10548408.2019.1665611

Feinerer, I., Hornik, K., & Meyer, D. (2008). Text mining infrastructure in R. *Journal of Statistical Software*, 25(5), 1–54. https://doi.org/10.18637/jss.v025.i05

Fuchs, M., Höpken, W., & Lexhagen, M. (2014). Big data analytics for knowledge generation in tourism destinations – A case from Sweden. *Journal of Destination Marketing & Management*, 3(4), 198–209.

Fuentes-Moraleda, L., Lafuente-Ibáñez, C., Muñoz-Mazón, A., & Villacé-Molinero, T. (2019). Willingness to pay more to stay at a boutique hotel with an environmental management system. A preliminary study in Spain. *Sustainability (Switzerland)*, 11(18), 5134. https://doi.org/10.3390/su11185134

Gil, J. (2019). Cambios en la producción y el consumo del turismo. El caso de Airbnb. In E. Cañada & I. Murray (Eds.), *Turistifación global. Perspectivas críticas en turismo* (Antrazyt 4, pp. 325–342). Icaria.

Gil, J., & Sequera, J. (2018). *Expansión de la ciudad turística y nuevas resistencias.* El Caso de Airbnb en Madrid Revista de Metodología de Ciencias Sociales. N°41 septiembre-diciembre, 15–32. https://doi.org/DOI/empiria.41. 2018.22602

Grün, B., & Hornik, K. (2011). Topicmodels: An R package for fitting topic models. *Journal of Statistical Software, 40*(13), 1–30. https://doi.org/10.18637/jss.v040.i13 https://doi.org/10.18637/jss.v040.i13

Gunter, U. (2018). What makes an Airbnb host a superhost? Empirical evidence from San Francisco and the Bay Area. *Tourism Management, 66*, 26–37. https://doi.org/10.1016/j.tourman.2017.11.003

Guo, Y., Barnes, S. J., & Jia, Q. (2017). Mining meaning from online ratings and reviews: Tourist satisfaction analysis using latent dirichlet allocation. *Tourism Management, 59*, 467–483. https://doi.org/10.1016/j.tourman.2016.09.009

Guttentag, D., Smith, S., Potwarka, L., & Havitz, M. (2018). Why tourists choose Airbnb: A motivation-based segmentation study. *Journal of Travel Research, 57*(3), 342–359. https://doi.org/10.1177/0047287517696980

Han, H., Hsu, L. J., & Lee, J. (2009). Empirical investigation of the roles of attitudes toward green behaviors, overall image, gender, and age in hotel customers' eco-friendly decision-making process. *International Journal of Hospitality Management, 28*(4), 519–528. https://doi.org/10.1016/j.ijhm.2009.02.004

Han, H., & Hyun, S. S. (2018). Eliciting customer green decisions related to water saving at hotels: Impact of customer characteristics. *Journal of Sustainable Tourism, 26*(8), 1437–1452. https://doi.org/10.1080/09669582.2018. 1458857

Han, H., Meng, B., & Kim, W. (2017). Emerging bicycle tourism and the theory of planned behavior. *Journal of Sustainable Tourism, 25*(2), 292–309. https://doi.org/10.1080/09669582.2016.1202955

Han, H., Olya, H. G. T., Cho, S., & Kim, W. (2018). Understanding museum vacationers' decision-making process: Strengthening the VBN framework. *Journal of Sustainable Tourism, 26*(6), 855–872.

Han, H., Ph, D., Hwang, J., Ph, D., Lee, S., & Ph, D. (2017). Cognitive, affective, normative, and moral triggers of sustainable intentions among convention-goers. *Journal of Environmental Psychology, 51*, 1–13. https://doi.org/10. 1016/j.jenvp.2017.03.003

Han, H., & Yoon, H. J. (2015). Hotel customers' environmentally responsible behavioral intention: Impact of key constructs on decision in green consumerism. *International Journal of Hospitality Management, 45*, 22–33. https://doi. org/10.1016/j.ijhm.2014.11.004

Higgins-Desbiolles, F. (2018). Sustainable tourism: Sustaining tourism or something more? *Tourism Management Perspectives, 25*(1), 157–160. https://doi.org/10.1016/j.tmp.2017.11.017

Hlee, S., Lee, H., & Koo, C. (2018). Hospitality and tourism online review research: A systematic analysis and heuristic-systematic model. *Sustainability (Switzerland), 10*(4), 1141. https://doi.org/10.3390/su10041141

Hu, Y. H., Chen, Y. L., & Chou, H. L. (2017). Opinion mining from online hotel reviews – A text summarization approach. *Information Processing and Management, 53*(2), 436–449. https://doi.org/10.1016/j.ipm.2016.12.002

Hunt, C. A., Gao, J., & Xue, L. (2014). A visual analysis of trends in the titles and keywords of top-ranked tourism journals. *Current Issues in Tourism, 17*(10), 849–855. https://doi.org/10.1080/13683500.2014.900000

Hurley, P. J. (2011). *A concise introduction to logic.* Cengage Learning.

Isen, A. M. (1987). Positive affect, cognitive processes, and social behavior. *Advances in Experimental Social Psychology, 20*(C), 203–253. https://doi.org/10.1016/S0065-2601(08)60415-3

Jockers, M. L. (2017). *Syuzhet: Extract sentiment and plot arcs from text.* R Package Version. https://github.com/ mjockers/syuzhet

Kim, Y., & Han, H. (2010). Intention to pay conventional-hotel prices at a green hotel - a modification of the theory of planned behavior. *Journal of Sustainable Tourism, 18*(8), 997–1014. https://doi.org/10.1080/09669582.2010.490300

Kim, A. E., Hansen, H. M., Murphy, J., Richards, A. K., Duke, J., & Allen, J. A. (2013). Methodological considerations in analyzing Twitter data. *Journal of the National Cancer Institute. Monographs, 2013*(47), 140–146. https://doi.org/ 10.1093/jncimonographs/lgt026

Kim, E., & Klinger, R. (2018). A survey on sentiment and emotion analysis for computational literary studies. *arXiv: 1808.03137v1.* pp. 1-26.

Lahno, B. (2001). On the emotional character of trust. *Ethical Theory and Moral Practice, 4*(2), 171–189. https://doi. org/10.1023/A:1011425102875

Lee, K. H., & Kim, D. H. (2019). A peer-to-peer (P2P) platform business model: The case of Airbnb. *Service Business, 13*(4), 647–669. https://doi.org/10.1007/s11628-019-00399-0

Li, G., Law, R., Vu, H., Rong, J., & Zhao, X. (2015). Identifying emerging hotel preferences using emerging pattern mining technique. *Tourism Management, 46*, 311–321.

Liu, B. (2015). *Sentimental analysis: Minings opinions, sentiments, and emotions.* Cambridge University Press.

Liu, X., Andris, C., Huang, Z., & Rahimi, S. (2019). Inside 50,000 living rooms: An assessment of global residential ornamentation using transfer learning. *EPJ Data Science, 8*(1),4. https://doi.org/10.1140/epjds/s13688-019-0182-z

Liu, S., Law, R., Rong, J., Li, G., & Hall, J. (2013). Analyzing changes in hotel customers' expectations by trip mode. *International Journal of Hospitality Management, 34*, 359–371.

Li, J., Xu, L., Tang, L., Wang, S., & Li, L. (2018). Big data in tourism research: A literature review. *Tourism Management, 68,* 301–323. https://doi.org/10.1016/j.tourman.2018.03.009

Li, H., Ye, Q., & Law, R. (2013). Determinants of customer satisfaction in the hotel industry: An application of online review analysis. *Asia Pacific Journal of Tourism Research, 18*(7), 784–802. https://doi.org/10.1080/10941665.2012.708351

Luo, Y., & Tang, R. L. (2019). Understanding hidden dimensions in textual reviews on Airbnb: An application of modified latent aspect rating analysis (LARA). *International Journal of Hospitality Management, 80,* 144–154. https://doi.org/10.1016/j.ijhm.2019.02.008

Meng, B., Chua, B., Ryu, H., & Han, H. (2020). Volunteer tourism (VT) traveler behavior: Merging norm activation model and theory of planned behavior. *Journal of Sustainable Tourism, 28*(12), 1947–1969. https://doi.org/10.1080/09669582.2020.1778010

Midgett, C., Bendickson, J. S., Muldoon, J., Solomon, S. J. (2017). The Sharing Economy and Sustainability: A case of AirBnB. *Small Business Institute Journal, 13*(2), pp. 51–71.

Mohammad, S. M., & Turney, P. D. (2013). *NRC emotion lexicon.* National Research Council Canada. https://doi.org/10.4224/21270984

Nadler, S. (2014). *The sharing economy: What is it and where is it going?* Massachusetts Institute of Technology.

Olteanu, A., Castillo, C., Diaz, F., & Kıcıman, E. (2019). Social data: Biases. *Frontiers in Big Data, 2,* 13. https://doi.org/10.3389/fdata.2019.00013

Pizam, A. (2009). Green hotels: A fad, ploy or fact of life? *International Journal of Hospitality Management, 28*(1), 1. https://doi.org/doi:https://doi.org/10.1016/j.ijhm.2008.09.001

Plutchik, R. (1994). *The psychology and biology of emotion.* Harper Collins.

Presenza, A., Petruzzelli, A. M., & Natalicchio, A. (2019). Business model innovation for sustainability. Highlights from the tourism and hospitality industry. *Sustainability, 11*(1), 212. https://doi.org/10.3390/su11010212

Quaglieri-Domínguez, A., & Sánchez Bergara, S. (2019). Alojamiento turístico y «economía colaborativa»: una revisión crítica de los discursos y las respuestas normativas. In E. Cañada & I. Murray (Eds.), *Turistifación global. Perspectivas críticas en turismo* (Antrazyt. 4, 343–366). Icaria.

Serrano, L., Sianes, A., & Ariza-Montes, A. (2019). Using bibliometric methods to shed light on the concept of sustainable tourism. *Sustainability (Switzerland), 11*(24), 6964. https://doi.org/10.3390/SU11246964

Sigala, M. (2018). New technologies in tourism: From multi-disciplinary to anti-disciplinary advances and trajectories. *Tourism Management Perspectives, 25,* 151–155. https://doi.org/10.1016/j.tmp.2017.12.003

Solomon, M. (2009). *Consumer behavior* (8th ed.). Upper Saddle River, NJ: Prentice Hall.

Song, Z., Xing, L., & Chathoth, P. K. (2015). The effects of festival impacts on support intentions based on residents' ratings of festival performance and satisfaction: A new integrative approach. *Journal of Sustainable Tourism, 23*(2), 316–337. https://doi.org/10.1080/09669582.2014.957209

Sthapit, E., & Jiménez-Barreto, J. (2018). Exploring tourists' memorable hospitality experiences: An Airbnb perspective. *Tourism Management Perspectives, 28,* 83–92. https://doi.org/10.1016/j.tmp.2018.08.006

Toni, M., Renzi, M. F., & Mattia, G. (2018). Understanding the link between collaborative economy and sustainable behaviour: An empirical investigation. *Journal of Cleaner Production, 172,* 4467–4477. https://doi.org/10.1016/j.jclepro.2017.11.110

Tussyadiah, I. P., & Park, S. (2018). When guests trust hosts for their words: Host description and trust in sharing economy. *Tourism Management, 67,* 261–272. https://doi.org/10.1016/j.tourman.2018.02.002

Tussyadiah, I. P., & Zach, F. J. (2015, April 10). Hotels vs. peer-to-peer accommodation rentals: Text analytics of consumer reviews in Portland. *Proceedings of 2015 TTRA Conference, Portland:OR.* https://doi.org/10.2139/ssrn.2594985

Wang, H., Lu, Y., Zhai, C. (2011). Latent aspect rating analysis without aspect keyword supervision. In *Proceedings of the ACM International Conference on Knowledge Discovery and Data Mining* (pp. 618–626). ACM. https://doi.org/10.1145/2020408.2020505

Wang, H., Lu, Y., & Zhai, C. (2010, July 24–28). *Latent aspect rating analysis on review text data: A rating regression approach* [Paper presentation]. Proceedings of the 16th ACM SIGKDD International Conference on Knowledge Discovery and Data Mining (pp. 783–792). Washington, DC. https://doi.org/10.1145/1835804.1835903

Wu, X., & Zhi, Q. (2016). Impact of shared economy on urban sustainability: From the perspective of social, economic, and environmental sustainability. *Energy Procedia, 104,* 191–196. https://doi.org/10.1016/j.egypro.2016.12.033

Xiang, Z., Du, Q., Ma, Y., & Fan, W. (2017). A comparative analysis of major online review platforms: Implications for social media analytics in hospitality and tourism. *Tourism Management, 58,* 51–65. https://doi.org/10.1016/j.tourman.2016.10.001

Xiang, Z., Magnini, V. P., & Fesenmaier, D. R. (2015a). Information technology and consumer behavior in travel and tourism: Insights from travel planning using the internet. *Journal of Retailing and Consumer Services, 22,* 244–249. https://doi.org/10.1016/j.jretconser.2014.08.005

Xiang, Z., Schwartz, Z., Gerdes, J. H., & Uysal, M. (2015b). What can big data and text analytics tell us about hotel guest experience and satisfaction? *International Journal of Hospitality Management, 44,* 120–130. https://doi.org/10.1016/j.ijhm.2014.10.013

Xu, F., Nash, N., & Whitmarsh, L. (2020). Big data or small data? A methodological review of sustainable tourism. *Journal of Sustainable Tourism, 28*(2), 144–166. https://doi.org/10.1080/09669582.2019.1631318

Yadav, R., Balaji, M. S., & Jebarajakirthy, C. (2019). How psychological and contextual factors contribute to travelers 'propensity to choose green hotels? *International Journal of Hospitality Management, 77*, 385–395. https://doi.org/10.1016/j.ijhm.2018.08.002

Yang, Y., & Mao, Z. (2019). Welcome to my home! An empirical analysis of Airbnb supply in US cities. *Journal of Travel Research, 58*(8), 1274–1287. https://doi.org/10.1177/0047287518815984

Ye, Q., Zhang, Z., & Law, R. (2009). Sentiment classification of online reviews to travel destinations by supervised machine learning approaches. *Expert Systems with Applications, 36*(3), 6527–6535. https://doi.org/10.1016/j.eswa.2008.07.035

Young, C. A., Corsun, D. L., & Xie, K. L. (2017). Travelers' preferences for peer-to-peer (P2P) accommodations and hotels. *International Journal of Culture, Tourism and Hospitality Research, 11*(4), 465–482. https://doi.org/10.1108/IJCTHR-09-2016-0093

Zhang, D., Zhou, L., Kehoe, J. L., & Kilic, I. Y. (2016). What online reviewer behaviors really matter? Effects of verbal and nonverbal behaviors on detection of fake online reviews. *Journal of Management Information Systems, 33*(2), 456–481. https://doi.org/10.1080/07421222.2016.1205907

Zhao, Y., Xu, X., & Wang, M. (2019). Predicting overall customer satisfaction: Big data evidence from hotel online textual reviews. *International Journal of Hospitality Management, 76*, 111–121. https://doi.org/10.1016/j.ijhm.2018.03.017

Zheng, W., & Ye, Q. (2009, November 21–22). Sentiment classification of Chinese traveler reviews by support vector machine algorithm. In *2009 Third International Symposium on Intelligent Information Technology Application* (Vol. 3, pp. 335–338). IEEE. https://doi.org/10.1109/IITA.2009.457

# An application of Delphi method and analytic hierarchy process in understanding hotel corporate social responsibility performance scale

Antony King Fung Wong(iD), Seongseop (Sam) Kim(iD), Suna Lee and Statia Elliot

**ABSTRACT**

This study aims to identify the essential indicators of hotel Corporate Social Responsibility (CSR) performance measurement with a standardized and composite CSR performance measurement index for the hotel industry. Employing both Delphi and Analytic Hierarchy Process (AHP) methods, three stakeholder groups are surveyed: academicians; hotel managers; and, hotel customers. Results reveal that three traditional CSR domains (legal, ethical, and social/philanthropic) are primary contributors to CSR performance, followed by two new environmental domains (room and restaurant; other general areas), and financial/economic domains as secondary contributors. This study shows the high level of consistency in the responses from stakeholder groups, supporting the effectiveness of the scale as a valuable tool to measure hotel CSR performance. Notably, domain weighted scores do differ slightly by respondent characteristic, indicating that the impacts of CSR are sensitive to respondent diversity.

## Introduction

At no other time in the history of commerce have the pressures of societal concerns for natural resource preservation, fair trade compliance, employee support, and the general welfare of our world been so critically important for business to address (Font & Lynes, 2018). Various and often vocal stakeholders in today's society call upon firms to assume additional responsibilities for the benefit of community and environment through what society sees as entitled business power. In response to this call, the hotel industry has gradually adopted "corporate social responsibility" (CSR) activities with varying degrees of effort (González-Rodríguez et al., 2019; Holcomb et al., 2007; Wong et al., 2019). Practices and performances range, with hotels applying a diversity of methods, measures, and approaches. This heterogeneity makes it difficult to compare CSR performance across hotels and highlights the need to develop a standardized measurement scale (De Grosbois, 2012; Lock & Seele, 2016; Madsen, 2009; Wong & Kim, 2020).

While the need to develop a standardized CSR performance measurement instrument in the hotel industry is recognized, efforts to identify hotel CSR performance measurement has been limited (Ko et al., 2019; Wong et al., 2019). Previous CSR researchers have adopted different scales that have led to diverse findings (Guzzo et al., 2019). In addition, most CSR scales have been applied in general business areas, thus lacking specificity to a hospitality context (Latif & Sajjad, 2018; Wong et al., 2019). Methodologically, there has been no effort to develop a validated hotel CSR scale using multiple hotel stakeholders to date (Serra-Cantallops et al., 2018). Yet previous research (Gond et al., 2017; O'Connor & Spangenberg, 2008; Turker, 2009) has argued for a diversity of stakeholder perspectives to be considered when developing CSR strategies and evaluating CSR performance. To address this gap, the current study aims to provide hotels with a multi-dimensional, multi-stakeholder method to measure and evaluate CSR performance by surveying three stakeholder groups (academics, hotel managers and hotel customers) and critically, encompassing a broader range of dimensions than current scales. CSR performance should be reviewed regularly given the changing needs and wants of stakeholders. This research seeks to provide a framework to support this process, thus contributing to the hotel industry insights into sector specific CSR dimensions to improve the evaluation of CSR measurement and performance.

This study has three research objectives: (1) identify the essential indicators of CSR performance measurement, (2) determine the perceived relative importance of CSR performance indicators, and (3) develop a standardized and composite CSR performance measurement index for the hotel industry. To achieve these objectives, this study employed both Delphi and analytic hierarchy process (AHP) techniques. Delphi can generate new, valuable and plausible ideas from the respondents, free from group intervention and strengthening the research validity by enabling a heterogeneity of panelists to contribute without the restriction of geographical distance. AHP is a mathematical technique for pairwise comparisons of multi-criteria, providing relative weights based on the importance of each measurement item to develop a standardized and composite CSR performance index.

## Literature review

### Existing CSR measurements

CSR is now a crucial aspect of business and society, with measurement an essential tool to ensure its validity and reliability (Carroll, 2000). CSR measurement includes the use of single- or multiple-issue indicators. One example of a single-issue indicator is the pollution control performance from the Council of Economic Priorities and Corporate Crime, which has been used in numerous studies (Baucus & Baucus, 1997; Davidson & Worrell, 1990). However, a uni-dimensional measurement approach has been shown to have substantial limitations in terms of exploring CSR and understanding the concept (Maignan & Ferrell, 2000). Another popular CSR measurement approach is the content analysis of annual corporate publications. This method is beneficial for deriving new measurement attributes for CSR (Abbott & Monsen, 1979). In recent years, CSR information has become considerably more accessible because of technological advancements and increased attention to CSR reporting (Gray et al., 1995). Ruf et al. (1998) claimed that the content analysis of corporate publications is an objective means of measuring CSR performance because the rating process is standardized after the selection of social attributes. However, Shnayder et al. (2015) argued that the information in annual corporate reports differs from tactual CSR performance. That is, companies may inadvertently mislead readers to gain a positive corporate image.

The next measurement involves using scales that evaluate CSR perception. Aupperle (1984) developed a widely accepted scale that follows Carroll's four-dimensional model to reflect managers' perceived CSR value. This CSR scale is the first attempt to grasp the multidimensional nature of CSR. However, Aupperle's (1984) scale is only applicable to the measurement of managers' perspectives. Another measurement was developed to measure managerial attitudes toward CSR by Quazi and O'brien (2000) who constructed a scale based on two-dimensional factors, namely, a range of outcomes of corporate social commitment and a span of corporate responsibility. Although this scale is advantageous to test a manager's motivation regarding and perception of CSR, the two-dimensional approach of this scale does not clearly define the composition of CSR, thereby limiting its explanatory ability.

The perceived role of ethics and social responsibility (PRESOR) scale was developed to measure the managerial perceptions of the role of social and ethical responsibility in achieving organizational effectiveness (Singhapakdi et al., 1996). Similar to the scale developed by Aupperle (1984), PRESOR focuses on measuring the managerial perceptions of CSR. However, PRESOR's lack of consideration of other perspectives limits its applicability. Moreover, Etheredge (1999) used PRESOR and conducted a replication study to analyze the perceived role of ethics and social responsibility from the managerial perspective. The results failed to depict the original factorial structure of the measurement.

Another well-known scale, developed by Pérez and Del Bosque (2013), aims to construct a reliable scale to measure customer perceptions of the CSR performance of their service providers. This scale adapts stakeholder theory, which encompasses customers, shareholders, employees, and society. This scale is suitable for investigating the socially responsible values of customers specifically in the banking industry, and is less suitable for measuring CSR in other industries because each has a unique business system and environment (Whitley, 1992).

Maignan and Ferrell (2000) developed an important scale to measure CSR and adopted the concept of corporate citizenship, which refers to the extent to which companies meet the standard of economic, legal, ethical, and discretionary responsibilities that influence various stakeholders. This approach integrates the four-dimensional CSR concept from Carroll's model (Carroll, 1979) and stakeholder theory, and has been tested and validated empirically in dissimilar cultural settings, such as the US and France. Maignan and Ferrell (2000) scale marks a substantial contribution to the CSR literature, yet failed to consider environmental impact as a CSR dimension.

Some studies (Alvarado-Herrera et al., 2017; Fatma et al., 2016) developed measurement scales to reflect consumers' perceptions of CSR, however, both studies have limitations. Alvarado-Herrera et al. (2017) employed a convenience sample of university students at an early stage of scale development thus they are not relevant to the practical interest of CSR in a company. While the two scales comprise three domains, namely, economic, environmental, and social, they fail to fully capture the specifics of the hotel industry. Moreover, their scales reflect only consumers' responses to CSR performance.

Table 1 summarizes existing studies of CSR scale development in terms of dimensionality, measurement, sample, setting, and noted weaknesses. Despite these advances in CSR scale development, no CSR study to date uses both multi-dimensional domains and multi-stakeholder samples to develop their scale. The current study addresses this gap by using three groups of hotel CSR stakeholders (academics, hotel managers, and hotel customers) as well as customized and expanded dimensions to develop a new sector-specific CSR measurement scale.

## CSR dimensions and measures

Although the exact nature and scope of CSR continues to be debated, there is some consensus as to its dimensional structure (Maignan & Ferrell, 2001; Smith, 2003; Wong & Kim, 2020). Carroll's (1998, 1999) CSR four-dimension conceptualization, inclusive of economic, legal, ethical, and philanthropic responsibilities, has been widely accepted and become most popular (Webb et al., 2008; Wood, 2010). Firstly, the financial/economic domain is the core requirement of sustaining business through seeking fair profit and attracting investors (Dahlsrud, 2008). The legal domain refers to the responsibility of an organization to comply with laws, regulations, and legal obligations (Ararat, 2008). The ethical domain reflects the responsibility of an organization to

**Table 1.** Summary of existing CSR performance scale development.

| No. | Author(s) | Theme | Dimensions | No. of items | Sample(s) | Research Setting | Weakness |
|---|---|---|---|---|---|---|---|
| 1 | Maignan and Ferrell (2000) | Scale development | 1. Economic 2. Legal 3. Ethical 4. Discretionary | 18 | 330 managers (210-US, 120-France) | Marketing | 1. Ignored environmental aspect 2. Single stakeholder group in sample (managers) |
| 2 | Quazi and O'brien (2000) | Cross-national testing CSR model | 1. Business of business is business 2. CSR is beneficial to business 3. Wider responsibility has its own benefits 4. Social responsibility costs money 5. Responsibility beyond regulation 6. Business must do more 7. Exercise CSR increases societal expectation | 20 | 320 Chief executive officers in food and textile manufacturers (102 - Australia, 218 - Bangladesh) | Food and textile | 1. Ignored environmental aspect 2. Single stakeholder group (Chief executive officers) |
| 3 | Singhapakdi et al. (1996) | Scale development | 1. Social responsibility and profitability 2. Long-term gains 3. Short-term gains | 13 | 442 Professional members of the American Marketing Association | General business | 1. Sub-domain of CSR is too general 2. Single stakeholder group in sample (Professional members) |
| 4 | Martínez et al. (2013) | Scale development | 1. Economy 2. Society 3. Environment | 17 | 1921 students, self-employed, worker, retired, pensioner, unemployed | Hospitality | 1. Ignored legal aspect 2. Unclear cut between ethical and social aspect 3. Single stakeholder group in sample (customers) |
| 5 | Alvarado-Herrera et al. (2017) | Scale development | 1. Social equity 2. Environmental protection 3. Economic development | 22 | 1147 Tourists | Tourism | 1. Ignored legal and ethical aspect 2. Single stakeholder group in sample (tourists) |
| 6 | Fatma et al. (2016) | Scale development | 1. Economic 2. Social 3. Environmental | 20 | 396 hotel guests | Tourism | 1. Ignored legal and ethical aspect 2. Single stakeholder group in sample (hotel guests) |
| 7 | Turker (2009) | Scale development | 1. CSR to society 2. CSR to employee 3. CSR to customers 4. CSR to government | 18 | 269 business professionals | General business | 1. Ignored legal and ethical aspect 2. Single stakeholder group in sample (business professionals) |

(continued)

**Table 1.** Continued.

| No. | Author(s) | Theme | Dimensions | No. of items | Sample(s) | Research Setting | Weakness |
|---|---|---|---|---|---|---|---|
| 8 | Pérez et al. (2013) | Scale development | 1. Customers 2. Shareholders 3. Employees 4. Society 5. General | 22 | 1124 customers | Banking | 1. Single stakeholder group in sample (customers) |
| 9 | Öberseder et al. (2014) | Scale development | 1. Community 2. Employee 3. Shareholder 4. Environmental 5. Societal 6. Customer 7. Supplier | 39 | Study 1 - 483 consumers, Study 2 – 1131 consumers | Manufacturer, fast-moving consumer goods company, and banking | 1. Single stakeholder group in sample (customers) |
| 10 | Skudiene and Auruskeviciene (2012) | Scale development | 1. Internal CSR- employees 2. External CSR – customers 3. External CSR - local communities 4. External CSR - business partners 5. External CSR – | 15 | 274 employees | General business | 1. Ignored legal, ethical and environmental aspect 2. Single stakeholder group in sample (employees) |
| 11 | Davenport (2000) | Scale development | 1. Ethical business behavior 2. Stakeholder commitment 3. Environmental commitment | 20 | 40 experts | General business | 1. Items to measure environmental aspect are too general and simplified (2 items only) 2. Single stakeholder group in sample (experts) |
| 12 | David et al. (2005) | Dual-Process Model of CSR and Purchase Intention | 1. Moral/ethical practices 2. Discretionary practices 3. Relational practices 4. Familiarity with CSR | 11 | 359 students | General business | 1. Ignored legal environmental aspect 2. Single stakeholder group in sample (students) |
| 13 | El Akremi et al. (2018) | Scale development | 1. Community-oriented CSR 2. Natural environment–oriented CSR 3. Employee-oriented CSR 4. Supplier-oriented CSR | 35 | 261 employees | Construction industry | 1. Single stakeholder group in sample (employees) |

(continued)

Table 1. Continued.

| No. | Author(s) | Theme | Dimensions | No. of items | Sample(s) | Research Setting | Weakness |
|---|---|---|---|---|---|---|---|
| | | | 5. Customer-oriented CSR<br>6. Shareholder-oriented CSR | | | | |
| 14 | Lee et al. (2016) | Scale development | 1. Usefulness<br>2. Credibility<br>3. Fairness | 21 | 253 Experts in communication | Media | 1. Specialize in media industry and unable to generalize to other industry<br>2. Single stakeholder group in sample (experts) |
| 15 | Pérez and Del Bosque (2013) | Scale development | 1. Customers<br>2. Shareholders and supervising boards<br>3. Employees<br>4. Society | 20 | 1124 banking service users | Banking | 1. Ignored legal and environmental aspect<br>2. Single stakeholder group in sample (customers) |
| 16 | Chow and Chen (2012) | Scale development | 1. Social development<br>2. Economic development<br>3. Environmental development | 15 | 314 managers | General business | 1. Ignored legal and ethical aspect<br>2. Single stakeholder group in sample (managers) |

exceed legal requirements by respecting the expectations of societal morals and ethical norms (Carroll, 1991). The social/philanthropic domain manifests the responsibility of an organization to support the fine and performing arts, and to partake in community service and volunteerism (Okoye, 2009).

In the context of hotels, Holcomb et al. (2007) performed content analysis on hotels' websites to identify their CSR patterns. The results showed that 80 percent of the hotel companies reported social activities, and classified the hotels' CSR activity patterns into four categories, namely, community, environment, marketplace vision and values, and workforce. De Grosbois (2012) investigated the CSR communication of leading global hotel companies. Hotels' CSR initiatives were categorized into four dimensions, namely, environmental goals, employment quality/ diversity and accessibility, society/community well-being, and economic prosperity. The results correspond to Carroll's (1998, 1999) four dimensions of CSR.

Though most CSR studies stem from Carroll's four dimensional CSR model, there are valid alternatives. David et al. (2005) proposed three domains of CSR - moral, discretionary, and relational aspects - aligning with Carroll's ethical and social/philanthropic domains. The relational domain is new, referring to practices that strive to build and maintain long-term relationships with related stakeholders. However, even though relational is considered to be a key domain of customer relationship management, it is not necessarily a CSR domain. Öberseder et al. (2014) and Turker (2009) proposed CSR dimensions based on stakeholders' positional perceptions, whether community, employee, shareholder, societal, customer, supplier, government, or environmental. Even though this approach has the advantage of covering a broad range of CSR aspects, definitions of the domains are vague and measurement items are unclear.

Alvarado-Herrera et al. (2017) also proposed a three-dimensional CSR model which manifests social equity, environmental protection and economic development. In their study social equity and economic development correspond to social/philanthropic and financial/economic domain, whereas environmental protection was considered to be the most important domain in CSR measurement. In addition, an increasing number of customers are keenly interested in environmental protection and hold positive attitudes toward eco-friendly products (Chan, 2001; Manaktola & Jauhari, 2007; Tsai & Tsai, 2008).

Items to measure environmental CSR performance should be different across hospitality businesses because environmental impacts-causing aspects are different. For example, the airline industry is a culprit of air pollution and the restaurant industry is a source of food waste, whereas the casino industry generates a negative image of a gambling and smoking culture. In addition, most previous studies (Kim et al., 2017; Song et al., 2015; Xu, 2014) have failed to adopt imminent environmental issues because they reflect realities of the 1990s.

As a result, this study will adopt Carroll's four-dimensional CSR concept and reinforce environmental dimensions through diving it into two domains such as room and restaurant area dimension, and other general areas dimension.

## Research methods

### Scale development through Delphi and AHP

This study aims to identify the important indicators of CSR performance in *a-priori* six dimensions and determine the weighted score of each dimension and indicator on the basis of their relative importance. The Delphi technique and AHP are adopted to achieve this objective. First, the Delphi technique is a widely used research method to collect and gather opinion from respondents within their domain of expertise (Hsu & Sandford, 2007). The Delphi technique provides anonymity to selected respondents, thereby minimizing the weakness of conventional means of collecting opinions from group interaction, such as the effect of dominant individuals, group pressure of conformity, and emotional noise from the discussion (Dalkey, 1969). Moreover, the

Delphi technique allows the use of electronic communication, such as e-mail and other smart communication tools, to solicit and exchange information that can help ensure research validity by obtaining heterogeneity within the population without the limitation of geographical distance (Maxwell, 2013). This study used three rounds of Delphi survey instead of four, as an approach to managing the decline in response rate when panel members are asked to stay involved for four-rounds (Keeney et al., 2001). Thus, this study used the first two rounds of the Delphi survey to investigate the essential indicators of hotel CSR performance measurement, and effectively used the third round to combine Delphi survey with AHP analysis to develop a standardized and composite hotel CSR performance measurement index. The main objective of Delphi, to reach response consensus, was met.

The second method adopted is AHP. Each industry has different characteristics and foci for CSR strategies. Current hospitality studies have indicated that the environmental aspect has become the center of attention compared with other dimensions (Kucukusta et al., 2013; Tsai et al., 2012). Relative importance weights should be applied when evaluating various CSR indicators. That is, important indicators should weigh more and less important indicators should weigh less. This measure can differ across industries, and the current study aims to explore the applicable and accurate CSR measures for the hotel industry. In a management decision-making process, Saaty (1987, 1994) developed the AHP model, which is a mathematical technique for multi-criteria decision-making based on pairwise comparisons between measurement criteria designed to obtain relative judgments.

AHP assists management to assess subjective and objective evaluation measures and provide a beneficial mechanism for checking the consistency of evaluation measures. AHP is used to construct an evaluation model and criterion weights and integrates various measures into a single overall score to rank decision alternatives. Moreover, applying this method often results in a multi-level hierarchical structure that simplifies a multiple-criterion problem (Chen, 2006). AHP has also been applied to a wide range of decisions and human judgments in the hospitality industry (Chen, 2006, 2014; Zaman et al., 2016). Tzeng et al. (2002) analyzed AHP in the decision-making process regarding a restaurant location selection. Chen (2006) tested AHP in critical criteria that affect decision-making in a convention site selection. The results showed that AHP is a beneficial tool in evaluating factors and enabling decision-makers to determine the relative importance of each influential factor.

### Scale development process

The literature review of CSR measurement studies has identified several weaknesses. First, the majority of CSR measures reflect general business, and not the hospitality industry. Limited CSR scales characterize the hotel industry. Given that the majority of CSR studies in a hotel context (Martínez & del Bosque, 2013; Su et al., 2015; Wong & Kim, 2020; Xiao et al., 2017; Xu, 2014) adopted CSR measurements developed for general business, the results of these studies generated diverse and incongruent findings, questioning their reliability and validity. Second, several CSR studies in the hotel industry adapted Carroll's (1991) measurement scale, developed over 25 years ago, thereby failing to represent the current situation. In addition, the hotel industry differs from other manufacturing businesses. For example, the hotel industry includes in its range, characteristics of luxurious and conspicuous consumption. Hotel facilities and services are involved with numerous non-environmentally friendly components in their processes of purchasing, cooking, discarding materials, maintaining facilities, preparing guest rooms, gardening, environmental resource management, and employee training and communication. In fact, most operations are related to environmentally sensitive activities, such as managing food and beverage, consuming water, washing linen, stocking amenities for bathrooms, using electricity, plus landscaping and gardening. Thus, hotel customers and employees actively co-create service elements of CSR while other industries may focus solely on the production process, highlighting the need to develop a sector specific CSR measurement scale.

Many conceptualizations of CSR fail to accommodate new industrial or consumer trends, customer preferences, new social paradigms, and advancements of new technologies. Hotel efforts to implement CSR are propelled by current social and industrial requirements and consumer concerns. For example, information or electrical technology development now facilitates the reduction of the amount of paper usage through e-mail, e-brochures, e-presentations, or e-check-in/out processes using tablets instead of distributing paper brochures or signing registration documents. To save on electricity, hotels use an automatic light on–off system, motion/heat sensor, LED light bulbs, and water-cooled and centralized air-conditioning systems. Hotel's CSR programs, such as donations or contributions to community causes, fair trade, eco-friendliness, sustainability, greening, or the sharing economy, reflect today's social needs. Therefore, a new CSR measurement instrument should consider these social and industrial trends.

To address these weaknesses, the current study reinforces the environmental domain by classifying it into two components, namely, (1) environmental (room and restaurant) domain and (2) environmental (other general areas) domain for the following reason. With more than 25 measurement items in the environmental domain, subdividing the domain allows for a more precise capturing of CSR dimensionality. Different departments in a hotel have different CSR policies, in particular, rooms and restaurants and other general areas. To verify this approach, four active CSR researchers and seven hotel managers were invited to a meeting to discuss the dimensionality of hotel CSR. Subsequently, a pre-test was carried out with 40 doctoral students majoring in hospitality and tourism management. As a consequence, this study adopted the six-dimensionality structure to measure hotel CSR as shown in Table 2.

To manifest social and industrial trends, this study included several new indicators from analyzing the content of 8 hotel and tourism sustainability and CSR reports, to derive items such as "extent to which the hotel excludes endangered species from the food menus", "extent to which the hotel recommends responsible drinking to customers," "extent to which the hotel fulfills the reuse/recycle program in guests' rooms," and "extent to which the hotel reduces the paper usage in operation (e.g., mobile check in, electronic invoice)". In total, 69 indicators were derived from 23 journal articles and 8 hotel and tourism sustainability and CSR reports (e.g., Knowles et al., 1999; Kroger, 2018; Maignan, 2001; Maignan et al., 1999; Ricaurte, 2011; Shangri-La Hotels and Resorts, 2016).

The Delphi method aims to achieve a consensus of opinions from experts or professionals regarding a specific topic (Ludwig, 1994). Online panel surveys were conducted to garner responses from the three stakeholder groups - academics, hotel managers, and hotel customers-

Table 2. Definition of the six dimensions of hotel CSR.

| Domain | Definition |
| --- | --- |
| Financial/Economic domain | Responsibility that an organization should fulfill in the financial or economic aspect, such as financial sustainability, operation efficiency, and profitability to enable long-term business success and survival. |
| Legal domain | Responsibility that an organization should fulfill in the legal aspects, such as compliance with laws, regulations, and legal obligations. |
| Ethical domain | Responsibility that an organization should fulfill in the ethical aspects; that is, going beyond the legal aspect, such as respecting norms and meeting the expectation of societal mores and ethical norms. |
| Social/Philanthropic domain | Responsibility that an organization should fulfill in the social/philanthropic aspect, such as providing assistance for the fine and performing arts and participating in community services and volunteerism. |
| Environmental (room and restaurant) | Responsibility that an organization should fulfill in the specific environmental measures in the hotel room and restaurant. |
| Environmental (other general areas) | Responsibility that an organization should fulfill in the specific environmental measures in hotel general areas other than the hotel room and restaurant. |

to develop the CSR scale. To ensure validity, the participants met several requirements in order to participate. Academics were required to have research or teaching experience in CSR or sustainable tourism. Hotel managers were required to have at least three years of managerial work experience in a hotel, and basic CSR knowledge. The panel of hotel customers had stayed at a hotel at least twice within the past 12 months and had a basic understanding of CSR.

### Analytical technique

This study adopted the mean value, median value, and Content validity ratio (CVR) as criteria to eliminate and refine the CSR indicators. If the indicators' mean values in the first survey were below 5.50 or the CVR values were below 0, then they were excluded. CVR indicates content validity by calculating the number of respondents who answered 6.00 or above on a 7-point Likert scale. The CVR value is a linear transformation of a proportional level of agreement on the number of respondents within the panel based on the indicators' appropriateness in computing CSR (Wilson et al., 2012).

Additional indicators were added on the basis of the respondents' responses to the open-ended questions. In the second survey, the indicators were excluded if their mean value or median value was below 5.50. In addition, if the indicators' CVR values were below 0.29 (Lawshe, 1975), then they were excluded.

The reason for demonstrating the different criteria of the CVR values between the first and second surveys is attributed to the fact that the first survey was deemed unable to cover all essential measurement indicators that were extracted as a result of the thorough review of the previous literature (Miller, 2001). The second survey represented the final decision on whether indicators were included or excluded, thereby requiring a high consensus in showcasing the appropriateness of indicators within each domain. Lawshe (1975) determined that when the sample size is at least 40, the minimum CVR value in reaching a consensus is 0.29. By contrast, Wilson et al. (2012) recommended that the minimum CVR value is 0.23 in the case that the number of respondents is 44. To reach a high consensus from the three stakeholder groups, this study adopted a critical CVR value of 0.29, as suggested by Lawshe (1975).

To ensure the consensus of the Delphi technique in this study, changes in the coefficient of variation (COV) were used to test the stability of the indicators. Table 3 shows that the average COV was 0.20 in the first survey, which decreased to 0.19 in the second survey, thereby indicating the acceptable consensus of the current study (Heiko, 2012; Heiko & Darkow, 2010).

After computing the CVR and COV values of the indicators in each domain in the first and second surveys, the third survey was conducted to determine the level of the relative importance of each domain and indicators through pairwise comparative judgments. This process enabled the panelists to assign relative priority when comparing two indicators or domains (Deng et al., 2002). The respondents could express their preference between every two indicators or domains and translate these preferences into numerical ratings of 1, 3, 5, 7, and 9 and 2, 4, 6, and 8 as intermediate values. Deng et al. (2002) explained that the comparative priority from the pairwise comparisons may lead to a certain degree of inconsistency. Therefore, the inconsistency ratio (IR) was used to check the consistency and reliability of the panelists' judgments. Furthermore, the consistency index (CI) and ratio should be considered. The CI can be calculated to measure each participant's consistency in the pairwise comparison.

Saaty (1980) developed the random index (RI), which refers to a constant value that corresponds to the mean random CI value according to $n$ (see Table 4). A Consistency ratios (CR) value between 0.10 and 0.20 is acceptable, whereas that below 0.10 represents a high consistency of responses (Saaty, 1980). If the respondents are non-experts, then the threshold value of CR is flexible at 0.20 or below (Ho et al., 2005). After analyzing the CR value, the relative weights

**Table 3.** Level of agreement of the Delphi studies (Coefficient of Variation).

| Domain | Indicator | Round 1 COV | Round 2 COV | Change COV |
|---|---|---|---|---|
| Financial / Economic domain | Length of the hotel's survival and long-term success | 0.27 | 0.28 | ↑ |
| | Proportion of hires that are local residents | 0.18 | 0.24 | ↑ |
| | Degree of the hotel's honesty in informing its shareholders of its economic situation | 0.21 | 0.26 | − |
| | Extent to which the hotel makes continuous improvements in product quality | 0.22 | 0.26 | ↑ |
| | Degree to which the hotel monitors employees' productivity | 0.28 | N/A | N/A |
| | Use of customer satisfaction as an indicator of the hotel's business performance | 0.21 | 0.22 | ↓ |
| | The hotel's level of improvement in financial performance | 0.25 | N/A | N/A |
| | Extent to which the hotel strictly monitors whether operating costs are properly spent | 0.24 | N/A | N/A |
| | Growth of the occupancy rate | 0.27 | N/A | N/A |
| | Growth rate of RevPAR | 0.28 | N/A | N/A |
| | Growth rate of return on assets | 0.23 | N/A | N/A |
| | Extent to which a hotel gains the high possible profit | 0.33 | N/A | N/A |
| | Extent to which the hotel uses local materials/products (e.g., food, flower, furniture) | N/A | 0.23 | N/A |
| Environmental domain | Extent to which the hotel donates leftover food to the community | 0.22 | 0.16 | ↓ |
| | Extent to which the hotel utilizes food waste (e.g., conversion to fertilizer) | 0.15 | 0.15 | − |
| | Extent to which the hotel excludes endangered species from the food menu (e.g., shark's fin soup) | 0.19 | 0.18 | ↓ |
| | Extent to which the hotel ensures food safety and hygiene | 0.20 | 0.21 | ↑ |
| | Extent to which the hotel recommends responsible drinking to customers | 0.30 | N/A | N/A |
| | Extent to which the hotel provides nutritional information on its menu | 0.22 | N/A | N/A |
| | Extent to which the hotel reduces water usage per available room | 0.21 | 0.19 | ↓ |
| | Extent to which the hotel reduces energy usage per available room | 0.20 | 0.18 | ↓ |
| | Extent to which the hotel reduces greenhouse gas emission per available room | 0.18 | 0.17 | ↓ |
| | Extent to which the hotel reduces solid waste per available room | 0.15 | 0.19 | ↑ |
| | Extent to which the hotel reduces bathroom amenities per available room (e.g., disposable shampoo, soap) | 0.22 | 0.25 | ↑ |
| | Extent to which the hotel reduces surplus towels per available room | 0.23 | N/A | N/A |
| | Extent to which the hotel implements an electronic management system in guests' rooms (e.g., motion sensors) | 0.24 | N/A | N/A |
| | Extent to which the hotel fulfills the reuse/recycle program in guests' rooms (e.g., reuse/recycle card reminder) | 0.14 | 0.17 | ↑ |

*(continued)*

**Table 3.** Continued.

| Domain | Indicator | Round 1 COV | Round 2 COV | Change COV |
|---|---|---|---|---|
| Environmental policy domain | Extent to which the hotel reduces natural resource consumption. | 0.22 | 0.25 | ← |
|  | Degree to which the hotel communicates with customers regarding its environmental practices | 0.17 | 0.17 | – |
|  | Extent to which the hotel uses renewable energy in a productive process that is environmentally friendly | 0.15 | 0.15 | – |
|  | Degree of the hotel's interest in protecting the natural environment | 0.16 | 0.18 | ←→ |
|  | Degree to which the hotel has a positive predisposition to use, purchase, or produce environmentally friendly goods | 0.18 | 0.16 | ←→ |
|  | Degree of a hotel customer's or employee's satisfaction with environmental effort | 0.19 | 0.21 | ← |
|  | Effort that that the hotel spends on environmental certification | 0.19 | 0.24 | ←← |
|  | Effort that the hotel spends on annual environmental audit | 0.16 | 0.21 | ←← |
|  | The amount of the hotel's average expenditure on and investment in environmental aspects | 0.19 | 0.21 | ← |
|  | Extent to which the hotel reduces paper usage in operation (e.g., mobile check in, electronic invoice) | 0.21 | 0.16 | → |
|  | Extent to which the hotel supports local and sustainable suppliers | 0.18 | 0.17 | →→ |
|  | Extent to which the hotel uses environmentally friendly equipment (e.g., LED light bulbs) | 0.17 | 0.13 | →→ |
| Legal Domain | Extent to which the hotel provides green training to employees | N/A | 0.15 | N/A |
|  | Extent to which the hotel's employees understand environmental law | 0.20 | 0.19 | →→ |
|  | Extent to which the hotel ensures that employees can fulfill their duty within the standards defined by local law | 0.15 | 0.21 | ← |
|  | Extent to which the hotel follows its contractual obligations | 0.17 | 0.16 | → |
|  | Extent to which the hotel avoids cheating on the law to improve performance | 0.21 | N/A | N/A |
|  | Extent to which the hotel complies with the principles defined by the business practice | 0.20 | 0.20 | – |
|  | Extent to which the hotel encourages workforce diversity (e.g., age, gender, race) | 0.17 | 0.18 | ← |
|  | Extent to which the hotel complies with all laws regulating hiring and employee benefits | 0.16 | 0.18 | ← |
|  | Extent to which the hotel meets legal standards for the product | 0.19 | 0.18 | → |
|  |  | 0.15 | 0.13 | → |

*(continued)*

**Table 3.** Continued.

| Domain | Indicator | Round 1 COV | Round 2 COV | Change COV |
|---|---|---|---|---|
| | Degree to which the hotel effectively implements internal policies to prevent discrimination in employees' compensation and promotion process | 0.20 | N/A | N/A |
| | Degree of the hotel's honesty in fulfilling its contractual obligations | | | |
| Ethical Domain | Extent to which the hotel does not compromise ethical standards to achieve corporate goals | 0.21 | 0.25 | ↑ |
| | Extent to which the hotel allows ethical problems that can negatively affect financial/economic performance | 0.25 | 0.19 | → |
| | Extent to which the hotel offers equal opportunities for promotion and hiring | 0.15 | 0.18 | ↑ |
| | Extent to which the hotel treats its employees fairly (without discrimination and abuse regardless of gender, race, origin, or religion) | 0.13 | 0.22 | ↑ |
| | Extent to which the hotel prioritizes ethical principles over economic performance | 0.21 | 0.21 | – |
| | Extent to which the hotel is committed to well-defined ethics and principles | 0.19 | 0.20 | ↑ |
| | Extent to which the hotel effectively implements confidential means for employees to report misconduct at work (e.g., stealing, sexual harassment) | 0.19 | 0.15 | → |
| | Extent to which the hotel provides accurate information to customers | 0.15 | 0.13 | → |
| | Extent to which the hotel follows a comprehensive code of conduct | 0.14 | 0.17 | ↑ |
| | Extent to which the hotel is recognized as a trustworthy company | 0.17 | 0.22 | ↑ |
| | Extent to which the hotel considers coworkers and business partners as an integral part of the employee evaluation process | 0.23 | 0.22 | → |
| | Level of effectiveness of procedures to respond to customer complaints | 0.23 | 0.14 | → |
| Social/Philanthropic Domain | Proportion of hotel's budget allocated for donations and social work to benefit poor people | 0.22 | 0.21 | → |
| | Extent to which the hotel allocates resources for philanthropic activities | 0.20 | 0.16 | → |
| | Effort that the hotel makes in society beyond profit generation | 0.25 | N/A | N/A |
| | Extent to which the hotel is committed to improving the welfare of the community | 0.18 | 0.15 | → |
| | Extent to which the hotel participates in managing public affairs | 0.24 | N/A | N/A |
| | Extent to which the hotel helps to solve social problems | 0.31 | N/A | N/A |
| | Extent to which the hotel participates in community services and volunteerism | 0.21 | 0.16 | → |
| | Extent to which the hotel actively sponsors or finances local and social events (e.g., sport, music …) | 0.26 | 0.23 | → |
| | Overall | 0.20 | 0.19 | → |

**Table 4.** RI values for the different values of n.

| n | 1 | 2 | 3 | 4 | 5 | 6 | 7 | 8 | 9 | 10 | 11 | 12 | 13 | 14 | 15 |
|---|---|---|---|---|---|---|---|---|---|----|----|----|----|----|----|
| RI | 0.00 | 0.00 | 0.58 | 0.90 | 1.12 | 1.24 | 1.32 | 1.41 | 1.45 | 1.49 | 1.51 | 1.48 | 1.56 | 1.57 | 1.59 |

of each indicator and domain were integrated thereafter to develop the final weighted score to measure CSR in the hotel industry.

## Findings

### Profile of respondents

The profiles of three groups of panelists in the three rounds are described in Table 5.

### First Delphi survey

From the results of the first survey, 17 out of the 69 indicators were excluded from the scale for the following reasons. First, the mean or median values of each attribute were below 5.50. Second, the CVR values of these attributes were below 0. Accordingly, a CVR ratio below 0 means that less than half of the respondents perceived the scale as an appropriate indicator for measuring CSR (Lawshe, 1975). Third, the attributes were not entirely related to the CSR measurement. For example, the "extent to which a hotel gains the highest possible profit" is an indicator of financial performance but not necessarily a CSR measurement. Lastly, some attributes overlapped. For example, "The hotel allows the decrease of profitability because of its compliance with ethical standards" overlapped with the indicator "extent to which the hotel is committed to well-defined ethics and principles."

The panelists' comments suggested that some attributes were not understood or explicitly presented. Thus, adjustments were necessary to enhance the understanding of the attributes by including practical examples. After the content analysis of respondents' comments, conducted by four CSR scholars, two additional indicators were added to the scale (e.g., "extent to which the hotel uses local materials/products" and "extent to which the hotel provides ethical studies and best practices to employees"). Consequently, 54 attributes were measured in the second survey.

### Second Delphi survey

To ensure the validity of the Delphi method, only respondents who participated in the first survey were invited to join the second survey. Indicators were removed from an initial pool of 54 indicators if they failed to meet the following criteria: (1) CVR value below 0.29 (Lawshe, 1975), (2) mean value of below 5.50, and (3) median value below 5.50. From the initial pool of 54 attributes, 14 were eliminated, and 40 attributes remained. To identify the internal consistency of each indicator in each domain, a reliability test was conducted with Cronbach's alpha values ranging from 0.75 to 0.89. The Cronbach's alpha value of all 40 indicators was 0.95. Thus, this scale achieved high internal consistency of indicators within each domain (Nunnally, 1978). Moreover, the average CVR value increased by 17% in the second survey, thereby indicating that a higher level of consensus was achieved in the second round than in the first round.

### Third Delphi survey (AHP analysis)

Again, only the respondents who answered in the first two surveys were invited to participate in the third round survey. Figure 1 shows a hierarchical decision-making graph. Three hierarchical levels are proposed in this study, namely, organizational level, individual level, and indicators.

**Table 5.** Profile of the respondents.

| Demographic information | | First survey (n = 50) | | | Second survey (n = 44) | | | Third survey (n = 27) | | |
|---|---|---|---|---|---|---|---|---|---|---|
| | | Academicians | Hotel managers | Hotel customers | Academicians | Hotel managers | Hotel customers | Academicians | Hotel managers | Hotel customers |
| n | | 19 | 15 | 16 | 15 | 14 | 15 | 11 | 7 | 9 |
| Gender | Female | 36.8% | 53.3% | 43.7% | 40.0% | 57.1% | 53.3% | 54.5% | 42.8% | 55.6% |
| | Male | 63.2% | 46.7% | 56.3% | 60.0% | 42.9% | 46.7% | 45.5% | 57.2% | 44.4% |
| Age | 20–29 | / | / | 31.5% | / | / | 33.3% | / | / | 44.4% |
| | 30–39 | / | / | 36.8% | / | / | 40.0% | / | / | 55.6% |
| | 40–49 | / | / | 10.5% | / | / | 20.0% | / | / | / |
| | 50–59 | / | / | 5.3% | / | / | 6.7% | / | / | / |
| Origin | Asia | 63.2% | / | 31.6% | 73.3% | / | 60.0% | 72.7% | / | 55.6% |
| | North America | 26.3% | / | 15.8% | 20.0% | / | 20.0% | 18.2% | / | 11.1% |
| | Africa | 10.5% | / | 10.5% | / | / | 6.7% | / | / | / |
| | Europe and Oceania | 5.3% | / | 36.8% | 6.7% | / | 13.3% | 9.1% | / | 33.3% |
| Position | Professor | 26.3% | / | / | 20.0% | / | / | 18.2% | / | / |
| | Associate Professor | 36.8% | / | / | 46.7% | / | / | 45.5% | / | / |
| | Assistant Professor | 36.8% | / | / | 33.3% | / | / | 36.4% | / | / |
| Department | Front of House | / | 60.0% | / | / | 42.8% | / | / | 42.8% | / |
| | Back of House | / | 40.0% | / | / | 57.2% | / | / | 57.2% | / |
| Working experience | < 10 years | / | 33.3% | / | / | 42.9% | / | / | 28.6% | / |
| | 10–20 years | / | 46.7% | / | / | 42.9% | / | / | 42.8% | / |
| | > 20 years | / | 13.3% | / | / | 14.2% | / | / | 28.6% | / |

Organizational Level

CSR measurement scale

Individual Level

| Financial/ Economic domain | Legal domain | Ethical domain | Social/ Philanthropic domain | Environmental (Room and Restaurant) | Environmental (other general areas) |

| Index score 11.6% | Index score 23.03% | Index score 21.90% | Index score 17.42% | Index score 12.90% | Index score 13.15% |

Indicators

FIN_1: 2.18%
FIN_2: 3.82%
FIN_3: 3.71%
FIN_4: 1.89%

LEG_1: 4.31%
LEG_2: 3.04%
LEG_3: 3.35%
LEG_4: 4.10%
LEG_5: 3.73%
LEG_6: 4.51%

ETH_1: 2.44%
ETH_2: 2.53%
ETH_3: 2.84%
ETH_4: 2.57%
ETH_5: 2.49%
ETH_6: 2.51%
ETH_7: 2.08%
ETH_8: 2.35%
ETH_9: 2.11%

SOC_1: 5.49%
SOC_2: 5.47%
SOC_3: 6.46%

ENV_1: 0.85%
ENV_2: 0.96%
ENV_3: 1.42%
ENV_4: 3.42%
ENV_5: 1.27%
ENV_6: 1.23%
ENV_7: 1.24%
ENV_8: 1.18%
ENV_9: 1.34%

ENVPO_1: 1.78%
ENVPO_2: 1.03%
ENVPO_3: 1.31%
ENVPO_4: 1.54%
ENVPO_5: 1.48%
ENVPO_6: 1.46%
ENVPO_7: 1.08%
ENVPO_8: 1.43%
ENVPO_9: 2.04%

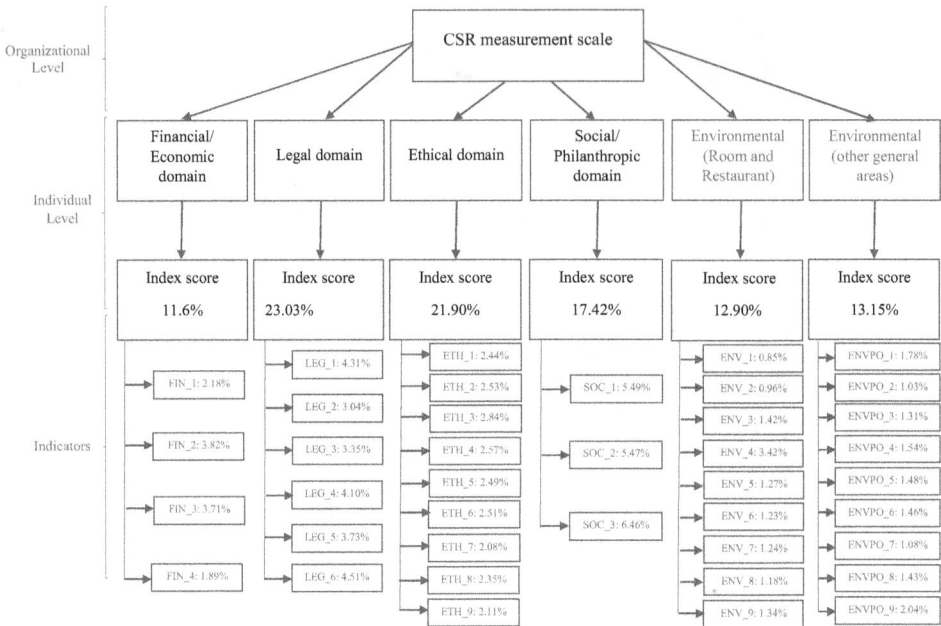

**Figure 1.** Results of the AHP hierarchy in the hotel CSR measurement scale.

Saaty (1980) suggested that the maximum number of analytic hierarchy levels is nine. Thus, this study's three-level hierarchical construct, with six second-order dimensions, is acceptable. In AHP, the appropriate solution is provided by two modes of synthesis, namely, distributive and ideal. The distributive mode normalizes the alternative score and creates dependency for rank reversal, whereas the ideal mode preserves ranks and explores relative importance by the score of the best alternative under each domain (Millet & Saaty, 2000). This study adopted the second model to assess the importance level of the domains and indicators.

Figure 1 presents the relative importance of the indicators at the individual level, which shows the financial/economic (11.60%), legal (23.03%), ethical (21.90%), social/philanthropic (17.42%), environmental (room and restaurant) (12.90%), and environmental (other general areas) (13.15%) domains. An overall CR value of 0.103 (Table 6) indicates that the responses have an acceptable consistency level. Table 6 provides details of the relative importance or weight among domains and within each domain. The responses were acceptably consistent with CR values in each domain ranging from 0.08 to 0.15. Table 6 shows the results of the relative importance of the indicators and their respective rankings.

Table 7 shows the comparison of the responses among three groups of stakeholders and overall respondents regarding the relative importance of the six major domains. Although the CVR and CR values in this study show the level of consistency in the responses of these cohorts, the differences in perceptions among stakeholders highlights some interesting incongruencies in their responses (AlWaer et al., 2008; Davis, 2014; Renfors, 2018).

The responses of academics regarding the relative importance of the major domains in measuring CSR in the hotel industry are most similar to the overall averages for the total sample. By contrast, the responses of hotel managers and customers differ from the response of academics, and from each other. This finding is attributed to the heterogeneous perspective toward CSR practices from various stakeholder groups (Chang et al., 2014). Hotel managers rank legal factors as most important, and have the lowest ranking for social factors in comparison to other stakeholders, whereas understandably, customers rank room and restaurant environment as being relatively more important and legal as less important than do other stakeholders. Thus, the current study reconfirms that various stakeholders can exhibit different perspectives of some hotel

**Table 6.** Analysis of the relative importance or weight.

| Domain | Items | Relative importance (Relative weight) | Rank within a domain | CVR | CR |
|---|---|---|---|---|---|
| Financial/Economic domain (11.60%) | Proportion of hiring local residents | 2.18% | 3 | 0.32 | 0.15 |
| | Degree of the hotel's honesty in informing its shareholders of its economic situation | 3.82% | 1 | 0.32 | |
| | Use of customer and employee satisfaction as an indicator of the hotel's business performance | 3.71% | 2 | 0.32 | |
| | Extent to which the hotel uses local materials/products (e.g., food, flower, furniture) | 1.89% | 4 | 0.36 | |
| Environmental domain (12.9%) | Extent to which the hotel donates leftover food to the community | 0.85% | 9 | 0.59 | 0.13 |
| | Extent to which the hotel utilizes food waste (e.g., conversion to fertilizer) | 0.96% | 8 | 0.68 | |
| | Extent to which the hotel excludes endangered species from the food menu (e.g., shark's fin soup) | 1.42% | 2 | 0.41 | |
| | Extent to which the hotel ensures food safety and hygiene | 3.42% | 1 | 0.32 | |
| | Extent to which the hotel reduces water usage per available room | 1.27% | 4 | 0.55 | |
| | Extent to which the hotel reduces energy usage per available room | 1.23% | 6 | 0.55 | |
| | Extent to which the hotel reduces greenhouse gas emission per available room | 1.24% | 5 | 0.55 | |
| | Extent to which the hotel reduces solid waste per available room | 1.18% | 7 | 0.41 | |
| | Extent to which the hotel fulfills the reuse/recycle program in guests' rooms (e.g., linen/towel reuse/recycle card reminder) | 1.34% | 3 | 0.50 | |
| Environmental policy domain (13.15%) | Extent to which the hotel reduces natural resource consumption. | 1.78% | 2 | 0.32 | 0.12 |
| | Degree to which the hotel communicates with customers regarding its environmental practices | 1.03% | 9 | 0.55 | |
| | Extent to which the hotel uses renewable energy in a productive process that is environmentally friendly | 1.31% | 7 | 0.32 | |
| | Degree of the hotel's effort in protecting the natural environment | 1.54% | 3 | 0.45 | |
| | Degree to which the hotel has use, purchase, or produce environmentally friendly goods | 1.48% | 4 | 0.50 | |
| | Extent to which the hotel reduces paper usage in operation (e.g., mobile check in, electronic invoice) | 1.46% | 5 | 0.32 | |
| | Extent to which the hotel supports local and sustainable suppliers | 1.08% | 8 | 0.45 | |
| | Extent to which the hotel uses environmentally friendly equipment (e.g., LED light bulbs) | 1.43% | 6 | 0.64 | |
| | Extent to which the hotel provides green training to employees | 2.04% | 1 | 0.50 | |
| Legal Domain (23.03%) | Extent to which the hotel ensures that employees can fulfill their duty within the standards defined by law | 4.31% | 2 | 0.45 | 0.07 |
| | Extent to which the hotel follows its contractual obligations | 3.04% | 6 | 0.32 | |
| | Extent to which the hotel encourages workforce diversity (e.g., age, gender, race) | 3.35% | 5 | 0.45 | |
| | Extent to which the hotel complies with all laws regulating hiring and employee benefits | 4.10% | 3 | 0.50 | |
| | Extent to which the hotel meets legal standards for the product | 3.73% | 4 | 0.45 | |
| | Degree to which the hotel effectively implements internal policies to prevent discrimination in employees' compensation and promotion process | 4.51% | 1 | 0.59 | |

(continued)

**Table 6.** Continued.

| Domain | Items | Relative importance (Relative weight) | Rank within a domain | CVR | CR |
|---|---|---|---|---|---|
| Ethical domain (21.9%) | Extent to which the hotel does not compromise ethical standards to achieve corporate goals | 2.44% | 6 | 0.36 | 0.09 |
| | Extent to which the hotel offers equal opportunities for promotion and hiring | 2.53% | 3 | 0.59 | |
| | Extent to which the hotel treats its employees fairly (without discrimination and abuse regardless of gender, race, origin, religion, disability and sexual orientation) | 2.84% | 1 | 0.68 | |
| | Extent to which the hotel is committed to well-defined ethics and principles | 2.57% | 2 | 0.50 | |
| | Extent to which the hotel effectively implements confidential means for employees to report misconduct at work (e.g., stealing, sexual harassment) | 2.49% | 5 | 0.50 | |
| | Extent to which the hotel provides accurate information to customers | 2.51% | 4 | 0.50 | |
| | Extent to which the hotel follows a comprehensive code of conduct | 2.08% | 9 | 0.64 | |
| | Extent to which the hotel is a trustworthy company | 2.35% | 7 | 0.41 | |
| | Extent to which the hotel provides ethical studies and best practices to employees | 2.11% | 8 | 0.45 | |
| Social Domain (17.42%) | Extent to which the hotel allocates resources for philanthropic activities | 5.49% | 2 | 0.32 | 0.08 |
| | Extent to which the hotel is committed to improving the welfare of the community | 5.47% | 3 | 0.41 | |
| | Extent to which the hotel participates in community services and volunteerism | 6.46% | 1 | 0.50 | |
| | Overall (Mean) | | | | 0.103 |

Table 7. Comparison of the responses among three cohorts.

| | Overall (n = 27) | | Academics (n = 11) | | Hotel Managers (n = 7) | | Hotel Customers (n = 9) | |
|---|---|---|---|---|---|---|---|---|
| Rank | Major domains | Relative importance | Major domains | Relative importance | Major domains | Relative importance | Major domains | Relative importance |
| 1 | Legal | 23.03% | Legal | 21.99% | Legal | 30.60% | Ethical | 21.85% |
| 2 | Ethical | 21.90% | Ethical | 20.72% | Ethical | 23.82% | Social | 19.15% |
| 3 | Social | 17.42% | Social | 19.81% | Financial/Economic | 12.07% | Legal | 18.41% |
| 4 | Environmental policy | 13.15% | Environmental policy | 14.40% | Social | 11.44% | Environmental | 14.97% |
| 5 | Environmental | 12.90% | Environmental | 12.30% | Environmental | 11.18% | Environmental policy | 13.37% |
| 6 | Financial/Economic | 11.60% | Financial/Economic | 10.77% | Environmental policy | 10.89% | Financial/Economic | 12.24% |

CSR policies, and share others. For example, hotel managers' perceived financial/economic domain is considered the third most important given that financial performance can directly affect managers' benefits, such as year-end bonuses and salary increases. By contrast, academics and hotel customers rated financial/economic domain as the least important because both groups are not directly impacted by hotel financial performance. However, the three stakeholder groups show a similar pattern in their perceived importance level of other domains. Thus, the result reached an acceptable consistency in terms of the panelists' responses. In summary, the importance of the three traditional CSR domains (i.e., legal, ethical, and social/philanthropic) was confirmed, while two new environmental domains were observed to be indispensable measurements of CSR in the hotel industry, notably ranking higher than the financial/economic domain overall.

## Discussion and implications

Results of this study reveal meaningful implications. First, the methodological approach to validate a hotel CSR scale through three rounds of a Delphi survey and AHP method is notable. The domains and indicators represent the unique characteristics of the hotel industry unlike any other CSR scale. CSR scholars in the hospitality field can adopt this standardized hotel CSR performance measurement instrument. Second, this study demonstrated that the various CSR domains have different relative importance weights when evaluating various CSR indicators. This significant finding reveals that important indicators or domains should weight more while less important indicators or domains should weigh less. With the examination of CSR effects so prevalent in the hospitality industry, future research should consider the relative importance of CSR effects and not assume that all domains have equal contribution.

Third, this study determined that the legal domain was perceived as the most important of the six domains in evaluating hotel CSR performance. Thus, indicators related to legal issues in evaluating hotel CSR performance should be included in the scale. This finding is slightly contradictory of Maignan's (2001) study, which indicated that American consumers highly value corporate economic responsibility. However, perceptions of CSR are likely to have changed over the last decade because of the increasing exposure of CSR practices through advanced technology and social media. Xiao et al. (2017) analyzed the ranking of consumers' perceived importance of Carroll's (1991) four CSR dimensions in the USA. Their results revealed that the legal domain was the most important, followed by ethical and social, whereas the economic domain was the least important. The results of this study supported Carroll's dimensional structure because the legal, ethical, and social/philanthropic domains were revealed as the most important domains, followed by the environment (room and restaurant; and, other general areas), and financial/economic domains. Legal CSR was perceived as a "must do" practice, whereas ethical CSR was perceived as the second foundation that goes beyond legal CSR. The social/philanthropic and the two environmental domains were perceived as additional practices that a hotel should opt to implement. The financial/economic domain was perceived as the least important domain, reflecting current sentiments that increasingly lean toward meaningful societal goals beyond profit maximization. Although this domain was the least important, it remains indispensable to the measurement of hotel CSR as a fundamental element of business existence.

Fourth, this study emphasized that commitment to environmental issues has become one of the most significant criteria when measuring CSR in the hotel industry. Two additional environmental domains that specify different areas provide a comprehensive view of the CSR effort within the hotel industry (Kucukusta et al., 2013; Tsai et al., 2012; Xiao et al., 2017). However, these domains were ranked as either fourth or fifth out of the six dimensions. This finding contradicts previous studies which have indicated that commitment to environmental issues is a top concern in measuring hotel CSR performance (Kucukusta et al., 2013; Tsai et al., 2012).

Importantly, both environmental domains surpass the financial/economic domain and in this regard are clearly important domains in CSR measurement.

Fifth, several common indicators in the financial/economic domain that were adopted in previous studies (e.g., Maignan, 2001; Maignan & Ferrell, 2001; Martínez & del Bosque, 2013) were removed from the current scale because these indicators were deemed unrelated to CSR. For example, "level of obtaining the greatest possible profits" did not contribute to CSR because financial/economic CSR does not require profits to be maximized. Instead, it must provide services or products that meet social expectations and needs at a reasonable price. In the results, four indicators remained within the financial/economic domain. The most important indicator was "degree of the hotel's honesty in informing its shareholders of its economic situation," which is consistent with the results of Mercer and Oskamp (2003), who emphasized honest communication to different stakeholders as a significant indicator of CSR. However, the second most important indicator contradicted the result of a previous study. Maignan et al. (1999) indicated that customer satisfaction as an indicator for measuring business performance is a non-significant indicator for measuring economic CSR. Nevertheless, "use of customer and employee satisfaction as an indicator of the hotel's business performance" was perceived as the second most important indicator. Given that the business and social environments have changed in the past 20 years, people are increasingly aware of customer and employee satisfaction rather than being concerned only with financial performance.

Sixth, the "extent to which the hotel ensures food safety and hygiene" was predictably perceived as the most important among the nine indicators in the environment domain. Food safety and hygiene help to avoid potentially severe health hazards, which is the first rule in the food and beverage industry. However, the "extent to which the hotel excludes endangered species from food menus (e.g., shark's fin soup)" was perceived as the second most important indicator in the environment domain. People are becoming aware of the importance of biodiversity because of the enhanced promotion of protecting endangered animals. Biological life on Earth is highly interdependent and humans are responsible for protecting endangered species and ensuring biodiversity.

Seventh, the "extent to which the hotel provides green training to employees" was revealed to be of utmost importance among the indicators in the hotel's environmental (other general areas) domain. This finding indicated that "soft skill" is a vital criterion in evaluating hotel CSR effort. Although the majority of previous studies (Costa & Menichini, 2013; De Grosbois, 2012; Manaktola & Jauhari, 2007; Martínez & del Bosque, 2013) have only considered "hard skill" when measuring CSR, other studies have analyzed the effects of employees' green training and supported the notion that such training improves the performance of green strategies and management (Teixeira et al., 2012, 2016). The exclusion of green training from the scales for assessing hotel CSR implies that hotels miss out on insights into their actual CSR performance.

Eighth, the "degree to which the hotel effectively implements internal policies to prevent discrimination in employees' compensation and promotion process," "extent to which the hotel ensures that employees can fulfill their duty within the standards defined by law," and "extent to which the hotel complies with all laws regulating hiring and employee benefits" are perceived as the most important indicators in the legal domain. Employees are key stakeholders of the legal CSR in the hotel industry (Chen & Hung-Baesecke, 2014), thereby justifying why indicators related to employee benefits and freedom from discrimination within legal requirements are perceived as the most important.

Ninth, indicators in the ethical domain are evenly distributed in the index. Similar to the legal domain, treating employees fairly and preventing discrimination are the most important indicators. This result reinforces the idea that employees are the most important stakeholders in legal and ethical CSR efforts. That is, the exclusion of employee consideration will lead to poor performance regarding the hotel's CSR effort.

Tenth, several indicators that were widely used as social CSR indicators in previous studies (Maignan, 2001; Maignan & Ferrell, 2001; Singh & del Bosque, 2008) were excluded, suggesting that, disappointingly, the hotel industry may not have made significant strides in community support efforts as other sectors. For example, the "extent to which the hotel helps to solve social problems" and the "proportion of hotel's budget allocated for donations and social work to benefit poor people" were dropped. Respondents may feel that the government is more accountable for handling these social issues than the hotel industry. The "effort that the hotel makes for philanthropic society beyond profit generation" was also excluded, however, the social responsibility of profit-making organizations to contribute to philanthropic endeavors, as times change, this perspective may too.

Eleventh, the results of this study add value to current scales by showing the specificity of CSR practices in the hotel industry. For example, the Sustainability Accounting Standards Board (SASB) developed a Materiality Map for 79 industries that includes the hotel industry. It shows that the hotel and lodging industry considers energy management, water and wastewater management, ecological impacts, labor practices and climate change as material issues (SASB, 2018). However, this study reveals the continued importance of legal and ethical issues as ranked by hotel CSR stakeholders, in addition to environmental domains that correspond to the materiality map but are measured at a more operational level (e.g. rooms and restaurants).

Lastly, the responses of academics, hotel managers, and hotel customers toward hotel CSR performance varied. This finding indicates that the respondents from different stakeholder groups are likely to produce different results. The results are sensitive to the diverse characteristics of the respondents, and their different levels of interest and benefits regarding hotel CSR efforts. Significantly, this study is the first to incorporate both multi-dimensions and multi-stakeholders' (employees, customers, and academics) opinions to develop and standardize a CSR scale specifically for the hotel industry by achieving consensus through three rounds of Delphi surveys. This method overcomes the weakness of past studies, which were limited to one stakeholder group as the research sample in their CSR scale development (Pérez & Del Bosque, 2013; Singhapakdi et al., 1996).

## Conclusion, limitations and future study

This study contributes to the understanding of essential indicators of hotel CSR performance measurement and develops a standardized and composite hotel CSR performance measurement index. Three traditional hotel CSR domains (legal, ethical, and social/philanthropic) are primary contributors to hotel CSR performance, followed by two new environmental domains (room and restaurant, and other general areas), and financial/economic domains as secondary contributors. Although the CVR and CR values show high levels of consistency in the responses, analysis by stakeholder group highlight some incongruences in perceptions, reflecting different levels of perceived interests and benefits.

As this study is an initial and creative endeavor, it is vulnerable to limitations. First, requiring respondents to participate in all rounds of the Delphi–AHP surveys was practically a challenge because longitudinal surveys result in lower participation. Thus, a future study is recommended to employ a semi-structured interview method despite its prolonged research duration. Second, this study is the first to combine the perspectives of various stakeholders in a CSR measurement scale for the hotel industry. Hence, future research should test this scale in different study contexts, such as countries or hotel brands, to enhance generalizability. Nevertheless, this new CSR scale for the hotel industry represents a timely and important step forward for both tourism research and practice.

## Disclosure statement

No potential conflict of interest was reported by the author(s).

## ORCID

*Antony King Fung Wong* (iD) http://orcid.org/0000-0001-5462-5397
*Seongseop (Sam) Kim* (iD) http://orcid.org/0000-0002-9213-6540

## References

Abbott, W. F., & Monsen, R. J. (1979). On the measurement of corporate social responsibility: Self-reported disclosures as a method of measuring corporate social involvement. *Academy of Management Journal, 22*(3), 501–515. https://doi.org/10.5465/255740

Alvarado-Herrera, A., Bigne, E., Aldas-Manzano, J., & Curras-Perez, R. (2017). A scale for measuring consumer perceptions of corporate social responsibility following the sustainable development paradigm. *Journal of Business Ethics, 140*(2), 243–262. https://doi.org/10.1007/s10551-015-2654-9

AlWaer, H., Sibley, M., & Lewis, J. (2008). Different stakeholder perceptions of sustainability assessment. *Architectural Science Review, 51*(1), 48–59. https://doi.org/10.3763/asre.2008.5107

Ararat, M. (2008). A development perspective for "corporate social responsibility": Case of Turkey. *Corporate Governance: The International Journal of Business in Society, 8*(3), 271–285. https://doi.org/10.1108/14720700810879169

Aupperle, K. E. (1984). An empirical measure of corporate social orientation. *Research in Corporate Social Performance and Policy, 6,* 27–54.

Baucus, M. S., & Baucus, D. A. (1997). Paying the piper: An empirical examination of longer-term financial consequences of illegal corporate behavior. *Academy of Management Journal, 40*(1), 129–151. https://doi.org/10.2307/257023

Carroll, A. B. (1979). A three-dimensional conceptual model of corporate performance. *Academy of Management Review, 4*(4), 497–505. https://doi.org/10.5465/amr.1979.4498296

Carroll, A. B. (1991). The pyramid of corporate social responsibility: Toward the moral management of organizational stakeholders. *Business Horizons, 34*(4), 39–48. https://doi.org/10.1016/0007-6813(91)90005-G

Carroll, A. B. (1998). The four faces of corporate citizenship. *Business and Society Review, 100–101*(1), 1–7. https://doi.org/10.1111/0045-3609.00008

Carroll, A. B. (1999). Corporate social responsibility: Evolution of a definitional construct. *Business & Society, 38*(3), 268–295. https://doi.org/10.1177/000765039903800303

Carroll, A. B. (2000). Ethical challenges for business in the new millennium: Corporate social responsibility and models of management morality. *Business Ethics Quarterly, 10*(1), 33–42. https://doi.org/10.2307/3857692

Chan, R. Y. (2001). Determinants of Chinese consumers' green purchase behavior. *Psychology and Marketing, 18*(4), 389–413. https://doi.org/10.1002/mar.1013

Chang, K., Kim, I., & Li, Y. (2014). The heterogeneous impact of corporate social responsibility activities that target different stakeholders. *Journal of Business Ethics, 125*(2), 211–234. https://doi.org/10.1007/s10551-013-1895-8

Chen, C. A. (2014). Suitable festival activities for Taiwan's tourism and nation branding with the application of the PR AHP program. *Asia Pacific Journal of Tourism Research, 19*(12), 1381–1398. https://doi.org/10.1080/10941665.2013.866579

Chen, C. F. (2006). Applying the analytical hierarchy process (AHP) approach to convention site selection. *Journal of Travel Research, 45*(2), 167–174. https://doi.org/10.1177/0047287506291593

Chen, Y. R. R., & Hung-Baesecke, C. J. F. (2014). Examining the internal aspect of corporate social responsibility (CSR): Leader behavior and employee CSR participation. *Communication Research Reports, 31*(2), 210–220. https://doi.org/10.1080/08824096.2014.907148

Chow, W. S., & Chen, Y. (2012). Corporate sustainable development: Testing a new scale based on the mainland Chinese context. *Journal of Business Ethics*, 105(4), 519–533. https://doi.org/10.1007/s10551-011-0983-x

Costa, R., & Menichini, T. (2013). A multidimensional approach for CSR assessment: The importance of the stakeholder perception. *Expert Systems with Applications*, 40(1), 150–161. https://doi.org/10.1016/j.eswa.2012.07.028

Dahlsrud, A. (2008). How corporate social responsibility is defined: an analysis of 37 definitions. *Corporate Social Responsibility and Environmental Management*, 15(1), 1–13. https://doi.org/10.1002/csr.132

Dalkey, N. (1969). An experimental study of group opinion: the Delphi method. *Futures*, 1(5), 408–426. https://doi.org/10.1016/S0016-3287(69)80025-X

Davenport, K. (2000). Corporate citizenship: A stakeholder approach for defining corporate social performance and identifying measures for assessing it. *Business & Society*, 39(2), 210–219. https://doi.org/10.1177/000765030003900205

David, P., Kline, S., & Dai, Y. (2005). Corporate social responsibility practices, corporate identity, and purchase intention: A dual-process model. *Journal of Public Relations Research*, 17(3), 291–313. https://doi.org/10.1207/s1532754xjprr1703_4

Davidson, W. N., & Worrell, D. L. (1990). A comparison and test of the use of accounting and stock market data in relating corporate social responsibility and financial performance. *Akron Business and Economic Review*, 21(3), 7.

Davis, K. (2014). Different stakeholder groups and their perceptions of project success. *International Journal of Project Management*, 32(2), 189–201. https://doi.org/10.1016/j.ijproman.2013.02.006

De Grosbois, D. (2012). Corporate social responsibility reporting by the global hotel industry: Commitment, initiatives and performance. *International Journal of Hospitality Management*, 31(3), 896–905. https://doi.org/10.1016/j.ijhm.2011.10.008

Deng, J., King, B., & Bauer, T. (2002). Evaluating natural attractions for tourism. *Annals of Tourism Research*, 29(2), 422–438. https://doi.org/10.1016/S0160-7383(01)00068-8

El Akremi, A., Gond, J. P., Swaen, V., De Roeck, K., & Igalens, J. (2018). How do employees perceive corporate responsibility? Development and validation of a multidimensional corporate stakeholder responsibility scale. *Journal of Management*, 44(2), 619–657. https://doi.org/10.1177/0149206315569311

Etheredge, J. M. (1999). The perceived role of ethics and social responsibility: An alternative scale structure. *Journal of Business Ethics*, 18(1), 51–64. https://doi.org/10.1023/A:1006077708197

Fatma, M., Rahman, Z., & Khan, I. (2016). Measuring consumer perception of CSR in tourism industry: Scale development and validation. *Journal of Hospitality and Tourism Management*, 27, 39–48. https://doi.org/10.1016/j.jhtm.2016.03.002

Font, X., & Lynes, J. (2018). Special issue: Corporate social responsibility for sustainable tourism. *Journal of Sustainable Tourism*, 26(7), 1027–1289. https://doi.org/10.1080/09669582.2018.1488856

Gond, J. P., El Akremi, A., Swaen, V., & Babu, N. (2017). The psychological microfoundations of corporate social responsibility: A person-centric systematic review. *Journal of Organizational Behavior*, 38(2), 225–246. https://doi.org/10.1002/job.2170

González-Rodríguez, M., Carmen Martín-Samper, R., Köseoglu, M., & Okumus, F. (2019). Hotels' corporate social responsibility practices, organizational culture, firm reputation, and performance. *Journal of Sustainable Tourism*, 27(3), 398–419. https://doi.org/10.1080/09669582.2019.1585441

Gray, R., Kouhy, R., & Lavers, S. (1995). Corporate social and environmental reporting: a review of the literature and a longitudinal study of UK disclosure. *Accounting, Auditing & Accountability Journal*, 8(2), 47–77. https://doi.org/10.1108/09513579510146996

Guzzo, Renata F, Abbott, JéAnna, & Madera, Juan M. (in press). A Micro-Level View of CSR: A Hospitality Management Systematic Literature Review. *Cornell Hospitality Quarterly*, https://doi.org/10.1177/1938965519892907.

Heiko, A. (2012). Consensus measurement in Delphi studies: review and implications for future quality assurance. *Technological Forecasting and Social Change*, 79(8), 1525–1536.

Heiko, A., & Darkow, I. L. (2010). Scenarios for the logistics services industry: A Delphi-based analysis for 2025. *International Journal of Production Economics*, 127(1), 46–59. https://doi.org/10.1016/j.ijpe.2010.04.013

Ho, D., Newell, G., & Walker, A. (2005). The importance of property-specific attributes in assessing CBD office building quality. *Journal of Property Investment & Finance*, 23(5), 424–444. https://doi.org/10.1108/14635780510616025

Holcomb, J. L., Upchurch, R. S., & Okumus, F. (2007). Corporate social responsibility: what are top hotel companies reporting? *International Journal of Contemporary Hospitality Management*, 19(6), 461–475. https://doi.org/10.1108/09596110710775129

Hsu, C. C., & Sandford, B. A. (2007). The Delphi technique: making sense of consensus. *Practical Assessment, Research & Evaluation*, 12(10), 1–8.

Keeney, S., Hasson, F., & McKenna, H. P. (2001). A critical review of the Delphi technique as a research methodology for nursing. *International Journal of Nursing Studies*, 38(2), 195–200. https://doi.org/10.1016/s0020-7489(00)00044-4

Kim, H. L., Rhou, Y., Uysal, M., & Kwon, N. (2017). An examination of the links between corporate social responsibility (CSR) and its internal consequences. *International Journal of Hospitality Management*, 61, 26–34. https://doi.org/10.1016/j.ijhm.2016.10.011

Knowles, T., Macmillan, S., Palmer, J., Grabowski, P., & Hashimoto, A. (1999). The development of environmental initiatives in tourism: responses from the London hotel sector. *International Journal of Tourism Research*, *1*(4), 255–265. https://doi.org/10.1002/(SICI)1522-1970(199907/08)1:4<255::AID-JTR170>3.0.CO;2-8

Ko, A., Chan, A., & Wong, S. (2019). A scale development study of CSR: Hotel employees' perceptions. *International Journal of Contemporary Hospitality Management*, *31*(4), 1857–1884. https://doi.org/10.1108/IJCHM-09-2017-0560

Kroger. (2018). *Sustainability Report 2018*. http://sustainability.kroger.com/Kroger_CSR2018.pdf

Kucukusta, D., Mak, A., & Chan, X. (2013). Corporate social responsibility practices in four and five-star hotels: Perspectives from Hong Kong visitors. *International Journal of Hospitality Management*, *34*, 19–30. https://doi.org/10.1016/j.ijhm.2013.01.010

Latif, K. F., & Sajjad, A. (2018). Measuring corporate social responsibility: A critical review of survey instruments. *Corporate Social Responsibility and Environmental Management*, *25*(6), 1174–1197. https://doi.org/10.1002/csr.1630

Lawshe, C. H. (1975). A quantitative approach to content validity 1. *Personnel Psychology*, *28*(4), 563–575. https://doi.org/10.1111/j.1744-6570.1975.tb01393.x

Lee, C. G., Sung, J., Kim, J. K., Jung, I. S., & Kim, K. J. (2016). Corporate social responsibility of the media: Instrument development and validation. *Information Development*, *32*(3), 554–565. https://doi.org/10.1177/0266666914559856

Lock, I., & Seele, P. (2016). The credibility of CSR (corporate social responsibility) reports in Europe. Evidence from a quantitative content analysis in 11 countries. *Journal of Cleaner Production*, *122*, 186–200. https://doi.org/10.1016/j.jclepro.2016.02.060

Ludwig, B. G. (1994). Internationalizing extension: An exploration of the characteristics evident in a state university Extension system that achieves internationalization [Doctoral dissertation]. The Ohio State University.

Madsen, P. M. (2009). Does corporate investment drive a "race to the bottom" in environmental protection? A reexamination of the effect of environmental regulation on investment. *Academy of Management Journal*, *52*(6), 1297–1318. https://doi.org/10.5465/amj.2009.47085173

Maignan, I. (2001). Consumers' perceptions of corporate social responsibilities: A cross-cultural comparison. *Journal of Business Ethics*, *30*(1), 57–72. https://doi.org/10.1023/A:1006433928640

Maignan, I., & Ferrell, O. C. (2000). Measuring corporate citizenship in two countries: The case of the United States and France. *Journal of Business Ethics*, *23*(3), 283–297. https://doi.org/10.1023/A:1006262325211

Maignan, I., & Ferrell, O. C. (2001). Corporate citizenship as a marketing instrument-Concepts, evidence and research directions. *European Journal of Marketing*, *35*(3/4), 457–484. https://doi.org/10.1108/03090560110382110

Maignan, I., Ferrell, O. C., & Hult, G. T. M. (1999). Corporate citizenship: Cultural antecedents and business benefits. *Journal of the Academy of Marketing Science*, *27*(4), 455–469. https://doi.org/10.1177/0092070399274005

Manaktola, K., & Jauhari, V. (2007). Exploring consumer attitude and behaviour towards green practices in the lodging industry in India. *International Journal of Contemporary Hospitality Management*, *19*(5), 364–377. https://doi.org/10.1108/09596110710757534

Martínez, P., & del Bosque, I. R. (2013). CSR and customer loyalty: The roles of trust, customer identification with the company and satisfaction. *International Journal of Hospitality Management*, *35*, 89–99. https://doi.org/10.1016/j.ijhm.2013.05.009

Martínez, P., Pérez, A., & Rodríguez del Bosque, I. (2013). Measuring corporate social responsibility in tourism: Development and validation of an efficient measurement scale in the hospitality industry. *Journal of Travel & Tourism Marketing*, *30*(4), 365–385. https://doi.org/10.1080/10548408.2013.784154

Maxwell, J. (2013). *Qualitative research design: An interactive approach* (3rd ed.), Applied social research methods series. Sage.

Mercer, J., & Oskamp, S. (2003). *Corporate social responsibility and its importance to consumers* [ProQuest dissertations and theses].

Miller, G. (2001). The development of indicators for sustainable tourism: Results of a Delphi survey of tourism researchers. *Tourism Management*, *22*(4), 351–362. https://doi.org/10.1016/S0261-5177(00)00067-4

Millet, I., & Saaty, T. L. (2000). On the relativity of relative measures–accommodating both rank preservation and rank reversals in the AHP. *European Journal of Operational Research*, *121*(1), 205–212. https://doi.org/10.1016/S0377-2217(99)00040-5

Nunnally, J. (1978). *Psychometric theory* (2nd ed.), McGraw-Hill series in psychology. McGraw-Hill.

Öberseder, M., Schlegelmilch, B. B., Murphy, P. E., & Gruber, V. (2014). Consumers' perceptions of corporate social responsibility: scale development and validation. *Journal of Business Ethics*, *124*(1), 101–115. https://doi.org/10.1007/s10551-013-1787-y

O'Connor, M., & Spangenberg, J. H. (2008). A methodology for CSR reporting: assuring a representative diversity of indicators across stakeholders, scales, sites and performance issues. *Journal of Cleaner Production*, *16*(13), 1399–1415. https://doi.org/10.1016/j.jclepro.2007.08.005

Okoye, A. (2009). Theorising corporate social responsibility as an essentially contested concept: is a definition necessary? *Journal of Business Ethics*, *89*(4), 613–627. https://doi.org/10.1007/s10551-008-0021-9

Pérez, A., & Del Bosque, I. R. (2013). Measuring CSR image: three studies to develop and to validate a reliable measurement tool. *Journal of Business Ethics*, *118*(2), 265–286. https://doi.org/10.1007/s10551-012-1588-8

Pérez, A., Martínez, P., & Del Bosque, I. R. (2013). The development of a stakeholder-based scale for measuring corporate social responsibility in the banking industry. *Service Business, 7*(3), 459–481. https://doi.org/10.1007/s11628-012-0171-9

Quazi, A. M., & O'brien, D. (2000). An empirical test of a cross-national model of corporate social responsibility. *Journal of Business Ethics, 25*(1), 33–51. https://doi.org/10.1023/A:1006305111122

Renfors, S. M. (2018). Developing the curriculum content of coastal and maritime tourism: Stakeholders' perspective of the sector-specific skills and knowledge in Finland. *Tourism in Marine Environments, 13*(2), 109–119. https://doi.org/10.3727/154427318X15265996581008

Ricaurte, E. (2011). Developing a sustainability measurement framework for hotels: Toward an industry-wide reporting structure. *Cornell Hospitality Report, 11*(13), 6–30.

Ricks, J. M. Jr. (2005). An assessment of strategic corporate philanthropy on perceptions of brand equity variables. *Journal of Consumer Marketing, 22*(3), 121–134. https://doi.org/10.1108/07363760510595940

Ruf, B. M., Muralidhar, K., & Paul, K. (1998). The development of a systematic, aggregate measure of corporate social performance. *Journal of Management, 24*(1), 119–133. https://doi.org/10.1177/014920639802400101

Saaty, R. W. (1987). The analytic hierarchy process—what it is and how it is used. *Mathematical Modelling, 9*(3–5), 161–176. https://doi.org/10.1016/0270-0255(87)90473-8

Saaty, T. (1980). *The analytic hierarchy process: Planning, priority setting, resource allocation.* McGraw-Hill International Book.

Saaty, T. (1994). *Fundamentals of decision making and priority theory with the analytic hierarchy process* (1st ed.), Analytic hierarchy process series. RWS Publications.

SASB. (2018). *SASB materiality map.* https://materiality.sasb.org/

Serra-Cantallops, A., Peña-Miranda, D. D., Ramón-Cardona, J., & Martorell-Cunill, O. (2018). Progress in research on CSR and the hotel industry (2006–2015). *Cornell Hospitality Quarterly, 59*(1), 15–38. https://doi.org/10.1177/1938965517719267

Shangri-La Hotels and Resorts. (2016). *Annual Report 2016.* http://ir.shangri-la.com/ir/en/reports/annualreports/2016/ar2016.pdf

Shnayder, L., Van Rijnsoever, F. J., & Hekkert, M. P. (2015). Putting your money where your mouth is: Why sustainability reporting based on the triple bottom line can be misleading. *PloS One, 10*(3), e0119036. https://doi.org/10.1371/journal.pone.0119036

Singh, J., & del Bosque, I. R. (2008). Understanding corporate social responsibility and product perceptions in consumer markets: A cross-cultural evaluation. *Journal of Business Ethics, 80*(3), 597–611. https://doi.org/10.1007/s10551-007-9457-6

Singhapakdi, A., Vitell, S. J., Rallapalli, K. C., & Kraft, K. L. (1996). The perceived role of ethics and social responsibility: A scale development. *Journal of Business Ethics, 15*(11), 1131–1140. https://doi.org/10.1007/BF00412812

Skudiene, V., & Auruskeviciene, V. (2012). The contribution of corporate social responsibility to internal employee motivation. *Baltic Journal of Management, 7*(1), 49–67. https://doi.org/10.1108/17465261211197421

Smith, N. C. (2003). Corporate social responsibility: Whether or how? *California Management Review, 45*(4), 52–76. https://doi.org/10.2307/41166188

Song, H. J., Lee, H. M., Lee, C. K., & Song, S. J. (2015). The role of CSR and responsible gambling in casino employees' organizational commitment, job satisfaction, and customer orientation. *Asia Pacific Journal of Tourism Research, 20*(4), 455–471. https://doi.org/10.1080/10941665.2013.877049

Su, L., Swanson, S. R., & Chen, X. (2015). Social responsibility and reputation influences on the intentions of Chinese Huitang Village tourists: Mediating effects of satisfaction with lodging providers. *International Journal of Contemporary Hospitality Management, 27*(8), 1750–1771. https://doi.org/10.1108/IJCHM-06-2014-0305

Teixeira, A. A., Jabbour, C. J. C., & de Sousa Jabbour, A. B. L. (2012). Relationship between green management and environmental training in companies located in Brazil: A theoretical framework and case studies. *International Journal of Production Economics, 140*(1), 318–329. https://doi.org/10.1016/j.ijpe.2012.01.009

Teixeira, A. A., Jabbour, C. J. C., de Sousa Jabbour, A. B. L., Latan, H., & De Oliveira, J. H. C. (2016). Green training and green supply chain management: evidence from Brazilian firms. *Journal of Cleaner Production, 116*, 170–176. https://doi.org/10.1016/j.jclepro.2015.12.061

Tsai, C. W., & Tsai, C. P. (2008). Impacts of consumer environmental ethics on consumer behaviors in green hotels. *Journal of Hospitality & Leisure Marketing, 17*(3–4), 284–313. https://doi.org/10.1080/10507050801984974

Tsai, H., Tsang, N. K., & Cheng, S. K. (2012). Hotel employees' perceptions on corporate social responsibility: The case of Hong Kong. *International Journal of Hospitality Management, 31*(4), 1143–1154. https://doi.org/10.1016/j.ijhm.2012.02.002

Turker, D. (2009). Measuring corporate social responsibility: A scale development study. *Journal of Business Ethics, 85*(4), 411–427. https://doi.org/10.1007/s10551-008-9780-6

Tzeng, G. H., Teng, M. H., Chen, J. J., & Opricovic, S. (2002). Multicriteria selection for a restaurant location in Taipei. *International Journal of Hospitality Management, 21*(2), 171–187. https://doi.org/10.1016/S0278-4319(02)00005-1

Webb, D. J., Mohr, L. A., & Harris, K. E. (2008). A re-examination of socially responsible consumption and its measurement. *Journal of Business Research, 61*(2), 91–98. https://doi.org/10.1016/j.jbusres.2007.05.007

Whitley, R. (1992). *Business systems in East Asia: Firms, markets and societies*. Sage.

Wilson, F. R., Pan, W., & Schumsky, D. A. (2012). Recalculation of the critical values for Lawshe's content validity ratio. *Measurement and Evaluation in Counseling and Development, 45*(3), 197–210. https://doi.org/10.1177/0748175612440286

Wong, A., & Kim, S. (2020). Development and validation of standard hotel corporate social responsibility (CSR) scale from the employee perspective. *International Journal of Hospitality Management, 87*(5), 102507. https://doi.org/10.1016/j.ijhm.2020.102507

Wong, A. K. F., Kim, S., & Lee, S. (in press). The evolution, progress, and the future of corporate social responsibility: Comprehensive review of hospitality and tourism articles. International *Journal of Hospitality & Tourism Administration*. https://doi.org/10.1080/15256480.2019.1692753

Wood, D. J. (2010). Measuring corporate social performance: A review. *International Journal of Management Reviews, 12*(1), 50–84. https://doi.org/10.1111/j.1468-2370.2009.00274.x

Xiao, Q., Heo, C., & Lee, S. (2017). How do consumers' perceptions differ across dimensions of corporate social responsibility and hotel types? *Journal of Travel & Tourism Marketing, 34*(5), 694–707. https://doi.org/10.1080/10548408.2016.1232671

Xu, Y. (2014). Understanding CSR from the perspective of Chinese diners: The case of McDonald's. *International Journal of Contemporary Hospitality Management, 26*(6), 1002–1020. https://doi.org/10.1108/IJCHM-01-2013-0051

Zaman, M., Botti, L., & Thanh, T. V. (2016). Weight of criteria in hotel selection: An empirical illustration based on TripAdvisor criteria. *European Journal of Tourism Research, 13*(1), 132–138.

# Comparing resident and tourist perceptions of an urban park: a latent profile analysis of perceived place value

Hwasung Song and Changsup Shim

**ABSTRACT**

Sustainable management of cities is only attainable when urban spaces are understood as spaces where residents and tourists can coexist even with different interests and priorities. Urban parks are a prime example of urban spaces where residents and tourists mingle; they are only sustainable if the different perceptions of place of the two groups are understood. Therefore, the current study examines 652 visitors at Gwanggyo Lake Park (GLP) in South Korea to compare tourist and resident perceptions. Employing Latent Profile Analysis of perceived place value, the current study identified three valid profiles for visitors of GLP: Relationship Seekers, Activity Seekers, and Environment Seekers. This analysis also found significant differences between profiles of residents and tourists visiting GLP in terms of demographic and behavioral characteristics. Based on the differences between resident and tourist perceptions, theoretical and practical implications are offered for sustainable management of urban parks and other attractions.

## Introduction

Urban areas are primarily spaces for residents to carry on their daily lives, although the growing popularity of urban tourism has shifted the focus of various urban spaces to increasingly serve as important tourist attractions (Novy & Colomb, 2017; UNWTO, 2012). Therefore, sustainable management of urban tourism now requires a deeper understanding than is required for non-urban settings of the interface between the urban setting, local residents, and tourists (Ashworth & Page, 2011; Beedie, 2005; Hinch, 1996). In particular, leisure spaces such as urban parks where residents and tourists closely mingle could not be sustainably managed without understanding the different use characteristics of residents and tourists (Bourdeau, De Coster & Paradis, 2001; Snepenger et al., 2003).

While the Boston Commons opened as the first open city space in 1634, and 16 other urban parks opened before 1800, including the National Mall in Washington, D.C., urban parks emerged in earnest in the 19th century and Central Park opened in New York City in 1857. Ever since, urban parks have been providing local residents natural spaces, so rare in cities, where residents can take walks in their spare time or enjoy recreational activities including sports, private social gatherings, or municipal events (Coley et al., 1997; Kaczynski, Potwarka & Saelens, 2008; Maas et al., 2009). Urban parks have also begun to assume an important role as tourist attractions for visitors to urban areas. Urban parks as tourist attractions symbolize the image and lifestyle of the city for visitors who can temporarily experience the life of local residents (Archer, 2006; Gobster, 2007; Masberg & Jamieson, 1999). For example, Hyde Park in London, Central Park in New York City, and Park Guell in Barcelona are tourist attractions that show the charming characteristics of each city very well, attracting millions of tourists each year. Therefore, city governments have increasingly put more effort into sustainable development and management of urban parks to provide diverse value, be it environmental, socio-cultural, educational, or recreational, to local residents and tourists alike (Crompton, 2007; Deng et al., 2010; Li, 2020).

The value assigned by visitors to a particular place influences their attitudes and behaviors (Homer & Kahle, 1988). Hence, the value assigned to a place must be accounted for in visitor segmentation and management to achieve the economic, environmental, and socio-cultural sustainability of the location. Previous studies have applied the concept of perceived value and have examined its effects in various tourism settings (Chen & Chen, 2010; Peña, Jamilena, & Molina, 2012). The perceived value of a place is not fixed or absolute but is relative: it is constructed based on individual and group characteristics or contextual factors (Relph, 1985; Tuan, 1977). This implies that different groups—for example, residents and tourists—may assign different value to urban parks. However, few empirical studies address the different values assigned to urban parks by the local residents and tourists who use them.

Accordingly, the current study examines the perceived place value of Gwanggyo Lake Park in South Korea and focuses on comparing tourist and resident perceptions. Located in a populated urban area (Suwon, South Korea), the park attracts over three million visitors annually, including both residents and tourists. Latent Profile Analysis, a novel classification procedure that is statistically more robust than cluster analysis, has been employed (Bergman & Trost, 2006; Magidson & Vermunt, 2002). More specifically, this study has the following three purposes. First, the study identifies sub-types of perceived place value assigned to the urban lake park both within the resident group and within the tourist group. Second, the study compares residents and tourists in the various sub-types by analyzing visit frequency, behavioral intentions, and demographic characteristics. Third and finally, theoretical and practical implications are offered for sustainable management of urban tourism space based on the differences between resident and tourist perceptions.

## Literature review

### Urban parks as tourist attractions

An urban park is a "bounded area of public open space that is maintained in a "natural" or semi-natural (landscaped) state and set aside for a designated purpose, usually to do with recreation" (Hilborn, 2009, p. 4). Urban parks are generally the result of planned urban development by a municipality or other governmental body. They are open to local residents and visitors alike, offering greenery, walking paths, and recreational as well as sporting facilities (Archer, 2006; Chiesura, 2004; Welch, 1991). Urban parks are perceived to be an essential part of urban planning in most developed countries, many of which have laws that provide for park space that is proportionate to the population (de Saz Salazar & Menéndez, 2007; Gobster, 1998; McCormack et al., 2010).

The role of urban parks has evolved over time. In the late 19th century, the purpose of parks was to provide natural spaces to urban residents who were tired of rapid industrialization and urbanization (Clanz, 1989). Since the middle of the 20th century, urban residents have more and more spare time and have increasingly turned to urban parks as spaces where they can enjoy sports and other recreational activities (Archer, 2006; Coley et al., 1997, Hayward, 1989). More recently, urban parks are expanding their role to serve as the center of various community activities that showcase culture and art, host public gatherings, and offer opportunities for experiential education (Crompton, 2007; Peters, Elands & Buijs, 2010; Xu et al., 2019). With the development of urban tourism, urban parks have broadened their appeal not only to urban residents, but also to all kinds of tourists and visitors to the city (Deng et al., 2010).

According to previous studies, tourists visit urban parks for several reasons. Most of all, urban parks are spaces where residents spend their spare time, and thus tourists can join the experience of a city's unique culture and forms of leisure. Hence, when visiting urban parks, tourists may enjoy feeling like city-dwellers, albeit only for a short while. In addition, as urban parks in world famous cities, like Central Park in New York City, have been represented by the media as symbols of the city in films, TV programs, and popular music, they are now perceived to be must-visit places. Urban parks are typically quite readily accessible to a large number of people since parks are generally centrally located; ready access and free entry are strong advantages of urban parks as tourist attractions since tourists can easily go there at little expense. The number of visitors to urban parks is also increasing because some offer leisure sports or other forms of outdoor recreation that are otherwise hard to find in the city (Donahue et al., 2018; Karanikola, Panagopoulos & Tampakis, 2017; Lee et al., 2017; Plunkett, Fulthorp & Paris, 2019; Qing, 2018; Wong & Domroes, 2005).

As urban parks have traditionally been perceived as spaces for residents to enjoy, studies related to urban parks have been conducted mainly in the fields of urban planning, leisure studies, and community studies rather than in tourism studies. Some studies on urban parks as tourist attractions have investigated urban ecotourism, pro-environmental behavior, the motivation to visit urban parks, and the effect of urban parks on the attractiveness of urban tourism (Deng et al., 2010; Lee, Quintal, & Phau, 2017). However, these previous studies have approached urban parks as one of many general tourist attractions, without specifically exploring the spatial context provided by cities for these parks. Moreover, each of these previous studies was conducted either on residents or tourists, largely ignoring differences between the two groups. Given the rapid development of urban tourism, a comparative analysis of the two groups is essential to examine sustainable development and future management of urban parks.

### Perceived place value

Perceived value is an individual's overall assessment of the worth or merit of a particular product. Perceived value has been shown to affect services and tourist locations (waterparks, festivals) as well as goods, and has been shown to affect not only the price people are willing to pay for something, but also customer satisfaction, loyalty, and behavioral intention. (Chen & Chen, 2010; Cronin et al., 2000; Jin, Lee & Lee, 2015). Perceived value of a specific place is subjective and can be formed directly through the senses through past experiences and impressions of that place or indirectly through images, symbols, myths, and the arts (Relph, 1985; Tuan, 1977). Relph explained that people develop a sense of place through the interaction of physical setting, activity, and meaning that allows the place to be culturally recognized (Relph, 1976). In other words, a sense of place, its personality so to speak, is created through a multitude of human interactions with the physical entity, the place, and in this process, each individual assigns different value to the same place depending on individual needs, desires, and socio-cultural background.

In addition, the perceived value of a place is affected by time and the environment (Shim & Santos, 2014; Song & Kim, 2018; Stedman, 2003). For example, McCleary, Weaver, and Hsu (2007) analyzed tourists who visited Hong Kong from seven geographic origins and found that tourists from the East including Mainland China, Taiwan, Malaysia and Singapore placed a higher overall value on Hong Kong than those from the U.S., Australia, and Western Europe. Lee, Lee, and Yoon (2009) compared first-time visitors and returning visitors to a particular festival, showing that the two groups assigned different value to different aspects of the same festival.

A few studies have compared perceptions of residents versus tourists toward the same place. For example, Simpson (1999) compared the perceptions of tourists and residents visiting the Historic Centre in Prague and showed that the two groups expected the place to develop in different directions according to their respective needs. In a study related to public beaches in South Carolina, USA, Oh, Draper, and Dixon (2010) found that tourists wanted to use public beaches for recreational purposes more than residents did. A study related to island tourism in Taiwan by Chao and Chao (2017) showed that tourists perceived the ecological value of the island to be high, while residents perceived its economic value to be high.

Perceived value is multi-dimensional and encompasses not only the intrinsic value of play, aesthetics, ethics, and spirituality, but also the extrinsic value of efficiency, excellence, status, and esteem (Holbrook, 1999). It has been suggested that the perception of place also assigns value based on a variety of aspects including the physical, ecological, and socio-cultural characteristics of particular places (Song & Kim, 2018; Yen & Teng, 2015). In addition, the value of a place to any particular individual or group of individuals is not fixed, but can be counted on to change since the value is affected by how the place is managed and consumed over time (Chen & Hu, 2010; Lee, Yoon & Lee, 2007; Peña et al., 2012). Previous studies have identified several dimensions according to which visitors perceive value in a place, including economic, educational, spiritual, cultural, aesthetic, biological, therapeutic, recreational, and existential values (Brown, 2004; Brown & Raymond, 2007; Rasoolimanesh, Dahalan & Jaafar, 2016).

Brown and Raymond (2007) examined differences in resident and tourist views concerning a particular tourist destination and showed that both groups ranked aesthetic and recreational value as very important; at the same time, they found economic value to be more important to residents while future value of the destination was more important to tourists. Zhu et al. (2010) found that residents primarily placed the highest value on aesthetic and recreational aspects of the destination while tourists highly valued wilderness and wildlife preservation features. On the other hand, Kim et al. (2015) studied visitors to Namhansanseong (i.e., mountain fortress), a UNESCO Cultural Heritage site in Korea, and found no differences in therapeutic, amenity, or atmosphere values although residents placed higher value than tourists on environmental factors including biological diversity, wilderness, and wildlife preservation.

Long ago, Tuan (1977) argued that tourism activities can create place recognition over a short period of time, resulting in various attempts to better understand place value. However, most studies have only focused on identification of the components of place value (e.g., Brown, 2004) or differences in the perception of place value based on differences in specific tourist destinations. Attempts to categorize various types of visitors according to their perceptions of place value have either simply divided groups based on differences in perception levels (e.g., Kaltenborn, 1998) or have created groups through an exploratory approach using qualitative methodologies (Hutson et al., 2010; Hutson & Montgomery, 2011; Wilson, 2005). However, the former approaches lump all aspects of sense of place together. Kaltenborn (1998), for example, relied on a previously identified sense of place index, classifying resident perceptions of sense of place as either at a high, middle, or low level on the sense of place index. Such an approach increases the risk that the results of the analysis of sense of place will be affected by variables with greater weight in the index. The limitation of this method is that it fails to differentiate various aspects of place value held by various groups within the population. In addition, although qualitative methods have the advantage of being able to show various relationships between places and users, they also have the primary weakness of being too dependent on subjective evaluation. A more comprehensive and holistic view of place value is needed, as well as a more realistic understanding of sense of place similarities and differences across different groups of visitors.

## Research method

### Study setting

Gwanggyo Lake Park(GLP), the setting of this study, houses South Korea's premier waterside and urban ecological park and is South Korea's largest lake park with a total area of 2,025,418 m$^2$

and lake area of 653,003 m$^2$. The construction took place from June 2010 to April 2013 at a cost of approximately 120 billion won. Of note, Suwon City, where GLP is located, also contains a UNESCO cultural heritage site, the Suwon Hwaseong Fortress. As a result, tourists visit Suwan Hwaseong Fortress and GLP together, the latter serving as an important tourism resource offering diverse forms of tourism in Suwon, a center of history and cultural resources, while promoting relaxation and health as well as functioning as a downtown neighborhood park.

Currently, there are various sports facilities in the park, including recreation facilities including auto camping grounds, water playgrounds, lawned squares, cultural facilities such as libraries, theaters, and exhibition halls, bird watching platforms, observation decks, and other facilities providing various cultural, ecological, and physical programs for residents and tourists to enjoy. GLP not only provides facilities, but offers various experiences to visitors, such as promotion of health, outdoor activities, festivals, and other events; it is utilized as a place of rest and exercise for residents and as a leisure and tourism site and complex cultural space for tourists.

The site on which GLP is located was extremely popular as an excursion and water play park due to its operation of Woncheon Amusement Park on its premises until August 2009. GLP thus has special value in South Korea as a place that conjures memories of the past for many visitors, distinguishing it from other parks in that its mere existence holds high value for long-time residents of Suwon. In summary, for residents and tourists, GLP is a waterside ecological park that holds ecological value as a natural resource, offers health, recreational, and cultural values of a park, and reminds visitors of the unique local value of a place where a well-known tourist spot was previously located. Due to its many functions, GLP presents a suitable research setting to examine various aspects of place and the differences between residents and tourists in their perceived value of the place.

### Data collection and measures

The study setting, GLP, has become a representative ecological green place as well as leisure place in Korea. GLP also evolving from an ordinary urban park to an eco-friendly leisure place through the provision of natural ecological and cultural programs. In this sense, based on prior studies (Beverly et al., 2008; Brown, 2004; Brown & Raymond, 2007; Park & Song, 2018; Zhu et al., 2010), questions about place value included eleven items addressing the value of human relationships, local culture, community, spirituality, diversity, wilderness, environmental learning, amenities, atmosphere, outdoor activities, and wellness (e.g., "GLP is valuable because I can learn about environment there") on a five-point Likert scale (1= not at all valuable, 3= moderately valuable, 5 = very valuable) to judge the value of each item. Other questionnaire items included number of visits, intention to revisit (1 = will not visit again, 3 = moderately, 5 = will visit again), and demographic characteristics (gender, age, place of residence, education, average monthly household income).

### Data analysis

Subtypes (profiles) of visitor perceptions of the place value of GLP were obtained using Latent Profile Analysis (LPA). LPA is a mixture model that identifies diversity within a group by extracting subtypes, or latent profiles, from a broader sample (Magidson & Vermunt, 2002). This methodology allows for the identification of diverse and multidimensional needs of visitors. The advantage of LPA is its ability to analyze a wide range of heterogeneous needs within a homogenous broader population, such as distinct types of travel motivations among international students (Song & Bae, 2018). LPA has the advantage of determining subtypes according to statistical criteria including fit indices, entropy, and the Lo-Mendell-Rubin likelihood ratio test (LMR-LRT), unlike cluster analysis or a researcher's arbitrary interpretations (Bergman & Trost,

2006). In addition, LPA is useful in describing the implications of each subtype since characteristics are grouped together with subtype-specific influence variables (Magidson & Vermunt, 2002). Accordingly, LPA has been widely used in the leisure/tourism field to examine cultural tourists (Pulido-Fernández & Sánchez-Rivero, 2010; Van der Ark & Richards, 2006), park visitors (Park & Song, 2018), and residents of locations with tourist attractions (Ven, 2016). In this study, LPA was used to classify residents and tourists according to their perceptions of place value. Differences in visit frequency, revisit intentions, and demographic factors were then be examined for each LPA profile using cross tabulation and one-way ANOVA analyses.

## Results

### Sample characteristics

The sample of 652 respondents included more women (n = 383, 58.7%) than men (n = 269, 41.3%). The average age of participants was 46.2 years (SD = 12.45). There were more visitors who were older than 50 (n = 238, 36.5%) than those who were in their 40s (n = 201, 30.8%) or 30s (n = 166, 25.5%). College attendance/graduation was the most common educational level (n = 359, 61.4%), and the average household income was seven thousand dollars a month, which indicated that a high proportion of participants were highly educated high-income visitors. The number of residents of Suwon who visited GLP was 301 (46.2%), whereas the number of tourists from outside of Suwon was 351 (53.8%).

### Selection of the latent profile model

The optimal model of latent profiles in LPA was selected based on statistical fitness, quality of classification, statistical significance, and classification ratio. The criteria to assess the explanatory power and parsimony of the model were the BIC (Baysian information criteria) and AIC (Akaike information criteria), with lower values indicating a better model (Nylund, Asparouhov, & Muthén, 2007). In addition, the quality of latent profile classification was based on an entropy value between 0 and 1, with values closer to 1 indicating greater accuracy of the latent profiles. However, statistical criteria from BIC, AIC, entropy, and p-values did not serve as absolute criteria dictating the final model to be used, but rather consideration of model parsimony, theoretical and practical interpretability, and variety of patterns were all considered in model selection (Magidson & Vermunt, 2002).

In order to obtain latent profiles of GLP visitors according to perceived place value, all 11 place value perception variables were included in the LPA for the entire sample. The same 11 variables were then included in a LPA of residents and tourists separately. Table 1 shows the AIC, BIC, entropy, and p-values for each model.

### Characteristics of the profiles

Table 2 shows the profiles obtained from LPA according to the perceptions of place value of all visitors. Values were based on the average of each profile on a five-point scale and the t-score value obtained from a comparison of profiles. The first profile consisted of 11.6% of all visitors, and the average obtained for this profile was generally below the average of all values, but the human relationship value was relatively high, and thus the profile was labeled Relationship Seekers (RS). The second profile, with 45.6% of the visitors, was labeled Activity Seekers (AS), and consisted of those who assigned high value to outdoor activities and wellness. The third profile, with 42.6% of the visitors, had the highest perceived value of the park of all three groups. This profile was labeled Environment Seekers (ES) because these visitors perceived urban parks to

**Table 1.** Goodness of fit model for LPA according to perceived place value.

|  | Number of profiles (k) | AIC | BIC | Entropy | p-value |
|---|---|---|---|---|---|
| Total | 2 | 20660.100 | 20812.422 | 0.867 | 0.000 |
|  | 3 | **19454.549** | **19660.631** | **0.897** | **0.000** |
|  | 4 | 19146.667 | 19406.51 | 0.859 | 0.079 |
| Residents | 2 | 9584.66 | 9710.70 | 0.878 | 0.005 |
|  | 3 | **8966.91** | **9137.44** | **0.916** | **0.001** |
|  | 4 | 8855.53 | 9070.54 | 0.924 | 0.199 |
| Tourists | 2 | 11102.339 | 11233.605 | 0.863 | 0.001 |
|  | 3 | **10526.601** | **10704.197** | **0.884** | **0.006** |
|  | 4 | 10313.192 | 10537.118 | 0.889 | 0.059 |

Notes: AIC = Akaike information criterion; BIC = Bayesian information criterion; LMR = Lo-Mendell-Rubin; LRT = Likelihood Radio Test (comparison with a (k-1) class model).

**Table 2.** GLP visitor latent profiles by perceived place value.

| Group | profile | Place Value (Item) | | | | | | | | | | |
|---|---|---|---|---|---|---|---|---|---|---|---|---|
|  |  | HR | LC | CO | SP | DI | WI | EL | AM | AT | OA | TH |
| AllVisitors | RS (11.9%) | 3.40 | 2.92 | 2.69 | 2.64 | 2.87 | 2.62 | 2.65 | 2.08 | 2.06 | 1.99 | 2.19 |
|  | AS (45.6%) | 2.17 | 2.52 | 2.84 | 2.85 | 2.86 | 3.25 | 3.16 | 4.07 | 4.07 | 4.19 | 4.33 |
|  | ES (42.6%) | 3.61 | 4.09 | 4.14 | 4.21 | 4.16 | 4.38 | 4.51 | 4.82 | 4.67 | 4.67 | 4.84 |
| Average |  | 3.06 | 3.18 | 3.22 | 3.23 | 3.30 | 3.41 | 3.44 | 3.65 | 3.60 | 3.62 | 3.79 |

Notes: RS = Relationship Seekers; AS = Activity Seekers; ES = Environment Seekers; HR = Human relationship, LC = Local culture; CM = Community; SP = Spirituality; DI = Diversity; WI = Wilderness; EL = Environmental learning; AM = Amenity; AT = Atmosphere; OA = Outdoor activity; TH = Therapeutic.

**Table 3.** Comparison between residents and tourists by profile rates.

| Number of profiles | Residents (%) | Tourists (%) |
|---|---|---|
| Relationship Seekers | 11.6 | 12.0 |
| Activity Seekers | 43.2 | 49.3 |
| Environment Seekers | 45.2 | 38.7 |

offer high value in terms of wellness, amenities, atmosphere, outdoor activity, environmental learning, and wilderness.

Visitors were then divided into two groups, residents and tourists, depending on whether they lived in or outside of Suwon where GLP is located. Each of the two groups was then subjected to LPA and, as shown in Table 3, the LPA for both groups indicated that there were fewer people who fit the RS profile than any other profile. Of the remaining two profiles, more residents fit into the ES profile (45.2%) whereas more tourists fit into the AS profile (49.3%).

## Demographic differences between residents and tourists

For a closer examination of the profiles of visitors to GLP, differences in demographic characteristics of the profiles of both residents and tourists were analyzed. There were no statistically significant differences in gender. In terms of marital status, although more residents were married than tourists, there were no statistically significant differences in perceived place value of married versus unmarried tourists between the various profiles.

On the other hand, there were statistically significant age differences for resident profiles: the RS profile consisted of residents with an average age of 47.51, with the highest proportion of individuals in their 40 s (48.6%); residents in the AS profile were an average of 42.80 years old, with the highest proportion of individuals in their 30 s (36.3%); and residents in the ES profile were an average of 50.38 years old, with the highest proportion of individuals in their 50 s (50.7.%). The LPA for tourists also showed statistically significant age differences. The average age of RS tourists was 50.40, with the highest proportion of individuals in their 50 s (45.2%); the

**Table 4.** Demographic differences between residents and tourists.

| | | Residents | | | | Tourists | | | |
| | | Latent Profile | | | | Latent Profile | | | |
| Socio-demographic | | RS | AS | ES | $\chi^2$/F | RS | AS | ES | $\chi^2$/F |
|---|---|---|---|---|---|---|---|---|---|
| Age(years) | | 47.5 | 42.8 | 50.4 | 14.089* | 50.4 | 42.1 | 48.6 | 14.537* |
| Gender(%) | Male | 31.4 | 46.9 | 38.2 | 3.631 | 45.2 | 43.9 | 37.0 | 1.805 |
| | Female | 68.6 | 53.1 | 61.8 | | 54.8 | 56.1 | 63.0 | |
| Marital status(%) | Single | 6.5 | 11.8 | 8.0 | 1.377 | 6.4 | 61.7 | 31.9 | 4.079 |
| | Married | 93.5 | 88.2 | 92.0 | | 13.1 | 46.7 | 40.1 | |
| Income/mo (Thousand dollars) | | 4.5 | 8.9 | 9.2 | 0.55 | 4.1 | 8.1 | 7.0 | 0.894 |

\*$p<.01$.
\*\*$p<.05$.

**Table 5.** Behavioral differences between residents and tourists.

| | | Residents | | | | Tourists | | | |
| | | Latent Profile | | | | Latent Profile | | | |
| Behavior Pattern | | RS | AS | ES | $\chi^2$/F | RS | AS | ES | $\chi^2$/F |
|---|---|---|---|---|---|---|---|---|---|
| Visit frequency | First visit, < 2 times/yr | 9.40% | 9.80% | 14.40% | 16.141* | 10.40% | 9.60% | 16.00% | 15.042* |
| | < 3 times/mo | 37.70% | 59.80% | 36.40% | | 53.00% | 32.90% | 53.00% | |
| | < 3 times/wk | 52.80% | 30.40% | 49.20% | | 36.60% | 57.50% | 31.00% | |
| Revisit intention | | 4.43 | 4.53 | 4.76 | 3.761** | 4.00 | 4.60 | 4.75 | 17.439* |

\*$p<.01$.
\*\*$p<.05$.

average age of AS tourists was 42.05, with the highest proportion of individuals in their 30 s (36.3%); and the average age of ES tourists was 48.57, with the highest proportion in their 50 s (47.8%). Overall, residents in the ES profile were oldest, whereas tourists in the RS profile were the oldest (see Table 4).

### Behavioral differences between residents and tourists

Table 5 shows differences in the number of visits and behavioral intentions of those in different perceived place value profiles. As to frequency of visits, the difference between residents and tourists was statistically significant ($\chi2 = 65.158$), with 48.7% of tourists having visited the park for the first time or having visited two times per year, with only 17.8% of residents having done so. Comparisons across profiles indicated that the RS profile of residents ($\chi2 = 16.141$) and the AS profile of tourists reflected the highest number of visits ($\chi2 = 15.042$).

An analysis of the differences in revisit intentions between the various GLP visitor profiles indicated that all profiles had high scores, with an average score of 4 or higher, with the score for RS < AS < ES. The RS profile showed the biggest difference between residents and tourists in terms of revisit intentions, with lower scores for tourists.

## Conclusion

### Discussion

In the era of the visitor economy (Law, 2002), sustainable management of cities is only attainable when urban spaces are understood as spaces where residents and tourists can coexist while fulfilling their own purposes. In particular, urban parks are a prime example of urban spaces where residents and tourists mingle, and the sustainability of these parks can improve if the different perceptions of place of the two groups are understood. Employing Latent Profile Analysis of perceived place value, the current study identified three valid profiles for visitors of GLP, namely

Relationship Seekers, Activity Seekers, and Environment Seekers. This analysis also found significant differences between residents and tourists in these profiles in terms of demographic and behavioral characteristics while visiting GLP. The current findings have several implications for sustainable development and management of urban parks.

First, GLP visitors were classified into Relationship Seekers, Activity Seekers, and Environment Seekers using LPA. Existing place value research has classified groups into high, middle, and low on an index that averaged sense of place components (e.g., Kaltenborn, 1998), or has taken an exploratory approach using place Q methodology to create groups (Hutson et al., 2010; Hutson & Montgomery, 2011; Wilson, 2005). These attempts are limited in that they overlook various aspects of place value or lead to the subjective evaluation that is inherent in qualitative research. By contrast, the present study used LPA to identify groups that reflect the various aspects of place value, providing a more meaningful analysis than Kaltenborn (1998)'s classification of high, middle, and low groups in the population based solely on overall averages of place value. The present study also extends Hutson et al. (2010)'s qualitative research that identified various aspects of place value by increasing the objectivity of the obtained groups by applying quantitative methods. Thus, LPA allowed an examination of the diverse aspects within each of two groups, residents and tourists, as well as a classification based on both specific statistical criteria and on sample-specific characteristics.

Second, urban parks in South Korea are currently perceived to be spaces with environmental and recreational value rather than spaces with socio-cultural value. In the history of urban parks, early urban parks were created to provide bleak cities with green spaces, which have since been utilized as spaces for various recreational activities and more recently have begun to be promoted as socio-cultural spaces for the community (Archer, 2006; Coley et al., 1997, Hayward, 1989). In this study, GLP was perceived to be low in socio-cultural value by both residents and tourists, with approximately 90% of the respondents classified as either Environment Seekers or Activity Seekers. This result may be attributed to the short history of the recently created GLP. Indeed, the use and popularity of urban parks is generally a more recent phenomenon in South Korea than in other developed countries, and South Korea appears to be progressing through the same developmental stages as those counterparts.

Third, residents assign the most value to the environmental aspects of urban parks while the highest proportion of tourists assign the most value to urban parks as spaces for recreational activities. Environment Seekers accounted for the highest proportion (45.2%) among residents, and the analysis of their age revealed that Environment Seekers were the oldest group (50.4 years old). This result shows that urban parks play a strong role as rare green spaces for urban residents, and in particular as green spaces that older people can easily find and use near their residences at a time in their lives when they may have more impediments to traveling away from home than younger people. On the other hand, Activity Seekers accounted for the highest proportion (49.3%) of tourists, and they were shown to be the youngest group (42.1 years old). In addition, the proportion of Environment Seekers among tourists (38.7%) was considerably lower than that of Environment Seekers (45.2%) among residents. This implies that the environmental value of urban parks is most attractive at the local level, but may not be attractive enough to lure visits by tourists from outside the city compared to the lure of natural resources such as mountains, rivers, lakes, or the ocean coastline, all of which would no doubt result in higher perceived environmental and ecological value.

Fourth, sustainable use of urban parks will require strengthening the role of urban parks as spaces for residents to spend time with friends and family and for tourists to promote their health and engage in recreational activities, both valued by Activity Seekers. Activity Seekers not only accounted for the highest proportion of tourists at 49.3%, but they were also the most frequent visitors to GLP. Among residents, Relationship Seekers accounted for only 11.6% of all residents and yet showed the highest frequency of visits compared to other groups of residents. This result shows that urban parks need to provide special activities to attract more tourists to those seeking social contact. If a park provides opportunities to experience activities that are not available in their nearby local park, tourists may readily visit the park despite the physical distance. On the other hand, residents will find increased utility in urban parks that are readily accessible, spaces where they can routinely spend time with friends and family. In other words, it is necessary that urban parks be perceived as a part of the everyday lives of residents so that locals feel free to visit even if there is no special purpose in terms of outdoor activities or wellness.

Together, these results indicate that there are strategies that can be used for GLP and urban parks throughout South Korea to achieve sustainable development while being used harmoniously by both residents and tourists. First, parks can be molded to emphasize different characteristics of the space: on weekdays they can be used mostly by residents, and on weekends they can be made inviting to tourists. On weekdays, it appears that urban parks need to faithfully play a role as green spaces desirable to residents, selling simple food and beverage and offering mellow music to help residents enjoy time with friends and family. On weekends, the same parks should offer an environment that enables various recreational activities that are not found elsewhere so that even tourists from other regions will be encouraged to come to the park. In the long term, given the developmental stages of urban parks, those in South Korea need to step up as spaces for residents to routinely visit and take advantage of community activities there. In this way, a virtuous circle of sustainable development can be achieved, one where urban parks also appeal to tourists as spaces for recreational activities as well as for socio-cultural activities that offer tourists a taste of culture, art, performances, and exhibitions and allow them to experience the unique local identity of a city.

### Limitations and future research

This study is subject to some limitations that may warrant future empirical research on urban parks as tourist attractions. First, the current study examined only one particular site, GLP in South Korea, and thus did not sufficiently consider the various types of urban parks (e.g., lake parks, playing fields, playgrounds) and other contextual factors affecting visitor perceptions of urban parks. Therefore, to understand urban parks more generally, future studies need to analyze more cases involving urban parks in many countries and their physical and historical characteristics. Second, the current results are limited because there is insufficient theoretical background upon which to define residents and tourists as park visitors. Tourists may not be adequately defined by simply determining that they live outside the administrative district where the attraction is located, but the definition of tourists may also need to take into account many other factors including travel distance, purposes, and behaviors. Therefore, more conceptual and scientific efforts should be given to defining tourists at various urban attractions in order to provide more practical implications for sustainable management of urban tourism.

## ORCID

*Changsup Shim* (iD) http://orcid.org/0000-0002-8000-863X

## References

Archer, D. (2006). Research note: urban parks and tourism. *Annals of Leisure Research, 9*(3-4), 277–282. https://doi.org/10.1080/11745398.2006.10816434

Ashworth, G., & Page, S. J. (2011). Urban tourism research: Recent progress and current paradoxes. *Tourism Management, 32*(1), 1–15. https://doi.org/10.1016/j.tourman.2010.02.002

Beedie, P. (2005). The adventure of urban tourism. *Journal of Travel & Tourism Marketing, 18*(3):37–48.

Bergman, L. R., & Trost, K. (2006). The person-oriented versus the variable-oriented approach: Are they complementary, opposites, or exploring different worlds? *Merrill-Palmer Quarterly, 52*(3), 601–632. https://doi.org/10.1353/mpq.2006.0023

Beverly, J. L., Uto, K., Wilkes, J., & Bothwell, P. (2008). Assessing spatial attributes of forest landscape values: an internet-based participatory mapping approach. *Canadian Journal of Forest Research*, *38*(2), 289–303. https://doi.org/10.1139/X07-149

Bourdeau, L., De Coster, L., & Paradis, S. (2001). Measuring satisfaction among festivalgoers: Differences between tourists and residents as visitors to a music festival in an urban environment. *International Journal of Arts Management*, *3*(2), 40–50.

Brown, G. (2004). Mapping spatial attributes in survey research for natural resource management: methods and applications. *Society & Natural Resources*, *18*(1), 17–39. https://doi.org/10.1080/08941920590881853

Brown, G., & Raymond, C. (2007). The relationship between place attachment and landscape values: Toward mapping place attachment. *Applied Geography*, *27*(2), 89–111. https://doi.org/10.1016/j.apgeog.2006.11.002

Chao, Y. L., & Chao, S. Y. (2017). Resident and visitor perceptions of island tourism: green sea turtle ecotourism in Penghu Archipelago. *Island Studies Journal*, *12*(2), 213–228. https://doi.org/10.24043/isj.27

Chen, C. F., & Chen, F. S. (2010). Experience quality, perceived value, satisfaction and behavioral intentions for heritage tourists. *Tourism Management*, *31*(1), 29–35. https://doi.org/10.1016/j.tourman.2009.02.008

Chen, P. T., & Hu, H. H. (2010). How determinant attributes of service quality influence customer-perceived value. *International Journal of Contemporary Hospitality Management*, *22*(4), 535–551. https://doi.org/10.1108/09596111011042730

Chiesura, A. (2004). The role of urban parks for the sustainable city. *Landscape and Urban Planning*, *68*(1), 129–138. https://doi.org/10.1016/j.landurbplan.2003.08.003

Coley, R. L., Kuo, F. E., & Sullivan, W. C. (1997). Where does community grow? The social context created by nature in urban public housing. *Environment and Behavior*, *29*(4), 468–494. https://doi.org/10.1177/001391659702900402

Cranz, G. (1989). *The politics of park design: A history of urban parks in America*. The MIT Press.

Crompton, J. L. (2007). The role of the proximate principle in the emergence of urban parks in the United Kingdom and in the United States. *Leisure Studies*, *26*(2), 213–234. https://doi.org/10.1080/02614360500521457

Cronin, J. J., Jr, Brady, M. K., & Hult, G. T. M. (2000). Assessing the effects of quality, value, and customer satisfaction on consumer behavioral intentions in service environments. *Journal of Retailing*, *76*(2), 193–218. https://doi.org/10.1016/S0022-4359(00)00028-2

del Saz Salazar, S., & Menéndez, L. G. (2007). Estimating the non-market benefits of an urban park: Does proximity matter? *Land Use Policy*, *24*(1), 296–305. https://doi.org/10.1016/j.landusepol.2005.05.011

Deng, J., Arano, K. G., Pierskalla, C., & McNeel, J. (2010). Linking urban forests and urban tourism: a case of Savannah, Georgia. *Tourism Analysis*, *15*(2), 167–181. https://doi.org/10.3727/108354210X12724863327641

Donahue, M. L., Keeler, B. L., Wood, S. A., Fisher, D. M., Hamstead, Z. A., & McPhearson, T. (2018). Using social media to understand drivers of urban park visitation in the Twin Cities. *Landscape and Urban Planning*, *175*, 1–10. https://doi.org/10.1016/j.landurbplan.2018.02.006

Gobster, P. H. (1998). Urban parks as green walls or green magnets? Interracial relations in neighborhood boundary parks. *Landscape and Urban Planning*, *41*(1), 43–55. https://doi.org/10.1016/S0169-2046(98)00045-0

Gobster, P. H. (2007). Urban park restoration and the" museumification" of nature. *Nature and Culture*, *2*(2), 95–114. https://doi.org/10.3167/nc.2007.020201

Hayward, J. (1989). Urban parks: research, planning and social change. In I. Altman & E. Zube (Eds.), *Public places and spaces*. Plenum Press.

Hilborn, J. (2009). *Dealing with crime and disorder in urban parks (No. 9)*. Office of Community Oriented Policing Services, U.S. Department of Justice.

Hinch, T. D. (1996). Urban tourism: perspectives on sustainability. *Journal of Sustainable Tourism*, *4*(2), 95–110. https://doi.org/10.1080/09669589608667261

Holbrook, M. B. (1999). *Consumer value. A framework for analysis and research*. Routledge.

Homer, P. M., & Kahle, L. R. (1988). A structural equation test of the value-attitude-behavior hierarchy. *Journal of Personality and Social Psychology*, *54*(4), 638–646. https://doi.org/10.1037/0022-3514.54.4.638

Hutson, G., Montgomery, D., & Caneday, L. (2010). Perceptions of outdoor recreation professionals toward place meanings in natural environments: A Q-method inquiry. *Journal of Leisure Research*, *42*(3), 417–442. https://doi.org/10.1080/00222216.2010.11950212

Hutson, G., & Montgomery, D. (2011). Demonstrating the value of extending qualitative research strategies into Q. *Operant Subjectivity*, *34*(4), 234–246.

Jin, N., Lee, S., & Lee, H. (2015). The effect of experience quality on perceived value, satisfaction, image and behavioral intention of water park patrons: New versus repeat visitors. *International Journal of Tourism Research*, *17*(1), 82–95. https://doi.org/10.1002/jtr.1968

Kaczynski, A. T., Potwarka, L. R., & Saelens, B. E. (2008). Association of park size, distance, and features with physical activity in neighborhood parks. *American Journal of Public Health*, *98*(8), 1451–1456. https://doi.org/10.2105/AJPH.2007.129064

Kaltenborn, B. P. (1998). Effects of sense of place on responses to environmental impacts: A study among residents in Svalbard in the Norwegian high Arctic. *Applied Geography*, *18*(2), 169–189. https://doi.org/10.1016/S0143-6228(98)00002-2

Karanikola, P., Panagopoulos, T., & Tampakis, S. (2017). Weekend visitors' views and perceptions at an urban national forest park of Cyprus during summertime. *Journal of Outdoor Recreation and Tourism, 17*, 112–121. https://doi.org/10.1016/j.jort.2016.10.002

Kim, H., & Song, H. (2017). Measuring hiking specialization and identification of latent profiles of hikers. *Landscape and Ecological Engineering, 13*(1), 59–68. https://doi.org/10.1007/s11355-016-0301-y

Kim, Y., Kim, H., & Paek, N. (2015). Comparison between hiker and non-hiker's sense of place value at Namhansanseong provincial park. *Journal of Korean Society of Rural Planning, 21*(4), 127–137. https://doi.org/10.7851/ksrp.2015.21.4.127

Law, C. M. (2002). *Urban tourism: The visitor economy and the growth of large cities.* Continuum.

Lee, J. S., Lee, C. K., & Yoon, Y. (2009). Investigating differences in antecedents to value between first-time and repeat festival-goers. *Journal of Travel & Tourism Marketing, 26*(7), 688–702.

Lee, C. K., Yoon, Y. S., & Lee, S. K. (2007). Investigating the relationships among perceived value, satisfaction, and recommendations: The case of the Korean DMZ. *Tourism Management, 28*(1), 204–214. https://doi.org/10.1016/j.tourman.2005.12.017

Lee, S., Quintal, V., & Phau, I. (2017). Investigating the push and pull factors between visitors' motivations of fringe and urban parks. *Tourism Analysis, 22*(3), 389–406. https://doi.org/10.3727/108354217X14955605216122

Li, C. L. (2020). Quality of life: The perspective of urban park recreation in three Asian cities. *Journal of Outdoor Recreation and Tourism, 29*, 100260. https://doi.org/10.1016/j.jort.2019.100260

Maas, J., Van Dillen, S. M., Verheij, R. A., & Groenewegen, P. P. (2009). Social contacts as a possible mechanism behind the relation between green space and health. *Health & Place, 15*(2), 586–595. https://doi.org/10.1016/j.healthplace.2008.09.006

Magidson, J., & Vermunt, J. (2002). Latent class models for clustering: A comparison with K-means. *Canadian Journal of Marketing Research, 20*(1), 36–43.

Masberg, B., & Jamieson, L. (1999). The visibility of public park and recreation facilities in tourism collateral materials: An exploratory study. *Journal of Vacation Marketing, 5*(2), 154–166. https://doi.org/10.1177/135676679900500204

McCleary, K. W., Weaver, P. A., & Hsu, C. H. (2007). The relationship between international leisure travelers' origin country and product satisfaction, value, service quality, and intent to return. *Journal of Travel & Tourism Marketing, 21*(2-3), 117–130.

McCormack, G. R., Rock, M., Toohey, A. M., & Hignell, D. (2010). Characteristics of urban parks associated with park use and physical activity: a review of qualitative research. *Health & Place, 16*(4), 712–726. https://doi.org/10.1016/j.healthplace.2010.03.003

Novy, J., & Colomb, C. (2017). Urban tourism and its discontents: An introduction. In C. Colomb, & J. Novy (Eds.). *Protest and resistance in the tourist city* (pp. 1–30). Routledge.

Nylund, K. L., Asparouhov, T., & Muthén, B. O. (2007). Deciding on the number of classes in latent class analysis and growth mixture modeling: A Monte Carlo simulation study. *Structural Equation Modeling: A Multidisciplinary Journal, 14*(4), 535–569. https://doi.org/10.1080/10705510701575396

Oh, C. O., Draper, J., & Dixon, A. W. (2010). Comparing resident and tourist preferences for public beach access and related amenities. *Ocean & Coastal Management, 53*(5-6), 245–251.

Park, C., & Song, H. (2018). Visitors' perceived place value and the willingness to pay in an urban lake park. *International Journal of Environmental Research and Public Health, 15*(11), 2518. https://doi.org/10.3390/ijerph15112518

Peña, A. I. P., Jamilena, D. M. F., & Molina, M. Á. R. (2012). The perceived value of the rural tourism stay and its effect on rural tourist behaviour. *Journal of Sustainable Tourism, 20*(8), 1045–1065. https://doi.org/10.1080/09669582.2012.667108

Peters, K., Elands, B., & Buijs, A. (2010). Social interactions in urban parks: Stimulating social cohesion? *Urban Forestry & Urban Greening, 9*, 93–100.

Plunkett, D., Fulthorp, K., & Paris, C. M. (2019). Examining the relationship between place attachment and behavioral loyalty in an urban park setting. *Journal of Outdoor Recreation and Tourism, 25*, 36–44. https://doi.org/10.1016/j.jort.2018.11.006

Pulido-Fernández, J. I., & Sánchez-Rivero, M. (2010). Attitudes of the cultural tourist: A latent segmentation approach. *Journal of Cultural Economics, 34*(2), 111–129. https://doi.org/10.1007/s10824-010-9115-1

Qing, W. U. (2018). Tourist satisfaction with urban parks: A case study of memorial Archway Park in Zhaoqing. *Journal of Landscape Research, 10*(6), 86–90.

Rasoolimanesh, S. M., Dahalan, N., & Jaafar, M. (2016). Tourists' perceived value and satisfaction in a community-based homestay in the Lenggong Valley World Heritage Site. *Journal of Hospitality and Tourism Management, 26*, 72–81.

Relph, E. (1976). *Place and placelessness.* Pion.

Relph, E. (1985). Geographical experiences and being-in-the-world: The phenomenological origins of geography. In D. Seamon & R. Mugerauer (Eds.), *Dwelling, place and environment: Toward a phenomenology of person and world.* Columbia University Press.

Shim, C., & Santos, C. A. (2014). Tourism, place and placelessness in the phenomenological experience of shopping malls in Seoul. *Tourism Management*, *45*, 106–114. https://doi.org/10.1016/j.tourman.2014.03.001

Simpson, F. (1999). Tourist impact in the historic centre of Prague: Resident and visitor perceptions of the historic built environment. *The Geographical Journal*, *165*(2), 173–183. https://doi.org/10.2307/3060415

Snepenger, D. J., Murphy, L., O'Connell, R., & Gregg, E. (2003). Tourists and residents use of a shopping space. *Annals of Tourism Research*, *30*(3), 567–580. https://doi.org/10.1016/S0160-7383(03)00026-4

Song, H., & Bae, S. Y. (2018). Understanding the travel motivation and patterns of international students in Korea: Using the theory of travel career pattern. *Asia Pacific Journal of Tourism Research*, *23*(2), 133–145. https://doi.org/10.1080/10941665.2017.1410193

Song, H., & Kim, H. (2018). Value-based profiles of visitors to a world heritage site: The case of Suwon Hwaseong fortress. *Sustainability*, *11*(1), 132–119. https://doi.org/10.3390/su11010132

Stedman, R. C. (2003). Is it really just a social construction?: The contribution of the physical environment to sense of place. *Society & Natural Resources*, *16*(8), 671–685. https://doi.org/10.1080/08941920309189

Tuan, Y. F. (1977). *Space and place: The perspective of experience*. University of Minnesota.

UNWTO. (2012). *Global report on city tourism. AM Reports: Volume Six*. UNWTO.

Van der Ark, L. A., & Richards, G. (2006). Attractiveness of cultural activities in European cities: A latent class approach. *Tourism Management*, *27*(6), 1408–1413.

Ven, S. (2016). Residents' participation, perceived impacts, and support for community-based ecotourism in Cambodia: A latent profile analysis. *Asia Pacific Journal of Tourism Research*, *21*(8), 836–861. https://doi.org/10.1080/10941665.2015.1075565

Welch, D. (1991). *The management of urban parks*. Longman Group UK Limited.

Wilson, I. (2005). *Person-place engagement among recreation visitors: A Q-method inquiry* [Unpublished dissertation]. Oklahoma State University.

Wong, K. K., & Domroes, M. (2005). The visual quality of urban park scenes of Kowloon Park, Hong Kong: Likeability, affective appraisal, and cross-cultural perspectives. *Environment and Planning B: Planning and Design*, *32*(4), 617–632. https://doi.org/10.1068/b31028

Xu, Y., Matarrita-Cascante, D., Lee, J. H., & Luloff, A. E. (2019). Incorporating Physical Environment-Related Factors in an Assessment of Community Attachment: Understanding Urban Park Contributions. *Sustainability*, *11*(20), 5603. https://doi.org/10.3390/su11205603

Yen, C. H., & Teng, H. Y. (2015). Celebrity involvement, perceived value, and behavioral intentions in popular media-induced tourism. *Journal of Hospitality & Tourism Research*, *39*(2), 225–244.

Zhu, X., Pfueller, S., Whitelaw, P., & Winter, C. (2010). Spatial differentiation of landscape values in the Murray River region of Victoria, Australia. *Environmental Management*, *45*(5), 896–911. https://doi.org/10.1007/s00267-010-9462-x

# Understanding backpacker sustainable behavior using the tri-component attitude model

Elizabeth Agyeiwaah ⓘD, Frederick Dayour, Felix Elvis Otoo ⓘD and Ben Goh

**ABSTRACT**

Research on sustainable practices of backpackers lacks a comprehensive model for understanding their sustainable behavior. This paper argues that the contribution of backpacker tourism to sustainable consumption can be achieved if backpackers' attitudes and behaviors are understood and managed. Predicated on the tri-component attitude model that conceptualizes attitude as a complex relationship among cognition (backpacker motivations and perceived impacts of backpacking), actions (backpacker sustainable and unsustainable behavior), and affection (backpacker satisfaction), this paper proposes a model for understanding backpacker sustainable behavior. We surveyed 400 backpackers in Ghana and tested 12 hypotheses using SPSS and AMOS software. The findings demonstrate that perceived positive impacts of backpacking predict backpacker sustainable behavior, suggesting that backpackers perceive their impacts positively, and hence engage in sustainable behaviors to reaffirm such perceptions. The results, however, reveal an insignificant relationship between the perceived positive impacts of backpacking and backpacker unsustainable behavior. The results also show that backpacker motivations explain the tendency to behave unsustainably. Based on these findings, we draw implications for promoting "intentional sustainability" by global organizations and destination management organizations.

## Introduction

Backpacker tourism is one of the niche markets that foster cross-cultural interaction, awareness, and promotion (Ooi & Laing, 2010; Scheyvens, 2002). This niche accounts for a considerable economic influence on overall tourist spending. Even at a 6% decline, the Australian backpacker tourism market, for example, accounted for AUS$4.1 billion in March 2019 according to Tourism Research Australia [TRA] (TRA, 2019). The niche also captures unique tourism trends, including "gap year" tourism and "begpack" both of which convey solipsism and novelty (Bernstein, 2019; Luzecka, 2016; Tolkach et al., 2019). Backpackers are an important tourist group in many developing countries, including Ghana, due to their interest in consuming local products and services in their attempt to spend little money at a destination and interact with host communities. Although they tend to be low spenders, their expenditure trickles down to the grassroots economy (Dayour et al., 2016; Hampton, 1998; Peel & Steen, 2007).

Beyond the economic and egalitarian value, backpacking is considered a key sustainability theme in tourism (Han et al., 2018; Iaquinto, 2015; Westerhausen & Macbeth, 2003). Achieving sustainable consumption as part of United Nations Sustainable Development Goal 12 (i.e., responsible consumption and production) requires that tourists change their unsustainable behaviors during travel and at destinations. Therefore, the sustainability practices of backpackers require closer industry and academic attention as tourist types continue to evolve in their traditional consumption characteristics and, more importantly, as the backpacker tourism enters into a stage of commercialization (Ooi & Laing, 2010). Increasingly, alternative tourism practices such as volunteer tourism and backpacking are becoming more demanding, more consumption-driven, and less sustainable (Abdulrazak & Quoquab, 2018; O'Reilly, 2006; Ooi & Laing, 2010).

Although backpacker tourism has received ample research and industry attention, there are some identifiable gaps in this literature. First, there is a global industry requirement to understand the extent to which backpacking can contribute to sustainable tourism development and practice in developing regions. It is now understood among researchers that underlying attitudinal issues among consumers prevent tourism businesses from adopting more sustainable practices. Failure to address the consumers' attitudes may result in unsustainable, and in some cases, unprofitable alternative tourism (Nok et al., 2017; Ooi & Laing, 2010).

Second, a comprehensive model for understanding the sustainability behavior of backpackers is unavailable. The lack of a comprehensive understanding of the backpacker market stems from the absence of a framework that captures the diverse constructs that influence a backpacker's decision to travel and the resulting perceived impact on behavior and experience evaluations (Nok et al., 2017; Pearce, 2007). Third, research that seeks to answer the question "how sustainable is sustainable tourism within the backpacker context?" is still nascent. Iaquinto and Pratt (2020) explored this question from a nationality perspective. They concluded that national differences contributed to a minor impact on the sustainability practices of backpackers and proposed that values, infrastructure, and corporeal abilities may offer better leverage to understand the sustainability behavior of backpackers. Berezan et al. (2013) also established the relative importance of green practices according to the nationality of the guest. While these studies further our understanding of national influences on sustainable tourism, they do not examine the complex set of factors that account for affective evaluation of visitor experiences.

Fourth, few studies identify a reverse paradigm where a sustainable path towards development is not always achieved with backpacker tourism (e.g., Bernstein, 2019). Some authors draw attention to the blurred line between backpacker tourism and mass tourism, and discuss the risks involved in promoting tourism practices that do not give serious attention to environmental issues (Ooi & Laing, 2010; Westerhausen, 2002; Westerhausen & Macbeth, 2003). As noted by Ooi and Laing (2010, p. 192), "the once 'off the beaten track' often becomes the path most trodden".

To address the gaps identified, the present study examines backpacker sustainability behavior with the aim of establishing the relationships among backpacker motivation, perceived impacts of backpacking, backpacker (un)sustainability behaviors, and their satisfaction. Specifically, the study investigates sustainable tourism development and practice within developing regions in order to provide a comprehensive framework for understanding the sustainability behavior of backpackers. The study further suggests ways by which backpacking can become more sustainable by utilizing a set of complex factors to evaluate the backpacker experience. The study holds implications for the implementation of strategies aimed at improving visitor behavior, resource efficiency, tourist behavior management, and sustainability practices. The rest of the paper reviews the relevant literature, presents the methods, results, as well as discussion, and ends with the conclusions of the study.

## Literature review

### Defining backpackers and their behavioral characteristics

Pearce (1990) defines backpackers as young travelers who are more likely to use budget accommodation facilities, meet other travelers, are independent, have a flexible travel schedule, stay longer at the destination, and focus on participatory holiday experiences. However, the description of backpackers as travelers who satisfy all the above characteristics has led some scholars to argue for a succinct approach that operationalizes backpackers as travelers who self-identify to be so (Cohen, 2011; Dayour et al., 2019). The present study uses this operationalization. Although this approach appears to have been favored by researchers in the last decade (see Hunter-Jones et al., 2008), it must be noted that the existing stereotypes associated with backpackers, including being 'noise makers', 'dopers', social miscreants, and disregarders of local norms, discourage some backpackers from disclosing their identity during travel (Dayour et al., 2017). The inherent weakness of the self-identification criterion implies that the current study may have missed out on backpackers who probably masked their identity during the data collection because of the above-mentioned stereotypes.

Backpacking is gradually becoming a popular phenomenon in the travel and tourism trade due to a wide range of economic, cultural, social, and environmental implications associated with it (Scheyvens, 2002). For instance, Visser (2004) notes that globally, backpackers' economic behaviors in South Africa contribute more towards foreign exchange earnings across a wider area within the local community compared to conventional travelers. This is because backpackers prefer to use rural or less traveled routes. Scheyvens (2002) submits that backpackers are sociable, interactive, and willing to learn local cultures and Nok et al. (2017) add that they show respect for local cultures and demonstrate low consumptive behaviors. These behaviors are further supported by residents' views. For example, residents of Lijiang in China think that backpackers are more environmentally friendly than conventional tourists given their choice of food, accommodation, and transportation (Luo et al., 2014). They are also cautious as far as the consumption of resources and energy and the disposal of garbage are concerned. Similarly, Dayour et al. (2016) note that backpackers are less acquisitive, less wasteful, and frugal in their expenditure when compared to mass tourists.

The positive traits of backpackers notwithstanding, the literature highlights some undesirable behaviors among backpackers and calls for a more controlled system to avoid possible negative impacts on destinations. Shaffer (2004) notes a bad image has been associated with backpackers because of unruly social activities such as loud 'moon parties' and use of drugs, which have been found to be common in popular backpacker enclaves in Goa, India, Chiang Mai, Thailand, and Vientiane, Laos (Howard, 2007; Scheyvens, 2002). This bad image has its roots in the activities of the hippie and beatnik subcultures of the 1960s and 1970s (Cohen, 1973). Therefore, the predominantly youthful character of backpackers (Hunter-Jones et al., 2008; Scheyvens, 2002) can lead to various social and environmental effects on destinations such as the disregard for local cultural norms, values, and principles as well as littering and other disruptive consumption behaviors. This argument finds expression in backpackers' negative association with visits to psychedelic enclaves and engagement in undesirable behaviors (Cohen, 2011). A recent study by Sroypetch et al. (2018) point to their appearance (e.g., scanty dressing) and other unruly behaviors such as over-consumption of alcohol, and overt sexual conduct that can offend host residents. Also, 'flashpackers' (i.e., high-end backpackers who often have more income than ordinary backpackers) contribute to economic leakages (Dayour et al., 2017).

Given the negative image associated with backpacking, certain destinations have discouraged visits from backpackers. For example, according to Scheyvens (2002), backpackers were discouraged from visiting the Maldives and barred from Bhutan because they were considered a threat

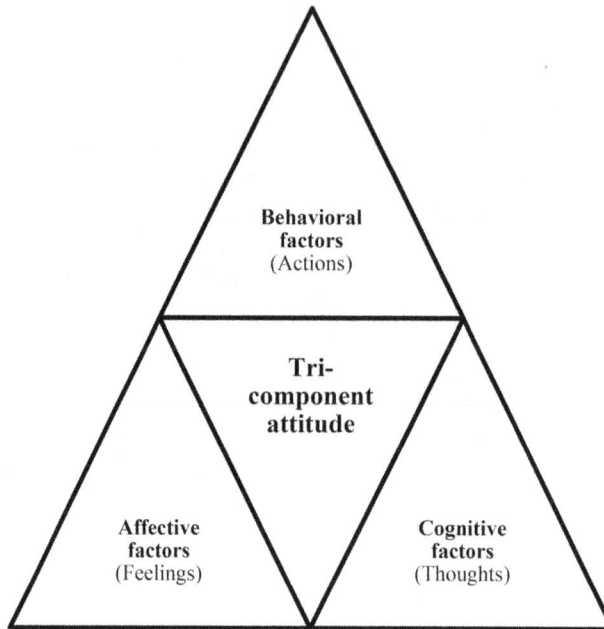

**Figure 1.** The tri-component attitude model adapted from Pickens Pickens (2005).

to the nation. However, the cognitive explanations of such (un)desirable behaviors are unknown and warrant empirical examination.

### Theoretical framework and hypotheses development

### The tri-component attitude model

Tourism scholars have explained the relationship between attitudes and behavior of visitors in various ways. Within backpacker research, for example, popular theoretical approaches include the theory of planned behavior that asserts that conscious attitudes and beliefs influence behaviors (Martin & McCurdy, 2009). Other theories, such as practice theory, overlook consciousness and emphasize agency (Iaquinto & Pratt, 2020) while social identity theory underscores how group identity influences individual behavior (Zhang et al., 2017). A limitation of these theories is the failure to recognize the complex relationships among individual thoughts, emotions, and actions. This limitation is addressed by the tri-component attitude theoretical model (see Figure 1) which defines how people consciously view situations and how they behave towards those situations (Howard & Sheth, 1969; Rosenberg & Hovland, 1960). The tri-component attitude model argues that attitude includes three components, namely cognition (a thought), an affect (a feeling), and behavior (action) (Pickens, 2005).

The cognitive component represents knowledge and information acquired from the consumer's self-knowledge, experience opinion, or learning which influences purchase or consumption desires. The affective component denotes emotional reactions or feelings towards a product or a situation. The behavioral component encompasses the psychological attitude to act towards a specific goal (Han et al., 2011; Yuan et al., 2008). Within the tourism and consumer behavior literature, motivations and perceived impacts of experiences constitute valuable constructs for understanding behavior (Prayag et al., 2018; Prayag & Brittnacher, 2014). However, backpacking studies tend to investigate these constructs separately (e.g., Chen et al., 2014; Nok et al., 2017). This study addresses this deficiency by considering backpacker sustainable behavior and backpacker satisfaction as an outcome of a combination of proximal antecedents such as backpacker motivations and perceived impacts of backpacking (Pickens, 2005; Trudel, 2019). The study develops a theoretical model comprising three components of cognition (backpacker motivations and perceived impacts of backpacking), behavior (backpacker sustainable and unsustainable behaviors), and affection (backpacker satisfaction). By making backpacker satisfaction an outcome of backpacker sustainable and unsustainable behavior, this study provides new theoretical insights into backpacker emotional response towards their (un)sustainable behavior.

### Backpacker motivation and perceived impacts of backpacking

This section discusses the link between backpacker motivations and perceived impacts of backpacking to support the proposed relationships in the model. It is noteworthy that the motivations and perceived impacts measured in this study are distinguishable from other studies as they focused on backpackers' behaviors at a destination rather than their motivations to travel. Motivation refers to a psychological need or want comprising integral forces that stimulate, direct, and join a person's behavior with an activity of any kind (Pearce, 1982). While in sociology and psychology, the definition of motivation focuses on cognitive and emotional motives (Ajzen & Fishbein, 1977) or internal and external motives (Gnoth, 1997), its definition in tourism centers on two factors: push and pull factors. These forces respectively define how tourists are pushed by intrinsic factors into making travel choices and pulled by extrinsic destination attributes (Uysal & Hagan, 1993). For instance, Dayour (2013) observed that backpackers who visited Ghana were motivated by intrinsic factors such as the need for escape, adventure and heritage tourism, and extrinsic factors such as cultural and ecological tourism. Similarly, Chen et al. (2014), in segmenting Chinese backpackers by travel motivations, found different motivations: self-actualizers, destination experiencers, and social seekers. A recent study by Nok et al. (2017) reports that the most pressing pull and push motivations for backpackers visiting Hong Kong are the unique food culture and the desire to learn new ideas respectively. This quest to learn about new cultures and countries has been previously documented by Pearce and Foster (2007). Nonetheless, not much is empirically known from these studies about the relationship between backpacker motivations and the perceived impact of backpacking.

Perceptions suggest that beliefs about tourism differ widely (Dogan, 1989). Backpacker perceived impacts in this study refer to their beliefs about the effects of their behaviors (good or bad) and/or activities at a particular location (Dogan, 1989; Weiler & Smith, 2009). Researchers and policymakers have been interested in understanding travel impacts to aid planning and the development of appropriate strategies to sustainably manage their destinations (Ap, 1992). From a mobility standpoint, both residents and tourists are aware of the impacts of tourism and their behaviors at destinations (Moscardo et al., 2013); hence, this can affect the consumption behavior of tourists (Prayag & Brittnacher, 2014). For instance, Iaquinto (2015; 2018) maintains that backpackers perform many sustainable practices that impact positively on the host, ranging from food sharing, car sharing, line-drying, purchase of local products, and services to ecotourism.

Despite the importance of tourist motivations in understanding the impact of visitations on destinations and attractions (Stone & Sharpley, 2008), the relationship between these two variables has hardly been examined within the tourism literature. For instance, Prayag et al. (2018) argue that tourists' perception of impacts at dark tourism attractions is a function of their motivations. Their study uncovered a positive relationship between tourist motivations and their perceived negative and positive impacts. More specifically, Biran et al. (2014) established that tourists who were motivated to visit dark attraction sites perceived both positive and negative. Bradt (1995) also found that backpackers who are motivated to discover novel destinations together with their predilection for authentic experiences damage the environment considerably. Similarly, Cooper et al. (2004) assert that backpackers can have a considerable impact on the environment because most of them want to have thrilling experiences, even if this will cause damage to the environment. These studies notwithstanding, backpacker tourism researchers, to date, have not examined the possible relationship between backpacker motivations and perceived impacts of backpacking. Therefore, the current study proposes that:

$H_{1a}$: Backpacker motivation has a positive effect on the perceived negative impact of backpacking.

$H_{1b}$: Backpacker motivation exerts a positive influence on the perceived positive impact of backpacking.

### Backpacker motivation and backpacker sustainable and unsustainable behavior

While the current study recognizes the contested definition of sustainability as development that meets the needs of the present without compromising future generation needs (Brundtland, 1987), it employs a triple bottom line conceptualization to reduce the ambiguity of this concept as comprising economic, socio-cultural, and environmental dimensions that mutually reinforce one another (Agyeiwaah et al., 2017; Elkington, 2013). Sustainability behaviors are responsible behaviors that contribute to triple bottom line sustainability while unsustainable behaviors do not contribute to triple bottom line sustainability (Trudel, 2019; Agyeiwaah et al. 2020).

The extant literature is replete with sustainable practices among backpackers. For instance, Nok et al. (2017) acknowledge that backpacking is an important conduit to the cultural and environmental sustainability of tourism since backpackers show respect for local cultures and have low consumptive behaviors. According to Scheyvens (2002), backpackers give advantages to local cultures by being sociable, interacting with local folks, and learning about new customs, thereby contributing to the rejuvenation and sustenance of local cultures. Interacting with local cultures also provides benefits to local economies through the patronage of local goods and services, which in turn benefits a wider spectrum of stakeholders. Iaquinto (2018) also argues that although backpackers may not be mindful of sustainability in their travel, sustainability exists in their practices.

Even though there appears to be limited empirical evidence in the literature to support the theoretical relationship between travel motivation and behavior, few studies speculate a possible correlation between the two variables, thereby necessitating the testing of this relationship in the current study. Eagles (1992) and Fodness (1994) are of the view that travel motivation is possibly the most important factor in comprehending tourist behavior at the destination and should be considered a significant driver in their behavioral analysis. Motivation can explain why a person or a group may act in a good or bad way to satisfy a need (Dann, 1981). For instance, a study by Abdulrazak and Quoquab (2018) which considered consumers' motivations for sustainable consumption, reported that respondents were motivated to engage in sustainable consumption based on the desire to feel good. However, sustainable consumption was found to be marginally affected by the need for competence among these respondents. In another study, Teo et al. (2014) observed that tourists interested in heritage sites were motivated to take action in support of sustainable ecotourism in order to show respect for local cultures. On undesired behaviors, Scheyvens (2002) reports that backpackers who desire to engage in psychedelic and unacceptable consumption behaviors are likely to have a negative impact on the destination and contribute to its unsustainability. Given the evidence in the literature in support of a possible relationship between backpacker motivation and behavior, backpacker motivation is hypothesized to have a positive impact on backpacker (un)sustainable behavior in this study.

$H_{2a}$: Backpacker motivation has a positive effect on backpacker sustainable behavior.

$H_{2b}$: Backpacker motivation has a positive effect on backpacker unsustainable behavior.

### Backpacker perceived impacts and behavior (sustainable and unsustainable)

Previous studies show that tourism development and tourists generate diverse positive impacts (e.g., construction projects, wildlife preservation, trickle-down effects, and improvement in the quality of life) as well as negative ones (e.g., traffic congestion, the surge in crime, alcoholism, and alteration of community identity) (Andereck et al., 2005; Ko & Stewart, 2002). Many such studies have focused on residents' perceptions at the expense of tourist perceptions, and have often concluded that residents who obtain beneficial outcomes perceive tourism positively. Generally, although an understanding of tourism perceptions and attitudes has increased in recent years, there is a limited understanding of how these perceptions of impacts result in sustainable and unsustainable behavior (Pietilä & Fagerholm, 2016; Prayag et al., 2018).

Among the research on tourists' perception of impacts, there is consensus that tourists' perceptions of impacts influence their behavior, including post-consumption behavior (Cottrell et al., 2004; Mainieri et al., 1997; Prayag et al., 2018; Prayag & Brittnacher, 2014). In the environmental psychology literature, Mainieri et al. (1997) found positive relationships between consumers' environmental concerns and their pro-environmental behavior using hierarchical multiple regression analysis. Cottrell et al. (2004) found that Germans were more concerned about the ecological impacts of tourism development as opposed to Dutch tourists and were more likely to promote local products. Zhang et al. (2015) also suggest that a perceived negative impact of haze pollution in China could influence the behavior of tourists. Further, it is expected that tourists who perceive beneficial outcomes of their travel will be more positive towards sustainability issues than those who perceive negative impacts (Han et al., 2018). From the preceding studies, it can be inferred that the more negative perceptions held by the backpacker, the less likely it is that s/he will engage in sustainable behavior and the more positive perceptions held, the higher the likelihood to engage in sustainable behavior. These inferences are presented in the following hypotheses:

$H_{3a}$: Perceived negative impact of backpacking has a negative influence on backpacker sustainable behavior.

$H_{3b}$: Perceived positive impact of backpacking has a positive influence on backpacker sustainable behavior.

$H_{4a}$: Perceived negative impact of backpacking has a positive impact on backpacker unsustainable behavior.

$H_{4b}$: Perceived positive impact of backpacking has a negative impact on backpacker unsustainable behavior.

### Perceived impacts of backpacking and backpacker satisfaction

Despite the contested definition of customer satisfaction, there is consensus in consumer behavior research that customer satisfaction is a comparison of what is expected and what is actually received (Oliver, 1993). In tourism research, tourist satisfaction is defined as an "individual's cognitive-affective state derived from a tourist experience" (del Bosque & San Martín, 2008, p. 553). A notable framework for examining satisfaction is the cognitive-affective-conative framework, which conceptualizes satisfaction as an emotional response to an experience (Han et al., 2011). There is evidence in the literature to suggest that there is a relationship between tourists' perceived impacts and satisfaction with tourism experience. For instance, previous studies within the dark tourism context affirm that while positive tourism impacts result in a strong statistically significant influence on satisfaction, negative perceived impacts result in an inverse but moderately significant effect on satisfaction (Prayag et al., 2018).

Similarly, Wang and Luo (2018) suggest that tourists' connection to destinations predisposes them to be highly sensitive to the impacts of tourism. The result is that such tourists tend to be more satisfied with their experiences. These relationships have been previously confirmed in studies on residents' attitudes that indicate that community satisfaction is positively related to perceived positive impacts whereas community satisfaction is negatively related to perceived negative impacts (Ko & Stewart, 2002). Despite the growing scholarly interest in backpacker tourism, the literature on perceived impacts-satisfaction relationship remains scant and usually adopts a qualitative approach (Wearing et al., 2002). Employing a quantitative approach to address this gap, the present study hypothesizes that:

$H_{5a}$: Perceived negative impact of backpacking has a negative influence on backpackers' satisfaction.

$H_{5b}$: Perceived positive impact of backpacking has a positive influence on backpackers' satisfaction.

### Backpacker sustainable and unsustainable behavior and backpacker satisfaction

The relationship between tourists' behavior and their evaluation of destination experiences is complex and mixed. Nonetheless, the conclusion that tourists with more sustainable behavior are more satisfied than those with less sustainable behavior is common (Berezan et al., 2013; Kastenholz et al., 2018; Nassani et al., 2013). For example, Berezan et al. (2013) tested the hypothetical relationship between green practices and visitors' overall satisfaction. They found a positive relationship between green practices on guests' satisfaction levels. In their investigation of tourists to Taiwan, Liu et al. (2016) tested whether sustainable tourism behavior is an antecedent of overall satisfaction and confirmed weak but statistically significant support for the hypothesis.

However, another stream of research holds an opposite view. For instance, Dodds et al. (2010) found that despite being generally happy during their vacation, the tourists in their study requested improvement in service through unsustainable and egotistic practices such as consistent power supply and freshwater showers, all of which overstretch communal resources, contribute to a surge in energy consumption and lead to material footprint per capita (Cohen, 2011; Cottrell et al., 2004; Scheyvens, 2002). Despite these contested findings coupled with the interest in backpacker research among sustainability scholars, empirical studies that test the relationship between backpackers' (un)sustainable behavior and satisfaction is absent in the backpacker literature (see Nash et al., 2006). Based on a recent study by Kastenholz et al. (2018) on rural tourists that show that clusters that exhibit more sustainable behavior possessed higher levels of satisfaction than those that exhibit less sustainable behavior, this paper assumes that backpackers that engage in sustainable behaviors will be more satisfied than those who exhibit unsustainable behavior. Consequently, it is hypothesized that:

$H_{6a}$: Backpacker sustainable behaviors have a positive influence on backpackers' satisfaction.

$H_{6b}$: Backpacker unsustainable behaviors have a negative influence on backpackers' satisfaction.

The hypothetical relationships are conceptually presented in Figure 2.

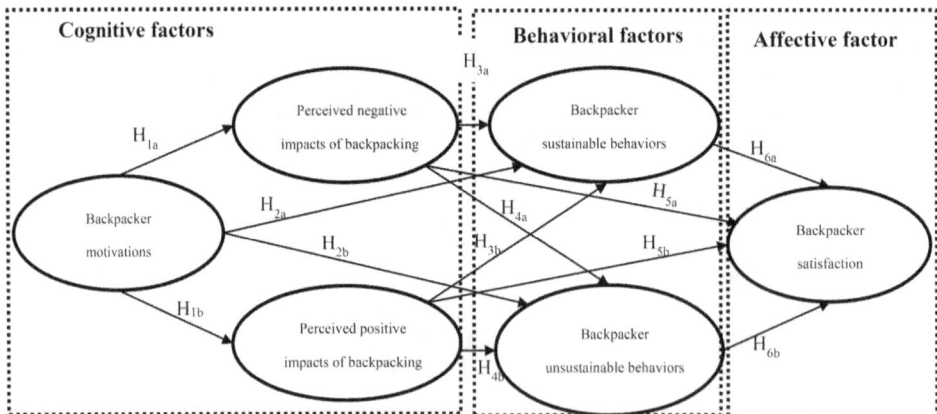

**Figure 2.** A proposed structural model.

## Methods

### Research setting

Ghana is recognized as one of the destinations that appeal to backpackers as far as the West African sub-region is concerned. This is because the country is recognized as a peaceful destination with rich cultural and historical resources for backpackers (Dayour et al., 2019). The study area, Cape Coast, is one of the popular tourist destinations in Ghana and it is home to the Cape Coast Castle, one of Ghana's flagship attractions popular for the infamous transatlantic slave trade. Also, the Oasis Beach Resort is currently the most popular enclave for backpackers and other tourists in Cape Coast (Dayour et al., 2019). Backpackers in Cape Coast predominantly come from North America and Europe (Adam, 2015). A typical backpacker in Cape Coast is commonly seen with their backpack and they prefer to live in backpacker hostels. They also prefer to interact with local residents and like to learn Ghanaian history and culture. Such behaviors have made their presence conspicuous to residents and researchers (Dayour, 2013).

### Measures

A structured questionnaire was used to collect data on backpacker motivations, perceived impacts of backpacking, backpacker (un)sustainable behaviors, and backpacker satisfaction as well as demographics and travel characteristics. The measures of constructs were generated from the literature and modified to suit the context of the present study. Backpacker motivation was assessed on a 7-point Likert scale [1 = Not at all important to 7 = Very important]. Perceived impacts of backpacking and backpacker satisfaction were also assessed based on a 7-point Likert scale [1 = Strongly disagree to 7 = Strongly agree]. However, backpacker (un)sustainable behaviors were measured on a 6-point Likert scale [1 = Very frequently to 6 = Never]. The questionnaire was organized into five parts. The first part comprised measures on backpackers' motivations such as asking backpackers to indicate whether they wanted "To learn about/experience another culture" (Nok et al., 2017; Pearce & Foster, 2007). These items were adopted from Pearce and Foster (2007, p. 1291)'s backpacker studies since the items reflect specific backpacker profiles. The second part examined backpacker sustainable behaviors (e.g., "Read the history of your destination" and "Buy and choose environmentally friendly accommodation") and backpacker unsustainable behaviors (e.g., "Smoking anywhere without considering those around them" and "causing congestion") in relation to the triple bottom line dimensions (Agyeiwaah et al., 2017; Iaquinto, 2015; Nok et al., 2017). The third part examined the perceived impact of backpacking, including both positive and negative aspects, based on previous impact studies (see Andereck et al., 2005; Ko & Stewart, 2002). The fourth part examined backpackers' satisfaction with the choice of destination (Agyeiwaah et al., 2016) and the final part examined demographics and travel characteristics such as gender, age, educational attainment, nationality, frequency of visit, travel party, and length of stay.

### Data collection and procedure

Following Chen and Huang (2017), face validity and content validity checks were carried out on the questionnaire before the actual survey. Doctoral students and academics in the field of tourism were consulted to ascertain whether the statements in the instrument were comprehensible. To further enhance the wording and relevance of statements, a pretesting of the questionnaire using 50 backpackers in Elmina (a nearby tourist destination around the study area) (Dayour et al., 2016) was carried out. Since most backpackers could read and write English, the questionnaire was administered in English. The data collection took place between August and

September 2019. The self-identification approach was used to select backpackers for the study (Adam, 2015; Hunter-Jones et al., 2008).

Informed by past studies (Adam, 2015; Dayour, 2013), we selected the Oasis Beach Resort and Cape Coast Castle as our research site. The former, according to the Ghana Tourism Authority's classification of hostels in Ghana, is a budget accommodation facility (Dayour et al., 2019). According to Dayour (2013), these locations are hotspots for backpackers in the city. Permission was sought from the managers of these facilities to conduct the survey. Two trained research assistants and one of the authors who was based in Ghana surveyed the two places. A convenience sampling technique was used to sample respondents; consequently, the researchers approached and introduced themselves to the guests, and explained the rationale for the study. The researchers then asked for guests' consent to be included in the survey. Those who consented were asked to indicate which of the following best described them: backpacker, traveler, or tourists. Those who identified as backpackers were handed a questionnaire to complete (Dayour et al., 2019). Persons who participated in the survey were not incentivized or coerced in any way. This is because surveys were conducted during checkout from the Cape Coast Castle, a time when most respondents are willing to respond to the questions while relaxing from their tour (Dayour et al., 2019). For those contacted at Oasis Beach Resort, the survey was conducted when respondents were waiting for their orders from the restaurant or simply relaxing at the beach. Despite categorizing the age range to capture the lowest age of backpackers based on the literature, we focused on only backpackers who were 18 years or older so that minors were excluded. A total of 420 questionnaires were administered of which 400 were found to be correctly completed, and therefore usable for the study. The response rate for the survey was 95.2%.

## Data analysis

A three-step approach was adopted in analyzing the data. First, an Exploratory Factor Analysis (EFA) was conducted to ensure that items were well loaded under their relevant constructs (Pallant, 2005). This was necessary because of the lack of a priori (confirmed) measures and constructs for the proposed model, thereby requiring an initial rotation of the items generated to determine their relevance for the model. As part of the EFA, one of the three items from the literature measuring perceived negative impacts of backpacking had a lower internal consistency of 0.436 Cronbach alpha (i.e., backpacker contributes to the increased cost of living). Deletion would increase the Cronbach alpha to 0.66 and improve the fit indices (Pallant, 2005); hence, the statement was deleted. Thus, the remaining two items measuring how backpacking contributes to prostitution and overcrowding in Ghana were used since they are statistically valid and reliable measures (see Table 4).

Second, a Confirmatory Factor Analysis (CFA) technique (in AMOS 22) was used to determine how well the proposed model fits the dataset before evaluating the proposed structural model (Hair et al., 2014). Here, validity and reliability analyses were conducted (Fornell & Larcker, 1981). Finally, the proposed theoretical model and hypotheses were tested using the maximum likelihood estimation method in covariance-based Structural Equation Modeling (SEM).

The demographic analysis revealed that of the 400 respondents surveyed, 45.5% were males and 54.5% were females consistent with previous backpacker studies (e.g., Iaquinto, 2015). The age groups of the respondents show that they are predominantly youthful travelers: 16-24 years old (38.5%), 25-34 years old (49.8%), 34-44 years old (8.0%), and 45 and above years old (3.7%). More than half (52.3%) of the respondents were college graduates, followed by those with postgraduate degrees (22.7%). The rest were high school graduates (16.8%), those with professional qualifications (7.5%), and others (0.7%). The top three nationalities were Germans (21.5%), British

**Table 1.** Demographic characteristics.

| Variable | Frequency | Percentage (%) |
|---|---|---|
| **Gender** | | |
| Male | 182 | 45.5 |
| Female | 218 | 54.5 |
| **Age** | | |
| 16-24 | 154 | 38.5 |
| 25-34 | 199 | 49.8 |
| 35-44 | 32 | 8.0 |
| 45 and above | 15 | 3.7 |
| **Level of education** | | |
| High school graduate or less | 67 | 16.8 |
| College graduate - undergraduate | 209 | 52.3 |
| Postgraduate degree | 91 | 22.7 |
| Professional qualification | 30 | 7.5 |
| Others | 3 | 0.7 |
| **Nationality** | | |
| German | 86 | 21.5 |
| British | 81 | 20.3 |
| American | 66 | 16.5 |
| Spanish | 56 | 14.0 |
| French | 47 | 11.8 |
| Australian | 38 | 9.5 |
| Others | 25 | 6.2 |
| South African | 1 | 0.2 |
| **Frequency of visits** | | |
| No previous visit | 45 | 11.3 |
| One time | 338 | 84.5 |
| Two times | 17 | 4.2 |
| **Travel party** | | |
| Friends | 185 | 46.3 |
| Alone | 182 | 45.5 |
| Organized tour | 15 | 3.8 |
| With your Spouse/Partner | 9 | 2.2 |
| With family members | 6 | 1.5 |
| Others | 3 | 0.7 |
| **Length of stay (in days)** | | |
| 1 to 5 | 246 | 61.5 |
| 6 to 10 | 78 | 19.5 |
| 11 to 15 | 68 | 17.0 |
| 16 days or more | 8 | 2.0 |

(20.3%) and Americans (16.5%), which are typical of Ghanaian backpacker generation regions of Europe and North America (Adam, 2015) (Table 1).

### Common methods bias (CMB)

Before presenting the study results, we examined CMB in the measures used in two ways. First, the correlation matrix in Table 4 was checked for any prospective high correlations and we found that the correlations among constructs were up to 0.22; hence, not considered high ($r > 0.90$) (Park & Tussyadiah, 2017). Second, based on Harman's single-factor (Podsakoff et al., 2003) approach, we noticed that the variance explained by the first (major) factor was 14.12% – less than the 50% cut-off point. This procedure confirmed the absence of common method errors in the results.

## Results

### Results of confirmatory factor analysis (CFA)

Before the CFA, initial results from the EFA helped to reduce the data into six factors that explain 65.5% variance. The EFA revealed a significant (*Sig. =0.000; df = 253*) Bartlett's test of sphericity

**Table 2.** Goodness-of-fit measures (N = 400).

| Stage | Chi-square (df) | P-value | RMSEA | SRMR | GFI | PCLOSE | AGFI | CFI | IFI | RFI |
|-------|----------------|---------|-------|------|-----|--------|------|-----|-----|-----|
| CFA | 309.27(209) | 0.000 | 0.04 | 0.06 | 0.94 | 1.00 | 0.92 | 0.98 | 0.98 | 0.91 |
| SEM | 311.56(212) | 0.000 | 0.03 | 0.06 | 0.94 | 1.00 | 0.92 | 0.98 | 0.98 | 0.91 |

Note: RMSEA (root mean square error of approximation), SRMR (standardized RMR), GFI (goodness-of-fit index), AGFI (adjusted GFI), CFI (comparative fit index), IFI (incremental fit index), and RFI (relative fit index).

and a Kaiser-Meyer-Olkin Measure of Sampling Adequacy ($KMO = 0.752$) greater than the recommended minimum of 0.6 (Pallant, 2005). Further results of a confirmatory factor analysis using the maximum likelihood estimation methods in AMOS revealed that the proposed measurement model fits the data well (Table 2). This was demonstrated through the multiple fit measures ($\chi^2$ =309.27; $df = 209$, $p < 0.001$, $\chi^2/df = 1.480$, CFI= 0.98, GFI = 0.94, RMSEA = 0.04 and PCLOSE 1.00) which represent the often applied fit indices for evaluating the measurement model (Dion, 2008). A detailed examination of the measurement model revealed predominantly significant loadings of measurement items of the constructs (Table 3). As suggested by Ursachi et al. (2015), measurement items must be simultaneously reliable in terms of their consistency in measuring a specific phenomenon and valid in terms of measuring what it is supposed to measure. Consequently, the internal consistency test of the measurement items using Cronbach's alpha revealed acceptable results of 0.6-0.9 (Ursachi et al., 2015). Following Brown's (2015) considerations about skewness (-3 to +3) and kurtosis (-10 to +10) when conducting SEM, the data was assessed to be symmetrically distributed as per the scores in Table 3. Further validity tests involved both convergent validity and discriminant validity to validate the measures of each construct. Following suggestions in the literature (e.g., Prayag et al., 2018) to inspect both composite reliability and average variance extracted to remove an item, convergent validity was established with all AVE values exceeding the recommended cut off point greater than or equal to 0.50 (Fornell & Larcker, 1981) except for the construct "backpacker unsustainable behaviors."

The construct, "backpacker unsustainable behaviors", was further validated following Fornell and Larcker (1981) argument that in instances where AVE is lesser than the recommended 0.5 but composite reliability is higher than 0.6, convergent validity is adequate for the construct (Table 4). Further composite reliability tests revealed adequate results, ranging from 0.7-0.92 (Hair et al., 2014). An additional test of discriminant validity showed that AVE values were greater than the squared correlation and maximum shared variance (Han et al., 2018). The assessment of the model fit, reliability tests, and construct validity of the measures of the various constructs was followed by the structural model analysis to either confirm or disconfirm the hypothesized relationship.

### Results of the structural model

Following the CFA, structural equation modeling was conducted to test the hypothesized relationships in the proposed model (Table 5). An ideal fit was found between the model and the data as summarized in the results ($\chi^2$ =311.56, $df = 212$, $p < 0.001$, $\chi^2/df = 1.470$, CFI = 0.98, GFI =0.94, RMSEA = 0.03 and PCLOSE 1.00). The results of the structural model supported six (H1$_b$, H2$_a$, H2$_b$, H3$_b$, H5$_b$, and H6$_a$) out of the twelve proposed relationships. For instance, backpacker motivation was found to exert a positive impact on perceived positive impacts of backpacking ($\beta = -0.22$; $p = 0.000$), backpacker sustainable behavior ($\beta = -0.12$; $p = 0.037$) and unsustainable behavior ($\beta = 0.19$; $p = 0.003$), thereby supporting H1$_b$, H2$_a$, and H2$_b$. These relationships substantiate the argument that behavior is an outcome of cognitive factors as backpacker sustainable and unsustainable behavior at the destination is explained by their motivations for travel (Pickens, 2005; Trudel, 2019). While backpacker motivations lead these travelers to perceive backpacker tourism as possessing positive impacts on community well-being, economy, and environment, these backpacker motivations do not influence their perceived negative impacts of

**Table 3.** Results of exploratory and confirmatory factor analysis.

| Constructs and items | Mean | EFA factor loadings | CFA Standardized factor loading | Cronbach's α | S.E. | t-values | Skewness | Kurtosis |
|---|---|---|---|---|---|---|---|---|
| *Backpacker motivations [BM]* | | | | 0.88 | | | | |
| To learn about / experience another culture | 6.29 | 0.915 | 0.96 | | 0.13 | 11.85*** | −1.36 | 2.12 |
| To learn about / experience another country | 6.30 | 0.884 | 0.93 | | 0.14 | 11.43*** | −1.23 | 0.86 |
| To interact with people of the host country | 6.22 | 0.883 | 0.75 | | 0.10 | 12.70*** | −0.99 | −0.09 |
| To explore and ask questions | 6.01 | 0.701 | 0.54 | | N/A | N/A | −0.94 | 0.60 |
| **Perceived negative impacts of backpacking [PNI]** | | | | 0.66 | | | | |
| Backpacking contributes to overcrowding | 3.44 | 0.859 | 0.60 | | 0.39 | 1.93 | −0.05 | −0.75 |
| Backpacking contributes to prostitution | 2.63 | 0.853 | 0.82 | | N/A | N/A | 0.46 | −0.91 |
| **Perceived positive impacts of backpacking [PPI]** | | | | 0.74 | | | | |
| Backpacking contributes to the preservation of natural areas | 5.21 | 0.894 | 0.90 | | 0.10 | 11.62*** | −0.64 | 0.33 |
| Backpacking contributes to the preservation of wildlife habitats | 5.22 | 0.857 | 0.80 | | N/A | N/A | −0.66 | 0.47 |
| Backpacking contributes to supporting the development of the local community | 5.10 | 0.605 | 0.45 | | 0.07 | 8.55*** | −0.72 | 0.61 |
| **Backpacker sustainable behaviors [SB]** | | | | 0.83 | | | | |
| Read the history of your destination | 1.71 | 0.807 | 0.65 | | N/A | N/A | 1.74 | 4.11 |
| Learn about Indigenous cultures | 1.73 | 0.796 | 0.69 | | 0.07 | 14.98*** | 1.42 | 1.87 |
| Interact with local residents | 1.70 | 0.782 | 0.78 | | 0.13 | 9.54*** | 1.43 | 2.05 |
| Ask permission before photographing | 1.78 | 0.742 | 0.69 | | 0.12 | 8.80*** | 1.26 | 1.28 |
| Buy and choose environmentally friendly accommodation | 1.79 | 0.676 | 0.58 | | 0.10 | 9.30*** | 1.42 | 2.06 |
| **Backpacker unsustainable behaviors [UB]** | | | | 0.69 | | | | |
| Smoking anywhere without considering those around them | 5.23 | 0.792 | 0.79 | | 0.19 | 8.12*** | −1.59 | 2.38 |
| Causing congestion or crowding problems because of their group behavior | 4.99 | 0.695 | 0.57 | | 0.16 | 7.50*** | −1.01 | 0.32 |
| Not respecting the religious or spiritual needs of others | 5.24 | 0.628 | 0.52 | | N/A | N/A | −1.64 | 2.67 |
| Expecting to be served before locals | 5.29 | 0.620 | 0.50 | | 0.14 | 6.92*** | −1.97 | 4.21 |
| Leaving TV, lights, and fan on always | 5.18 | 0.569 | 0.42 | | 0.14 | 6.15*** | −1.68 | 2.98 |

*(continued)*

**Table 3.** Continued.

| Constructs and items | Mean | EFA factor loadings | CFA Standardized factor loading | Cronbach's α | S.E. | t-values | Skewness | Kurtosis |
|---|---|---|---|---|---|---|---|---|
| **Backpacker satisfaction [SAT]** | | | | 0.91 | | | | |
| I truly enjoyed the experience of backpacking in Ghana. | 6.14 | 0.922 | 0.96 | | 0.08 | 16.10*** | −0.99 | 0.47 |
| I am satisfied with the decision to backpack in Ghana. | 6.18 | 0.896 | 0.83 | | 0.06 | 19.87*** | −1.56 | 3.37 |
| I feel good about the decision to backpack in Ghana. | 6.20 | 0.888 | 0.89 | | 0.07 | 16.19*** | −0.97 | 0.15 |
| I am sure it was the right thing to backpack in Ghana. | 6.11 | 0.820 | 0.70 | | N/A | N/A | −1.24 | 1.56 |

Note: S.E. = standard error for unstandardized coefficient; ***Significant at significant at p < 0.001; **Significant at significant at p < 0.01; *Significant at significant at p < 0.05.

**Table 4.** Validity measures and correlation matrix.

| | CR | AVE | MSV | Mean | SD | BM | PNI | PPI | SB | UB | SAT |
|---|---|---|---|---|---|---|---|---|---|---|---|
| BM | 0.92 | 0.79 | 0.04 | 6.20 | 0.83 | 1.00 | | | | | |
| PNI | 0.70 | 0.52 | 0.01 | 3.03 | 1.43 | −0.01 (0.00) | 1.00 | | | | |
| PPI | 0.77 | 0.55 | 0.03 | 5.18 | 1.09 | −0.19**(0.04) | −0.06(0.00) | 1.00 | | | |
| SB | 0.81 | 0.50 | 0.06 | 1.74 | 0.72 | −0.13*(0.02) | −0.01(0.00) | −0.14**(0.02) | 1.00 | | |
| UB | 0.70 | 0.33 | 0.04 | 5.18 | 0.74 | 0.17**(0.03) | 0.04(0.00) | −0.13*(0.02) | −0.01(0.00) | 1.00 | |
| SAT | 0.91 | 0.72 | 0.06 | 6.16 | 0.92 | 0.02(0.00) | −0.08(0.01) | 0.22**(0.05) | −0.21**(0.04) | 0.01(0.00) | 1.00 |

Note: Maximum Shared Variance (MSV), Composite Reliability (CR), Average Variance Extracted (AVE); **Correlation is significant at the 0.01 level and *0.05; Squared correlation is in parenthesis.

**Table 5.** Path coefficients of the structural equation model.

| Hypotheses | | β | S.E. | t-values | p | Supported? |
|---|---|---|---|---|---|---|
| H$_{1a}$: Backpacker motivation | Perceived negative impacts of backpacking | 0.02 | 0.15 | 0.36 | 0.721 | No |
| H$_{1b}$: Backpacker motivation | Perceived positive impact of backpacking | −0.22 | 0.11 | −3.72 | *** | Yes |
| H$_{2a}$: Backpacker motivation | Backpacker sustainable behavior | −0.12 | 0.06 | −2.08 | 0.037* | Yes |
| H$_{2b}$: Backpacker motivation | Backpacker unsustainable behavior | 0.19 | 0.06 | 2.97 | 0.003** | Yes |
| H$_{3a}$: Perceived negative impact of backpacking | Backpacker sustainable behavior | −0.00 | 0.02 | −0.07 | 0.943 | No |
| H$_{3b}$: Perceived positive impact of backpacking | Backpacker sustainable behavior | −0.21 | 0.04 | −3.31 | *** | Yes |
| H$_{4a}$: Perceived negative impact of backpacking | Backpacker unsustainable behavior | 0.05 | 0.02 | 0.69 | 0.490 | No |
| H$_{4b}$: Perceived positive impact of backpacking | Backpacker unsustainable behavior | −0.08 | 0.03 | −1.29 | 0.199 | No |
| H$_{5a}$: Perceived negative impact of backpacking | Backpacker satisfaction | −0.09 | 0.04 | −0.93 | 0.351 | No |
| H$_{5b}$: Perceived positive impact of backpacking | Backpacker satisfaction | 0.12 | 0.04 | 2.14 | 0.033* | Yes |
| H$_{6a}$: Backpacker sustainable behavior | Backpacker satisfaction | −0.21 | 0.08 | −3.53 | *** | Yes |
| H$_{6b}$: Backpacker unsustainable behavior | Backpacker satisfaction | 0.04 | 0.08 | 0.66 | 0.510 | No |

Note: S.E. = standard error for unstandardized coefficient; ***Significant at significant at p < 0.001; **Significant at significant at p < 0.01; *Significant at significant at p < 0.05.

backpacking. Also, $H_{3b}$, $H_{5b}$, and $H_{6a}$ were supported as perceived positive impacts of backpacking was found to influence backpacker sustainable behaviors ($\beta$=-0.21; $p$ = 0.000) and backpacker satisfaction ($\beta$ = 0.12; $p$ = 0.033). Additionally, backpacker sustainable behavior had a positive influence on backpacker satisfaction ($\beta$=-0.21; $p$ = 0.000). Further, positive perceptions of backpacking stimulated sustainable behavior among this group (Prayag & Brittnacher, 2014). Six hypotheses were not supported ($H_{1a}$, $H_{3a}$, $H_{4a}$, $H_{4b}$, $H_{5a}$, and $H_{6b}$). For example, backpacker motivations did not significantly influence the perceived negative impacts of backpacking ($\beta$ = 0.02; $p$ = 0.721) and perceived negative impacts of backpacking did not have a negative influence on sustainable behavior ($\beta$=-0.00; $p$ = 0.943) and backpacker satisfaction ($\beta$=-0.09; $p$ = 0.351). Moreover, the hypothesized relationship between perceived negative impacts of backpacking and backpacker unsustainable behavior was not significant ($\beta$ = 0.05; p = 0.490), perceived positive impacts of backpacking did not influence backpacker unsustainable behavior ($\beta$=-0.08; $p$ = 0.199) and backpacker unsustainable behavior equally did not have any influence on backpacker satisfaction ($\beta$ = 0.04; $p$ = 0.510). These insightful findings of predictors and outcomes of backpacker sustainable and unsustainable behaviors have been further discussed together with their implications for achieving sustainable consumption as part of United Nations Sustainable Development Goal 12 (United Nations [UN. , 2019). A summary of the results of the structural equation modeling is presented in Figure 3.

## Discussion

Employing the tri-component attitude model (Howard & Sheth, 1969; Rosenberg & Hovland, 1960), the present study has investigated the relationship between cognitive (backpacker motivations and perceived impacts of backpacking), behavioral (backpacker sustainable and unsustainable behavior), and affective (backpacker satisfaction) components of backpacking. This study addresses a gap in the literature with respect to the absence of a comprehensive model for examining the antecedents and outcomes of backpackers' sustainable and unsustainable behaviors.

The results demonstrate strong support for previous backpacker studies that stated that backpacker motivations can lead to both desirable and undesirable behaviors at the destination. For example, previous backpacker studies suggest that backpacker motivations are predominantly positive and comprise learning new cultures, social interactions, self-actualization, and personal development (Chen et al., 2014; Loker-Murphy, 1997; Pearce & Foster, 2007). While these positive motivations lead backpackers to perceive their impacts positively (Nok et al., 2017), positive motivations can lead some backpackers to engage in unsustainable behaviors when an extreme emphasis on the "youthful self" translate into irresponsible behaviors at the destination (Cohen, 2011; Scheyvens, 2002). This suggests the possibility of backpacker motivation to trigger divergent behavioral outcomes. For the most part, the two conflicting (un)sustainable behavioral outcomes of backpacker motivations in this study lay further claims that backpackers have varied levels of motivations as suggested in backpacker motivation-based cluster studies (Chen et al., 2014). It can also be argued from a mobility perspective that when backpackers are highly motivated to backpack at a destination, they become sensitive to the beneficial outcomes of their visits since sustainability exists in their practices as confirmed by studies in Hong Kong (Nok et al., 2017) and Australia (Iaquinto, 2018).

Perceived positive impacts of backpacking had a positive influence on both backpackers' sustainable behavior and backpacker satisfaction as reflected in both environmental psychology and tourism sustainability studies (Cottrell et al., 2004; Mainieri et al., 1997) as well as tourism impacts studies (Ko & Stewart, 2002; Prayag et al., 2018). Backpackers who engaged in sustainable behaviors are more satisfied, confirming the findings of previous studies (Kastenholz et al., 2018; Nassani et al., 2013). Nonetheless, contrary to the hypothesized relationships, perceived negative

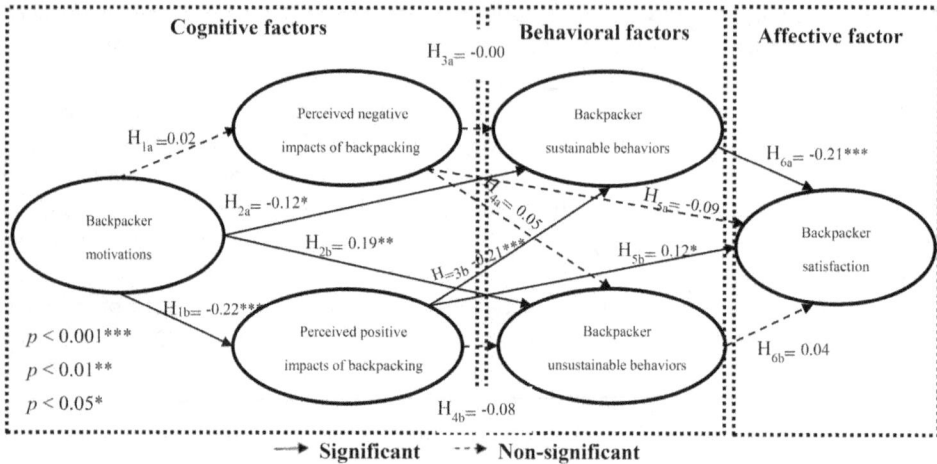

**Figure 3.** Structural model of backpacker sustainable behavior.

impacts of backpacking did not influence backpacker sustainable behavior, backpacker unsustainable behavior, and backpacker satisfaction. We propose a possible explanation that most backpackers do not perceive themselves as negative travelers and do not prefer to be identified as "tourists" as confirmed in most backpacker social identity theory studies (Zhang et al., 2018). Having such a positive mindset of themselves is reflected in their perceptions of their impacts on the destination, which subsequently influences how they behave and their emotional appraisal of the destination.

Moreover, backpacker unsustainable behaviors did not have a negative influence on backpacker satisfaction contrary to rural tourist studies that suggest that travelers who behave unsustainably are less satisfied (e.g., Kastenholz et al., 2018). This finding implies that there is a possibility for a positive relationship where backpackers attain satisfaction from unsustainable behaviors. This is because as experienced travelers, backpackers demonstrate their satisfaction by bargaining for cheap prices of local products (Riley, 1988). This partially explains why they have been banned from third world countries, including Bhutan and Maldives (Scheyvens, 2002).

Finally, backpacker motivations did not influence the perceived negative impacts of backpacking and perceived positive impacts of backpacking did not influence backpacker unsustainable behavior. Overall, the findings suggest that backpacker motivations have a wider variety of influence on behavior than the perceived impacts of backpacking within the tri-component attitude model (Abdulrazak & Quoquab, 2018). This suggests that backpackers are purposeful travelers seeking to fulfill different motivational needs by engaging in both sustainable and unsustainable actions (Ooi & Laing, 2010).

### Theoretical implications

This study adds to knowledge and understanding of how backpackers contribute to sustainable tourism development in developing destinations in three ways. First, backpackers have the tendency to act sustainably because of their motivations and perceived positive impacts of backpacking as many of these young travelers perceive backpacking as a one-time opportunity to experience foreign countries and cultures (O'Reilly, 2006). The overarching positive thoughts of backpackers imply that negative perceptions do not explain backpackers' sustainable behavior and satisfaction. The study, thus, builds on the existing literature by identifying two significant cognitive factors that underlie backpackers' sustainable actions.

Second, the tendency for backpackers to engage in unsustainable behaviors does not derive from the perceived negative impacts and perceived positive impacts of backpacking but rather because of the inclination to satisfy their positive internal and external motivations. Hence, positive internal and external motivations may not always translate into sustainable backpacker behaviors since backpackers are not mindful of sustainability. Therefore, this study sheds light on the need for a change from "inadvertent sustainability" to "intentional sustainability" by reconceptualizing backpacker motivations within a sustainable development framework. Third, the study contributes to an understanding of the complexity of backpackers' sustainable attitude and behavior as the findings demonstrate that backpacker satisfaction may stem from the perceived positive impacts of their travel and sustainable behaviors at the destination.

### Practical implications

The study holds two practical implications for destination management organizations and global tourism organizations towards achieving UN Sustainable Development Goal 12 (responsible consumption). First, the finding that backpackers who engage in sustainable behaviors become more satisfied implies that destination management organization should focus more attention on packaging and marketing services and recreational activities that will stimulate backpackers to behave sustainably and gain satisfaction. For example, destination management organizations can emphasize the packaging of eco-friendly tours, cultural displays, patronage of local products and services, etc. while discouraging visits to psychedelic enclaves, engaging in prostitution and drug trade and other overly consumptive behaviors.

Second, reconceptualizing backpacker motivations within a sustainable development framework, practically speaking, requires both global and local tourism organizations to develop awareness and advertisement campaigns that align backpacker motivations with backpacker sustainable behaviors. For example, backpackers could be encouraged by global tourism organizations' websites to buy and choose environmentally friendly accommodation in order to learn about the destination's environmental practices and contribute to local economy, read the history of their chosen destination in order to explain and ask relevant questions, and interact with local residents in order to understand and learn about the destination's culture. By aligning motivational needs with sustainable behaviors, unsustainable actions will be minimized and backpacker tourism will become more sustainable and satisfying.

### Limitations and suggestions for future research

The limitations of the study provide an opportunity for future research. First, the study sample is limited to one context (Cape Coast); hence, the results may not be generalizable to tourists in other settings. Therefore, future research can test the proposed model on other backpacker groups in different contexts. Second, although both backpacker (un)sustainable behaviors and perceived impact of backpacking measures, in this study, focused on major triple bottom line questions, these measures can be expanded in future studies. Finally, while we ensured direct reduction of social desirability bias through anonymity and confidentiality assurances, we acknowledge that the negative measures of perceived negative impacts of backpacking and backpacker unsustainable behavior may influence such biases.

## Conclusions

As part of achieving UN Sustainable Development Goal 12, destination management organizations and global tourism organizations need to understand how they can direct backpackers' attitudes and behaviors towards a more sustainable path. One way of achieving this objective is to investigate the complex relationships among cognitive, behavioral, and affective components of backpacker tourism. Against this backdrop, the present study examined the relationship among backpacker motivations, perceived impacts of backpacking, backpacker sustainable and unsustainable behaviors, and backpacker satisfaction to determine the influence of motivations and perceived impacts of backpacking on backpacker (un)sustainable behaviors and backpacker satisfaction. Overall, it was found that backpackers have positive perceptions of their impacts and frequently engage in sustainable behaviors than unsustainable behaviors. While both backpacker motivations and perceived positive impacts of backpacking influence sustainable behaviors, only backpacker motivation is a significant predictor of backpacker unsustainable behaviors. Moreover, the perceived negative impact of backpacking is an insignificant predictor of backpacker (un)sustainable behaviors and backpacker satisfaction. Finally, no adverse relationship was observed between backpacker unsustainable behavior and backpacker satisfaction. Strategies that instill "intentional sustainability" are crucial for sustainable consumption behaviors and backpacker satisfaction.

## ORCID

Elizabeth Agyeiwaah  http://orcid.org/0000-0002-8357-0011
Felix Elvis Otoo  http://orcid.org/0000-0003-2803-3835

## References

Abdulrazak, S., & Quoquab, F. (2018). Exploring consumers' motivations for sustainable consumption: A self-deterministic approach. *Journal of International Consumer Marketing*, *30*(1), 14–28. https://doi.org/10.1080/08961530.2017.1354350

Adam, I. (2015). Backpackers' risk perceptions and risk reduction strategies in Ghana. *Tourism Management*, *49*, 99–108. https://doi.org/10.1016/j.tourman.2015.02.016

Agyeiwaah, E., Adongo, R., Dimache, A., & Wondirad, A. (2016). Make a customer, not a sale: Tourist satisfaction in Hong Kong. *Tourism Management*, *57*, 68–79. https://doi.org/10.1016/j.tourman.2016.05.014

Agyeiwaah, E., McKercher, B., & Suntikul, W. (2017). Identifying core indicators of sustainable tourism: A path forward? *Tourism Management Perspectives*, *24*, 26–33. https://doi.org/10.1016/j.tmp.2017.07.005

Agyeiwaah, E., Pratt, S., Iaquinto, B. L., & Suntikul, W. (2020). Social identity positively impacts sustainable behaviors of backpackers. *Tourism Geographies*, 1–22. https://doi.org/10.1080/14616688.2020.1819401

Ajzen, I., & Fishbein, M. (1977). Attitude-behavior relations: A theoretical analysis and review of empirical research. *Psychological Bulletin*, *84*(5), 888–918. https://doi.org/10.1037/0033-2909.84.5.888

Andereck, K. L., Valentine, K. M., Knopf, R. C., & Vogt, C. A. (2005). Residents' perceptions of community tourism impacts. *Annals of Tourism Research*, *32*(4), 1056–1076. https://doi.org/10.1016/j.annals.2005.03.001

Ap, J. (1992). Residents' perceptions on tourism impacts. *Annals of Tourism Research*, *19*(4), 665–690. https://doi.org/10.1016/0160-7383(92)90060-3

Berezan, O., Raab, C., Yoo, M., & Love, C. (2013). Sustainable hotel practices and nationality: The impact on guest satisfaction and guest intention to return. *International Journal of Hospitality Management, 34*, 227–233. https://doi.org/10.1016/j.ijhm.2013.03.010

Bernstein, J. D. (2019). Begging to travel: Begpacking in Southeast Asia. *Annals of Tourism Research, 77*(C), 161–163. https://doi.org/10.1016/j.annals.2018.12.014

Biran, A., Liu, W., Li, G., & Eichhorn, V. (2014). Consuming post-disaster destinations: The case of Sichuan. *Annals of Tourism Research, 47*, 1–17. https://doi.org/10.1016/j.annals.2014.03.004

Bradt, H. (1995). Better to travel cheaply. *The Independent on Sunday, 12*, 49–50.

Brown, T. A. (2015). *Confirmatory factor analysis for applied research.* Guilford Publications.

Brundtland, G. H. (1987). What is sustainable development. In *Our common future* (pp. 8–9). Oxford University Press.

Chen, G., Bao, J., & Huang, S. (2014). Segmenting Chinese backpackers by travel motivations. *International Journal of Tourism Research, 16*(4), 355–367. https://doi.org/10.1002/jtr.1928

Chen, G., & Huang, S. (2017). Toward a theory of backpacker personal development: Cross-cultural validation of the BPD scale. *Tourism Management, 59*, 630–639. https://doi.org/10.1016/j.tourman.2016.09.017

Cohen, E. (1973). Nomads from affluence: Notes on the phenomenon of drifter-tourism1. *International Journal of Comparative Sociology, 14*(1-2), 89–103. https://doi.org/10.1177/002071527301400107

Cohen, S. A. (2011). Lifestyle travellers: Backpacking as a way of life. *Annals of Tourism Research, 38*(4), 1535–1555. https://doi.org/10.1016/j.annals.2011.02.002

Cooper, M., O'Mahony, K., & Erfurt, P. (2004). Backpackers: Nomads join the mainstream? An analysis of backpacker employment on the'harvest trail circuit'in Australia. In Richards, G, Wilson, J (Eds.), *The global nomad: Backpacker travel in theory and practice.* (pp. 180–195). Channel View Publications.

Cottrell, S., van der Duim, R., Ankersmid, P., & Kelder, L. (2004). Measuring the sustainability of tourism in Manuel Antonio and Texel: A tourist perspective. *Journal of Sustainable Tourism, 12*(5), 409–431. https://doi.org/10.1080/09669580408667247

Dann, G. M. S. (1981). Tourist motivation an appraisal. *Annals of Tourism Research, 8*(2), 187–219. https://doi.org/10.1016/0160-7383(81)90082-7

Dayour, F. (2013). Are backpackers a homogeneous group? A study of backpackers' motivations in the Cape Coast-Elmina conurbation, Ghana. *European Journal of Tourism, Hospitality and Recreation, 4*(3), 69–94.

Dayour, F., Adongo, C. A., & Taale, F. (2016). Determinants of backpackers' expenditure. *Tourism Management Perspectives, 17*, 36–43. https://doi.org/10.1016/j.tmp.2015.11.003

Dayour, F., Kimbu, A. N., & Park, S. (2017). Backpackers: The need for reconceptualisation. *Annals of Tourism Research, 66*, 191–193. https://doi.org/10.1016/j.annals.2017.06.004

Dayour, F., Park, S., & Kimbu, A. N. (2019). Backpackers' perceived risks towards smartphone usage and risk reduction strategies: A mixed methods study. *Tourism Management, 72*, 52–68. https://doi.org/10.1016/j.tourman.2018.11.003

del Bosque, I. R., & San Martín, H. (2008). Tourist satisfaction a cognitive-affective model. *Annals of Tourism Research, 35*(2), 551–573. https://doi.org/10.1016/j.annals.2008.02.006

Dion, P. A. (2008). Interpreting structural equation modeling results: A reply to Martin and Cullen. *Journal of Business Ethics, 83*(3), 365–368. https://doi.org/10.1007/s10551-007-9634-7

Dodds, R., Graci, S. R., & Holmes, M. (2010). Does the tourist care? A comparison of tourists in Koh Phi Phi, Thailand and Gili Trawangan, Indonesia. *Journal of Sustainable Tourism, 18*(2), 207–222. https://doi.org/10.1080/09669580903215162

Dogan, M. (1989). *Pathways to power: Selecting rulers in pluralist democracies.* Westview Press.

Eagles, P. F. J. (1992). The travel motivations of Canadian ecotourists. *Journal of Travel Research, 31*(2), 3–7. https://doi.org/10.1177/004728759203100201

Elkington, J. (2013). Enter the triple bottom line. In *The triple bottom line.* (pp. 23–38). Routledge.

Fodness, D. (1994). Measuring tourist motivation. *Annals of Tourism Research, 21*(3), 555–581. https://doi.org/10.1016/0160-7383(94)90120-1

Fornell, C., & Larcker, D. F. (1981). Evaluating structural equation models with unobservable variables and measurement error. *Journal of Marketing Research, 18*(1), 39–50. https://doi.org/10.1177/002224378101800104

Gnoth, J. (1997). Tourism motivation and expectation formation. *Annals of Tourism Research, 24*(2), 283–304. https://doi.org/10.1016/S0160-7383(97)80002-3

Hair, J. F., Gabriel, M., & Patel, V. (2014). AMOS covariance-based structural equation modeling (CB-SEM): Guidelines on its application as a marketing research tool. *Brazilian Journal of Marketing, 13*(2), 44–55.

Hampton, M. P. (1998). Backpacker tourism and economic development. *Annals of Tourism Research, 25*(3), 639–660. https://doi.org/10.1016/S0160-7383(98)00021-8

Han, H., Kim, Y., & Kim, E.-K. (2011). Cognitive, affective, conative, and action loyalty: Testing the impact of inertia. *International Journal of Hospitality Management, 30*(4), 1008–1019. https://doi.org/10.1016/j.ijhm.2011.03.006

Han, H., Yu, J., Kim, H.-C., & Kim, W. (2018). Impact of social/personal norms and willingness to sacrifice on young vacationers' pro-environmental intentions for waste reduction and recycling. *Journal of Sustainable Tourism, 26*(12), 2117–2133. https://doi.org/10.1080/09669582.2018.1538229

Howard, R. W. (2007). Five backpacker tourist enclaves. *International Journal of Tourism Research*, *9*(2), 73–86. https://doi.org/10.1002/jtr.593

Howard, J. A., & Sheth, J. N. (1969). *The theory of buyer behavior*. John Wiley & Sons, Inc.

Hunter-Jones, P., Jeffs, A., & Smith, D. (2008). Backpacking your way into crisis: an exploratory study into perceived risk and tourist behaviour amongst young people. *Journal of Travel & Tourism Marketing*, *23*(2-4), 237–247.

Iaquinto, B. L. (2015). I recycle, I turn out the lights": Understanding the everyday sustainability practices of back-packers. *Journal of Sustainable Tourism*, *23*(4), 577–599. https://doi.org/10.1080/09669582.2014.978788

Iaquinto, B. L. (2018). Backpacker mobilities: Inadvertent sustainability amidst the fluctuating pace of travel. *Mobilities*, *13*(4), 569–583. https://doi.org/10.1080/17450101.2017.1394682

Iaquinto, B. L., & Pratt, S. (2020). Practicing sustainability as a backpacker: The role of nationality. *International Journal of Tourism Research*, *22*(1), 100–107. https://doi.org/10.1002/jtr.2321

Kastenholz, E., Eusébio, C., & Carneiro, M. J. (2018). Segmenting the rural tourist market by sustainable travel behav-iour: Insights from village visitors in Portugal. *Journal of Destination Marketing & Management*, *10*, 132–142.

Ko, D.-W., & Stewart, W. P. (2002). A structural equation model of residents' attitudes for tourism development. *Tourism Management*, *23*(5), 521–530. https://doi.org/10.1016/S0261-5177(02)00006-7

Liu, C.-H., Horng, J.-S., Chou, S.-F., Chen, Y.-C., Lin, Y.-C., & Zhu, Y.-Q. (2016). An empirical examination of the form of relationship between sustainable tourism experiences and satisfaction. *Asia Pacific Journal of Tourism Research*, *21*(7), 717–740. https://doi.org/10.1080/10941665.2015.1068196

Loker-Murphy, L. (1997). Backpackers in Australia: A motivation-based segmentation study. *Journal of Travel & Tourism Marketing*, *5*(4), 23–45.

Luo, X., Brown, G., & Huang, S. (2014). Resident perceptions of backpackers' impacts: A case study from Lijiang, China. In *CAUTHE 2014: Tourism and Hospitality in the Contemporary World: Trends, Changes and Complexity*. School of Tourism, The University of Queensland.

Luzecka, P. (2016). Take a gap year!" A social practice perspective on air travel and potential transitions towards sustainable tourism mobility. *Journal of Sustainable Tourism*, *24*(3), 446–462. https://doi.org/10.1080/09669582. 2015.1115513

Mainieri, T., Barnett, E. G., Valdero, T. R., Unipan, J. B., & Oskamp, S. (1997). Green buying: The influence of environ-mental concern on consumer behavior. *The Journal of Social Psychology*, *137*(2), 189–204. https://doi.org/10.1080/00224549709595430

Martin, S. R., & McCurdy, K. (2009). Wilderness food storage in yosemite: Using the theory of planned behavior to understand backpacker canister use. *Human Dimensions of Wildlife*, *14*(3), 206–218. https://doi.org/10.1080/10871200902858993

Moscardo, G., Konovalov, E., Murphy, L., & McGehee, N. (2013). Mobilities, community well-being and sustainable tourism. *Journal of Sustainable Tourism*, *21*(4), 532–556. https://doi.org/10.1080/09669582.2013.785556

Nash, R., Thyne, M., & Davies, S. (2006). An investigation into customer satisfaction levels in the budget accommo-dation sector in Scotland: A case study of backpacker tourists and the Scottish Youth Hostels Association. *Tourism Management*, *27*(3), 525–532. https://doi.org/10.1016/j.tourman.2005.01.001

Nassani, A. M., Khader, J. A., & Ali, I. (2013). Consumer environmental activism, sustainable consumption behavior and satisfaction with life. *Life Science Journal*, *10*(2), 1000–1006.

Nok, L. C., Suntikul, W., Agyeiwaah, E., & Tolkach, D. (2017). Backpackers in Hong Kong–motivations, preferences and contribution to sustainable tourism. *Journal of Travel & Tourism Marketing*, *34*(8), 1058–1070.

O'Reilly, C. C. (2006). From drifter to gap year tourist: Mainstreaming backpacker travel. *Annals of Tourism Research*, *33*(4), 998–1017. https://doi.org/10.1016/j.annals.2006.04.002

Oliver, R. L. (1993). Cognitive, affective, and attribute bases of the satisfaction response. *Journal of Consumer Research*, *20*(3), 418–430. https://doi.org/10.1086/209358

Ooi, N., & Laing, J. H. (2010). Backpacker tourism: Sustainable and purposeful? Investigating the overlap between backpacker tourism and volunteer tourism motivations. *Journal of Sustainable Tourism*, *18*(2), 191–206. https://doi.org/10.1080/09669580903395030

Pallant, J. (2005). *SPSS survival manual: a step by step guide to data analysis using SPSS for windows*. Allen & Unwin.

Park, S., & Tussyadiah, I. P. (2017). Multidimensional facets of perceived risk in mobile travel booking. *Journal of Travel Research*, *56*(7), 854–867. https://doi.org/10.1177/0047287516675062

Pearce, P. L. (1982). *The social psychology of tourist behaviour*. Pergamon Press.

Pearce, P. L. (1990). *The backpacker phenomenon: Preliminary answers to basic questions*. James Cook University of North Queensland.

Pearce, P. L. (2007). Sustainability research and backpacker studies: Intersections and mutual insights. In K. Hannam, & I. Ateljevic (Eds.), *Backpacker tourism-concepts and profiles*. (pp. 38–53). Channel View Publications.

Pearce, P. L., & Foster, F. (2007). A "university of travel": Backpacker learning. *Tourism Management*, *28*(5), 1285–1298. https://doi.org/10.1016/j.tourman.2006.11.009

Peel, V., & Steen, A. (2007). Victims, hooligans and cash-cows: Media representations of the international backpacker in Australia. *Tourism Management*, *28*(4), 1057–1067. https://doi.org/10.1016/j.tourman.2006.08.012

Pickens, J. (2005). Attitudes and perceptions. In N. Borkowski (Ed.), *Organizational behavior in Health Care* (pp. 43–76). Jones and Bartlett Publishers.

Pietilä, M., & Fagerholm, N. (2016). Visitors' place-based evaluations of unacceptable tourism impacts in Oulanka National Park, Finland. *Tourism Geographies*, *18*(3), 258–279. https://doi.org/10.1080/14616688.2016.1169313

Podsakoff, P. M., MacKenzie, S. B., Lee, J.-Y., & Podsakoff, N. P. (2003). Common method biases in behavioral research: a critical review of the literature and recommended remedies. *The Journal of Applied Psychology*, *88*(5), 879–903. https://doi.org/10.1037/0021-9010.88.5.879

Prayag, G., & Brittnacher, A. (2014). Environmental impacts of tourism on a French urban coastal destination: Perceptions of German and British visitors. *Tourism Analysis*, *19*(4), 461–475. https://doi.org/10.3727/108354214X14090817031116

Prayag, G., Suntikul, W., & Agyeiwaah, E. (2018). Domestic tourists to Elmina Castle, Ghana: Motivation, tourism impacts, place attachment, and satisfaction. *Journal of Sustainable Tourism*, *26*(12), 2053–2070. https://doi.org/10.1080/09669582.2018.1529769

Riley, P. J. (1988). Road culture of international long-term budget travelers. *Annals of Tourism Research*, *15*(3), 313–328. https://doi.org/10.1016/0160-7383(88)90025-4

Rosenberg, M. J., & Hovland, C. I. (1960). Cognitive, affective, and behavioral components of attitude. In M. J. Rosenberg, Hovland, C.I., McGuire, W.J., Abelson, R.P. and Brehm, J.W. (Ed.), *Attitude Organization and Change: An Analysis of Consistency among Attitude Components* (pp. 1–14). Yale University Press.

Scheyvens, R. (2002). Backpacker tourism and third world development. *Annals of Tourism Research*, *29*(1), 144–164. https://doi.org/10.1016/S0160-7383(01)00030-5

Shaffer, T. S. (2004). Performing backpacking: Constructing" authenticity" every step of the way. *Text and Performance Quarterly*, *24*(2), 139–160. https://doi.org/10.1080/1046293042000288362

Sroypetch, S., Carr, N., & Duncan, T. (2018). Host and backpacker perceptions of environmental impacts of backpacker tourism: A case study of the Yasawa Islands, Fiji. *Tourism and Hospitality Research*, *18*(2), 203–213. https://doi.org/10.1177/1467358416636932

Stone, P., & Sharpley, R. (2008). Consuming dark tourism: A thanatological perspective. *Annals of Tourism Research*, *35*(2), 574–595. https://doi.org/10.1016/j.annals.2008.02.003

Teo, C. B. C., Khan, N. R. M., & Rahim, F. H. A. (2014). Understanding cultural heritage visitor behavior: The Case of Melaka as world heritage city. *Procedia - Social and Behavioral Sciences*, *130*, 1–10. https://doi.org/10.1016/j.sbspro.2014.04.001

Tolkach, D., Thuen Jørgensen, M., Pratt, S., & Suntikul, W. (2019). Encountering begpackers. *Tourism Recreation Research*, *44*(1), 17–32. https://doi.org/10.1080/02508281.2018.1511943

TRA (2019). *International visitors in Australia: Year ending march 2019*. Retrieved on 21/3/2020 from https://www.tra.gov.au/ArticleDocuments/185/IVS%20Summary%20March%202019.pdf.aspx.

Trudel, R. (2019). Sustainable consumer behavior. *Consumer Psychology Review*, *2*(1), 85–96.

UN. (2019). 5/3/2020). *Sustainable Development Goal 12: Ensure sustainable consumption and production patterns*. Retrieved 5/3/2020 from https://www.un.org/sustainabledevelopment/sustainable-consumption-production/.

Ursachi, G., Horodnic, I. A., & Zait, A. (2015). How reliable are measurement scales? External factors with indirect influence on reliability estimators. *Procedia Economics and Finance*, *20*, 679–686. https://doi.org/10.1016/S2212-5671(15)00123-9

Uysal, M., & Hagan, L. A. R. (1993). Motivation of pleasure travel and tourism. *Encyclopedia of Hospitality and Tourism*, *21*(1), 798–810.

Visser, G. (2004). The developmental impacts of backpacker tourism in South Africa. *GeoJournal*, *60*(3), 283–299. https://doi.org/10.1023/B:GEJO.0000034735.26184.ae

Wang, J., & Luo, X. (2018). Resident perception of dark tourism impact: The case of Beichuan County. *Journal of Tourism and Cultural Change*, *16*(5), 463–481. https://doi.org/10.1080/14766825.2017.1345918

Wearing, S., Cynn, S., Ponting, J., & McDonald, M. (2002). Converting environmental concern into ecotourism purchases: A qualitative evaluation of international backpackers in Australia. *Journal of Ecotourism*, *1*(2-3), 133–148. https://doi.org/10.1080/14724040208668120

Weiler, B., & Smith, L. (2009). Does more interpretation lead to greater outcomes? An assessment of the impacts of multiple layers of interpretation in a zoo context. *Journal of Sustainable Tourism*, *17*(1), 91–105. https://doi.org/10.1080/09669580802359319

Westerhausen, K. (2002). *Beyond the beach: An ethnography of modern travellers in Asia*. White Lotus Press.

Westerhausen, K., & Macbeth, J. (2003). Backpackers and empowered local communities: Natural allies in the struggle for sustainability and local control? *Tourism Geographies*, *5*(1), 71–86. https://doi.org/10.1080/1461668032000034088

Yuan, J., Morrison, A. M., Cai, L. A., & Linton, S. (2008). A model of wine tourist behaviour: A festival approach. *International Journal of Tourism Research*, *10*(3), 207–219. https://doi.org/10.1002/jtr.651

Zhang, J., Morrison, A. M., Tucker, H., & Wu, B. (2018). Am I a backpacker? Factors indicating the social identity of Chinese backpackers. *Journal of Travel Research*, *57*(4), 525–539. https://doi.org/10.1177/0047287517702744

Zhang, J., Tucker, H., Morrison, A. M., & Wu, B. (2017). Becoming a backpacker in China: A grounded theory approach to identity construction of backpackers. *Annals of Tourism Research*, *64*, 114–125. https://doi.org/10.1016/j.annals.2017.03.004

Zhang, A., Zhong, L., Xu, Y., Wang, H., & Dang, L. (2015). Tourists' perception of haze pollution and the potential impacts on travel: Reshaping the features of tourism seasonality in Beijing. *Sustainability*, *7*(3), 2397–2414. https://doi.org/10.3390/su7032397

# Index

Note: **Bold** page numbers refer to tables; *italic* page numbers refer to figures.

Abdulrazak, S. 178
action for climate change (ACC) 65, 67–68, 70, 72–74
Agyeiwaah, E. 15
Ahn, J. 85
Aiken, L. S. 33
Airbnb 14, 114–128
Ajzen, I. 4–5
Albarracin, D. 25
Almaguer, J. 67
Altinay, L. 13, 54
Alvarado-Herrera, A. 135, 139
anchoring effect 13, 63–64, 66, 68–70, 74
Anderson, J. C. 68, 70
Anderson, K. 64
anticipated emotions 5, 7, 11, 14, 79–80, 82–86, 89–92
anticipated pride and guilt 8, 11, 83
arbitrary coherence 64
Ariely, Dan 64
ARIMA model 99, 101–102, 104–105, 107
aspect-based sentiment analysis 123
augmented Dickey–Fuller (ADF) tests 104
Aupperle, K. E. 134–135
aviation green tax 13–14, 62–65, 68–69, 74

Böcker, L. 115
backpackers 15, 173–175, 177–178, 180–182, 187–190; motivation 15, 174, 177–178, 181, 184, 187–190; sustainable behavior 15, 173, 176, 178–181, 184, 187–189; unsustainable behaviors 180–181, 184, 188, 190
back-translation method 50
Baek, J. 100
Bagozzi, R. P. 11, 79, 83, 86
Banerjee, B. 67
Baumgartner, H. 82, 85
Becken, S. 65
behavioral characteristics 15, 168, 175
Behavioral differences 167

behavioral intentions 4–5, 36, 68–69, 79–83, 85, 161–162, 167
Belk, R. 117
Berezan, O. 174, 180
big data 14, 117–118, 121
Bilgihan, A. 83
Biran, A. 177
Blei, D. M. 119
Bloor, M. 68
Bohlen, G. 90
Bourke, J. G. 47
Bozkurt, V. 67
Bradley, G. L. 81
Bradt, H. 177
brand equity 26–29, 31–34
Brown, G. 163
Brown, T. A. 184
Bullock, E. V. 84

Caber, M. 49
Cai, X. 13
Carroll, A. B. 135, 139–140, 152
Casaló, L. V. 49
Chaminda, J. W. D. 48
Chan, E. S. W. 10
Chao, S. Y. 163
Chao, Y. L. 163
Chen, C. F. 140
Chen, G. 177, 181
Chen, S. Y. 26
Chen, X. 66
Cheng, M. 114–115, 117
Chi, G-q. 49
Chi, J. 100
Chiang, Y. T. 84
Choe, Y. J. 14, 100
Cleveland, M. 79, 82, 86
climate change risk perception (CCRP) 65–68, 70, 72–74
Colebrook-Claude, C. 82
common methods bias (CMB) 183

common method variance (CMV) 32
configurational model testing 52
confirmatory factor analysis (CFA) 50–51, 68, 70, **71**, 86, **88**, 182–184
connectedness 8, 12
Cooper, M. 177
corporate social responsibility (CSR) 52, 134–135, 139, 146, 152–153
Cottrell, S. 179
customer loyalty 47–49

data analysis 50, 85–86, 92, 118–119, 164, 182
data collection 50, 68, 86, 118, 164, 175, 181
David, P. 139
Dawson, S. 67
Dayour, F. 175, 177, 182
De Grosbois, D. 139
Del Bosque, I. R. 135
Delphi technique 134, 139–140, 142, 154
demographic differences 166
demographic profiles 69, 92
Deng, J. 142
De Pelsmacker, P. 83
descriptive social norm 8, 10
destination loyalty intention 25, 28–29, 32, 34, 37
Dixon, A. W. 163
Dodds, R. 180
Dono, J. 82
Draper, J. 163
drone food delivery services 14, 78–82, 84–86, 89–92

Eagles, P. F. J. 178
economic sustainability measures 48
Elliot, S. 14
emotional polarity 125–126
emotions 79, 82–85, 89, 92, 115, 119–120, 122, 124–125, 127, 176
energy efficiency 25, 29–31, 33
environmental corporate social responsibility 8, 11
environmental knowledge 7–10
environmental performance 29–31, 33–34, 37
environmental sustainability measures 48
environmental tax reform (ETR) 63
epidemics 14, 97–101, 107–109
Etheredge, J. M. 135
exploratory factor analysis (EFA) 182–183

familiarity 13, 43–45, 47–48, 50–56
Farmaki, A. 13
Felix, R. 67
Ferrell, O. C. 135
Fielding, K. S. 81
Figliozzi, M. A. 79
first Delphi survey 146
five-point Likert scale 68, 164

Fodness, D. 178
Fornell, C. 184
Foster, F. 177, 181

Gannon, M. 45
Gerbing, D. W. 68, 70
Giger, J. C. 83
goal-directed behavior 4–5, 30, 85
Goh, C. 101
Gravante, T. 84
green consumerism 116
green equity 13, 24–34, 37
green image 7–9
green-induced tourist equity model 13, 23–25, 26, 34
green initiatives 24–28, 31
green perceived value (GPV) 65, 67–68, 70, 72–74
green practices 27–28, 30, 37, 84–85, 174, 180
green product attachment 10
green tourists 65, 115
green users 14, 114, 118, 122
green value 8, 12
Guagnano, G. A. 81
Gursoy, D. 13, 48–49, 54

Ha, Y. 85
Han, H. 6, 8, 11–12, 66, 83–84
health, safety, environment, and quality (HSEQ) 81
hedonism 66–68, 71, 73
hierarchical linear modeling (HLM) 33
Holcomb, J. L. 139
Howard, J. A. 6
Hsu, C. H. 162
Huang, G. 13
Huang, S. 181
Hui, T. K. 32
Hurley, P. J. 118
Hutson, G. 168
Hwang, J. 14, 78, 80–81, 86, 90
hypotheses development 27, 176
Hyun, S. S. 13, 66, 90

Iaquinto, B. L. 174, 178
internal environmental locus of control (INELOC) 14, 79–85, 89–90
International Air Transport Association (IATA) 63–64, 74

Jacobson, R. P. 4
Jang, Y. J. 84
Jin, X. 114–115, 117
Jockers, M. L. 125

Kahneman, D. 64
Kals, E. 79, 85
Kaltenborn, B. P. 163, 168

Kara, A. 47
Kastenholz, E. 47, 180
Kenebayeva, A. 13
Kim, E. 119
Kim, H. 13, 78
Kim, J. J. 80–81
Kim, S. B. 14, 99
Kim, Y. 163
Klinger, R. 119
knowledge gaps 24
Koenig-Lewis, N. 83
Korea Culture and Tourism Institute 99
Korea Tourism Organization 99–101
Krajhanzl, J. 2
Kwon, J. 85
Kwon, O. 49

Lahno, B. 122
Laing, J. H. 174
Lange, F. 64
Larcker, D. F. 184
latent profile model 165
Law, R. 101
Lawshe, C. H. 142
Lee, C. K. 162
Lee, J. S. 162
Lee, S. 14
Lee, Y. 49
Lemon, K. N. 27
Lewis, C. D. 105
Li, K. X. 99
Liu, C.-H. 180
Liu, Y. V. 100
locus of control 14, 78–79, 81, 84
Loureiro, S. M. C. 47
loyalty 13, 25–29, 33, 44–48, 50–56, 66, 162
Luo, X. 179
Luo, Y. 124, 127
Lyu, S. O. 86

Maditinos, Z. 100
Maes, J. 79, 85
Maignan, I. 135, 152–153
Mainieri, T. 179
managerial implications 37, 127
materialism 66–68, 71–74
McCleary, K. W. 162
McGraw, A. P. 82, 84
McKeage, K. 67
measurement instruments 68
measurement items 68, 85–86, 139, 141, 184
measurement model testing 51
measurement scales 48, 140
Meelen, T. 115
Mellers, B. A. 82, 84
Mercer, J. 153
Middle East Respiratory Syndrome coronavirus
   (MERS) 14, 97–101, 104–105, 107–109

Midgett, C. 115
Modica, P. D. 54
Moon, S. 4
Moons, I. 83
Mussweiler, T. 64

Nadler, S. 117
Nelson, K. M. 64
Nok, L. C. 175, 177–178
norm activation theory 4, 6–7
nudge theory 13, 62–64, 69, 74

Oberseder, M. 139
O'brien, D. 135
Oh, C. O. 163
Okumus, B. 83
Oliver, R. L. 47
Olya, H. G. 13, 54, 66
Onwezen, M. C. 6, 83
Ooi, N. 174
Orel, F. D. 47
Oskamp, S. 153

Park, J. 80
Pearce, P. L. 175, 177, 181
Peng, Q. 66
perceived effectiveness 8, 11
perceived place value 160–162, 165–167
perceived value 12, 66, 161–164
Perera, L. C. R. 48
Pérez, A. 135
Perugini, M. 11, 79
Piçarra, N. 83
Plewa, C. 48
5-point Likert-type scale 49
Poma, A. 84
Powell, D. M. 84
Pratt, S. 174
Prayag, G. 177
price elasticity 64
pro-environmental behaviors 2–11, 37, 64–65,
   81–82, 84, 89, 92, 116, 162, 179

Quazi, A. M. 135
questionnaire design 48
Quoquab, F. 178

Raymond, C. 163
reasoned action theory 4–5, 81
regional environmental performance 23, 32
relationship equities 13, 24, 26–29, 31–33, 35
Rezvani, Z. 83
Richins, M. L. 67–68
Rosenbaum, M. S. 26, 28, 31
Rotter, J. B. 81
Ru, X. 66
Ruan, W. 13
Ruf, B. M. 134

Rust, R. T. 31–32
Ryley, T. 64

Saaty, R. W. 140
Saaty, T. 142, 148
sample characteristics 68, 165
satisfaction 13, 15, 25, 28, 44–48, 50–56, 115–118, 127, 153, 174, 176–177, 179–181, 187–190
Satta, G. 37
scale development process 135, 139–140
Scheyvens, R. 175, 178
Schwartz, S. H. 6, 66
Schwarz value survey (SVS) 66
second Delphi survey 146
self-enhancement value 66
self-identification approach 182
self-identity 65
sentiment analysis 118–120, 122–123, 126
Serrano, L. 14
sharing economy 14, 114–118, 126–127, 141
Shi, W. 99
Shim, C. 15
Shnayder, L. 134
Simpson, F. 162
Smedby, N. 65
Sohn, T. H. 100
Song, H. 15
Song, J. 99
Sonnenschein, J. 65
source market 13, 24–25, 29–31, 33–34, 36–38
Sparks, B. A. 81
Sroypetch, S. 175
Steg, L. 3
stepwise autoregressive model 14, 98–99, 103, 107
Stern, P. C. 6
Stolaroff, J. K. 79
Strack, F. 64
structural equation modeling (SEM) 44–45, 48, 50–51, 68, 70, 72, 87–88, 119, 121–122, 161, 164, 182, 184, 187, 189
structural invariance model 71
structural model 51, 73, 182, 184
Suh, H. 99
sustainability practices 13–16, 43–48, 51, 53–56, 79–83, 89–90, 97–98, 114–119, 122–124, 126–127, 178, 187, 189; in hotel industry 45
sustainable tourism 13–14, 28, 31, 34, 36–37, 62, 74, 97–98, 108, 142, 174; consumption 14

Türkel, S. 47
Tang, R. L. 124, 127
Teo, C. B. C. 178

Tepeci, M. 47
third Delphi survey 146
tourism 2–3, 7, 9, 11–13, 15–16, 24–25, 27, 36, 65, 97–98, 100–101, 108–109, 117–118, 174, 176–179
tourist attractions 108, 161–162, 165, 169
tourist equity 13, 24–27, 29–30, 35, 37–38; drivers 25, 27, 29, 37; model 26
tri-component attitude model 15, 173, 176, 187–188
Tuan, Y. F. 163
Turker, D. 139
Tversky, A. 63
Tzeng, G. H. 140

urban parks 15, 108, 160–162, 167–169
Ursachi, G. 184

value-belief-norm (VBN) theory 4, 6–7, 81
value equity 27, 29, 31–34
value proposition 31
Van der Linden, S. 66
Vargas, P. 25
Vassiliadis, C. 100
Visser, G. 175
Vlek, C. 3
Vogel, V. 27, 32

Wang, J. 179
Wang, S. 31
Ward, D. O. 65
Weaver, P. A. 162
West, S. G. 33
WHO 99
Wilson, F. R. 142
winters model 103, 105
Wong, A. K. F. 14
Wong, I. 13
Wong, I. A. 26–28, 31
Woodside, A. G. 54
Wu, H. C. 65

Xiao, Q. 152
Xie, C. 84
Xu, X. 48–49, 54

Yim, E. S. 100
Yoon, Y. 162

Zeithaml, V. A. 12, 86
Zhang, A. 179
Zhao, H. 31
Zhu, X. 163

For Product Safety Concerns and Information please contact our EU
representative  GPSR@taylorandfrancis.com
Taylor & Francis Verlag GmbH, Kaufingerstraße 24, 80331 München, Germany

www.ingramcontent.com/pod-product-compliance
Lightning Source LLC
Chambersburg PA
CBHW082033230326
41598CB00081B/6289

9 781032 187952